Administration
and
Management
of
Physical Education
and
Athletic Programs

Fourth Edition

Clayne R. Jensen
Brigham Young University

Steven J. Overman
Jackson State University

CBS

CBS Publishers & Distributors Pvt. Ltd.

New Delhi • Bengaluru • Chennai • Kochi • Mumbai • Pune
Hyderabad • Kolkata • Nagpur • Patna • Vijayawada

WAVELAND

PRESS, INC.
Long Grove, Illinois

Administration and Management of Physical Education and Athletic Programs

Waveland ISBN: 978-1-57766-272-3

Copyright © 2003 by Clayne R. Jensen & Steven J. Overman
Copyright © 1992 by Clayne R. Jensen

The worldwide edition of this book is published by Waveland Press, Inc., 4180, IL, Route 83, Suite 101, Long Grove, Illinois 60047, United States of America

All rights reserved. No part of this book may be reproduced or transmitted in any form or by any means, electronic or mechanical, including photocopying, recording, or any information storage and retrieval system without permission, in writing, from the author and publishers.

This edition has been published in India with the permission of Waveland Press, Inc. for sale in India, Bangladesh, Myanmar, Pakistan, Sri Lanka, Maldives, Bhutan and Nepal.

CBS Reprint: 2015

CBS ISBN: 978-81-239-2661-2

Published by:
Satish Kumar Jain for CBS Publishers & Distributors Pvt. Ltd.,
4819/XI Prahlad Street, 24 Ansari Road, Daryaganj, New Delhi - 110002
delhi@cbspd.com, cbspubs@airtelmail.in • www.cbspd.com
Ph.: 23289259, 23266861, 23266867 • Fax: 011-23243014

Corporate Office: 204 FIE, Industrial Area, Patparganj, Delhi - 110 092
Ph: 49344934 • Fax: 011-49344935
E-mail: publishing@cbspd.com • publicity@cbspd.com

Branches:
• *Bengaluru:* 2975, 17th Cross, K.R. Road, Bansankari 2nd Stage, Bengaluru - 70
  Ph: +91-80-26771678/79 • Fax: +91-80-26771680
  E-mail: cbsbng@gmail.com, bangalore@cbspd.com
• *Chennai:* No. 7, Subbaraya Street, Shenoy Nagar, Chennai - 600030
  Ph: +91-44-26681266, 26680620 • Fax: +91-44-42032115
  E-mail: chennai@cbspd.com
• *Kochi:* 36/14, Kalluvilakam, Lissie Hospital Road, Kochi - 682018
  Ph: +91-484-4059061-65 • Fax: +91-484-4059065
  E-mail: cochin@cbspd.com
• *Mumbai:* 83-C, Dr. E. Moses Road, Worli, Mumbai - 400018
  Ph: +91-9833017933, 022-24902340/41 • E-mail: mumbai@cbspd.com
• *Pune:* Bhuruk Prestige, Sr. No. 52/12/2+1+3/2,
  Narhe, Haveli (Near Katraj-Dehu Road Bypass), Pune - 411041
  Ph: +91-20-64704058/59, 32342277 • E-mail: pune@cbspd.com

Representatives:

• Hyderabad: 0-9885175004
• Nagpur: 0-9021734563
• Vijayawada: 0-9000660880
• Kolkata: 0-9831437309, 0-9051152362
• Patna: 0-9334159340

Printed at: India Binding House, Noida, U.P.

Contents

iii

Part III
Program Administration 257

Appendix A: Professional Resources 369

Appendix B: AAHPERD Code of Ethics 387

Appendix C: Facility Planning Checklist 389

Appendix D: Criteria for Evaluating Physical Education Curriculum, Grades 9–12 395

Appendix E: Position Statements 399

Preface

The last decade has witnessed some dramatic developments in education, especially in physical education and athletics. The place of public schools in American society has been challenged as never before with concerns regarding accountability, competition from the private sector, and a decrease in financial support. Concern about the health and fitness of young Americans has been linked to declining requirements for and enrollment in physical education programs. At the same time, a resurgence in intramural recreation on college campuses has been accompanied by a building boom in recreational facilities. During the same period, the number of young women participating in interscholastic and intercollegiate sports has increased dramatically, while the number of women coaches declined. Concurrently, college athletics, particularly the men's football and basketball programs, have undergone increasing scrutiny because of overemphasis, commercialism, and poor graduation rates. All of these developments have implications for administration.

Technological advances now influence how we teach in the classroom and in the gymnasium, and how instructional, recreational, and athletics programs are managed. Teachers, coaches, directors, and administrators all make use of new hardware and software in order to perform more efficiently.

Management theory also has evolved. Paradigms are shifting and new management models emerging. A trend toward site-based management is changing the way school districts function. Heightened security concerns and liability issues impact facility management. Equal employment opportunity continues to be an issue that engages human-resource managers. Effective public relations has never been more crucial for schools. The fourth edition addresses the administrative implications of these issues.

Changes in emphasis in this revision have been accompanied by structural changes. The fourth edition is reorganized into sixteen chapters under three major headings: Part I, The Art of Administering; Part II, Management Techniques; and Part III, Program Administration. Material from several chapters of the previous edition are combined, some chapters are renamed, and chapters are renumbered. Reference material within chapters—including that which made up chapter 20 in the third edition—is relocated in Appendices.

The fourth edition also gives readers the opportunity to access information on the Internet. Within the text, the reader will find frequent references to such documents as mission statements, strategic plans, and policy handbooks accessible online. A list of Online Sources at the end of each chapter refers the reader to professional organizations and other pertinent sites. Appendix A, Professional Resources, provides an exhaustive catalog of professional associations and service organizations, journals, and newsletters with contact information, including Internet addresses.

This book is aimed toward improving physical education and athletics programs at all school levels by helping those who manage them to implement better content, methods, and procedures. Its main focus is on effective planning, organization, and management. The book was written specifically for upper division or graduate students who have particular interest in administrative work. But the content will also serve well those preparing for or serving in teaching and coaching positions, because their jobs include some elements of management, and they must know how to work effectively with the school administrators.

A fourth edition would not be possible without those who contributed their efforts to Dr. Jensen's third edition. Special appreciation is expressed to the following individuals for contributing to the book and for their critical reviews of certain sections: Linda Carpenter, Joyce Harrison, Bruce Holley, Boyd Jarman, Mary Bee Jensen, Jay Naylor, Betty Vickers and Pete Witbeck.

Finally, grateful acknowledgment is extended to the following individuals who contributed or critically evaluated material for the fourth edition: Lucy Chaffin, Doyice Cotten, Maurie Denney, Julie Flaherty, Mary Fry, James Gavette, Anthony Lovell, Todd Miner, and Elizabeth Overman. Special thanks go to Dr. Clayne Jensen for his wisdom, generosity, and support in the fourth revision of this book, and to my editor, Gayle Zawilla, for her guidance and assistance.

Steven J. Overman

Part I

The Art of
Administering

1 Principles and Philosophy of Administration

THE STUDY OF ADMINISTRATION

Why study administration? It is important at the outset to discuss the rationale for doing so and to establish certain meanings attached to the topic. Furthermore, we need to examine the nature of administration and how it functions in the educational system, and to describe its basic tools.

Administration has been described in a general way as the art of getting things done. More precisely, it means getting the right things done the right way for the right reasons. The practical activities of administration involve both deciding and doing, and those two tasks—deciding and doing—are closely integrated processes that pervade the entire organization.

One might ask why college students should study this topic when administrative assignments are not imminent, and many of them have no intention of pursuing careers in administration. Certainly, the study of administration has a more direct and immediate value for those preparing to become administrators as well as current administrators who want to improve their skills. The fact is that all teachers and coaches need to have a basic understanding of administrative theory and practice because they have administrative responsibilities, such as managing student records, selecting and ordering class materials, and utilizing facilities. All else being equal, a teacher or coach who has knowledge and insight about administrative processes will be able to work more effectively with those in administrative positions.

Physical educators are more likely to have administrative responsibilities than most classroom teachers, because they teach through the medium of human movement. This type of teaching involves complex logistics and diversified equipment and facilities, as well as oversight of out-of-class programs such as intramurals. Coaching encompasses even greater administrative dimensions. Coaches must be able to arrange transportation, manage sports facilities, and work with school administrators, the media, sports officials, parents of athletes, and the general public. Athletics includes responsibilities pertaining to finances and public relations that do not exist to the same degree in the assignments of most teachers.

3

Increased emphasis on participation by women makes good administrative procedures in physical education and athletics even more important. Women and men now share facilities more frequently, which escalates the need to use those facilities more efficiently. The expansion of women's programs increases budgeting requirements and underscores the need for better financial management. In addition, on the heels of several decades of school consolidation, physical education and athletics departments have become larger and more complex, requiring more efficient administrators. The administrative responsibilities inherent in teaching physical education and coaching may explain the high proportion of professionals in this area who eventually pursue administrative careers outside their own discipline. Many go on to become department chairs, directors of athletics, deans, school principals, and superintendents. Surveys indicate that in some states almost half of all school principals were former physical education teachers or coaches.

Administration is not for everyone. Many teachers and coaches are promoted to administrative positions and finish out their careers at that level; others discover over time that they prefer the classroom or the athletic field and return to these assignments. Administration is no different than other careers in that it involves tradeoffs. Following is a list of advantages and disadvantages of holding an administrative position (Horine, 1999).

Advantages

- Increased compensation
- Prestige in one's profession and the community
- Professional challenges
- Opportunity to exert power positively
- Feelings of accomplishment
- Opportunity to effect positive change
- Association with others who are leaders

Disadvantages

- Long work hours and insufficient time to accomplish tasks
- Changes in personal relationships with staff and students
- Less time for teaching and doing research
- Pressure of responsibilities for programs, personnel, and facilities
- Public scrutiny of decisions and results

Meaning and Context of Administration

The need for effective administration of educational programs is obvious when one considers the important role that education plays in society and the tremendous amount of public and private resources devoted to it. However, the exact meaning of administration is not so obvious. Let's analyze what the term *administration*, and the related term, *management*, imply.

Administration, the broader of the two terms, can be defined as the universal process of efficiently organizing people and directing their activities toward common goals and objectives. Administration is both an art and a science (an inexact one)—and arguably a craft, because administrators are not judged by their education or training, but by their performance on the job. Administration encompasses leadership skills such as vision and influence (see chapter 3). Administrators must visualize the "big picture," establish the mission and clarify the direction of an organization, help the organization determine what goals are appropriate, rally the workers behind these goals, and set strategies for meeting the goals (Northouse, 1997)

A prevailing view has been that administrative skills are generic, that they are transferable across organizations (O'Connor, 2001). Similar skills apply whether we are discussing the administration of a school, a city recreation department, a law-enforcement agency, or a hospital. The fact that college athletics directors and head coaches (who are largely administrators) successfully move back and forth between positions in intercollegiate and professional sports reinforces this view. An alternate conception is that organizations are becoming more complex, technology dependent, and differentiated, requiring management skills that are situation specific. This implies that sports managers must consider how public schools differ from organizations in the private sector.

Management is defined as the aspect of administration associated with the day-to-day operations of various elements within the organization. Management brings order and consistency to an organization; it provides what is needed for the organization to run effectively and efficiently. It focuses on providing a structure in which to work, on the physical context of work, and on relationships between workers. Management means placing people in the right jobs and developing rules and procedures regarding how work is to be performed. Managers develop work incentives and systems for monitoring work. Management skills, which tend to be specialized, include managing records, technology, finances, public relations, facilities and equipment, and human resources (Northouse, 1997). Subsequent chapters in this book address these specific areas of management. Every organization must be managed well if it is to be successful.

Background of School Administration

The complexity of administration depends on the size and nature of the organization. Let's relate this idea to the educational system. A teacher in a one-room school would engage in teaching as the primary responsibility. However, even in this basic educational setting, there would be certain administrative duties. The facility would have to be provided and cared for, supplies and equipment kept on hand, student records maintained, and student information disseminated. A budget would have to be worked out to cover the teacher's salary and the expenses of the facility and equipment. There would be public-relations considerations involving parents and local citizens. The school would operate under some form of local authority, such as a committee, council, or board, and this would involve an administrative relationship between the teacher and the higher authority. Therefore, it is apparent that even in the basic setting of the one-room school, the teacher, being the sole employee, would need to devote a portion of time to the managerial functions inherent in the situation.

As one-room schools became less prevalent and several teachers were employed, educators decided that most of the administrative tasks should be handled by one of the teachers and that a portion of that teacher's time should be given for this purpose. This designated person came to be known as the school principal. As the educational system became even more complex, each state created a state department of education with a state superintendent in charge. States subdivided into school districts with a district superintendent appointed to administer the educational programs of the local area.

In a similar fashion, colleges evolved from small operations into large and complex universities. The first college presidents were faculty members who continued to teach part time. As small colleges grew into universities, they were organized into separate colleges, schools, and departments with an administrative head over each unit. Each of these steps in development resulted in administrative positions that previously had not existed. In a general sense, the areas of administrative duties are the same under today's large and complex institutions of higher learning as they were in the one-teacher school, but the complexity of and level of involvement in each aspect of administration has increased immensely.

Current Status of Administration

Even in our present educational arrangement, no school employee is totally free of administrative tasks and procedures. Although a teacher cannot separate him- or herself totally from management responsibilities, such involvement should be carefully controlled in order to facilitate effective teaching. Those in administrative positions should handle most administrative tasks. This division of labor assumes that those given administrative assignments will handle these tasks effectively, thus relieving full-time teachers of these assignments so that they are better able to teach.

Even though many of the responsibilities of an administrator are managerial in nature, an educational administrator is certainly more than a manager. In addition to the direct day-to-day management of the organization, he or she must be a leader who will provide: (1) enthusiasm and direction to the organization; (2) guidance on both curricular and noncurricular matters; (3) broad and far-reaching planning of budget, personnel, and facilities; (4) help in defining and interpreting goals and continuing support toward achieving them; (5) opportunities for improvement and enrichment of staff; and (6) direction concerning public relations.

In fact, some administrative positions are largely managerial, focusing on the efficiency of the daily operation, whereas others are much broader in scope. For example, the chair of a physical education division in a high school has primarily managerial responsibilities. The principal of the same school is also involved in management, but on a broader basis. The principal must give adequate attention to the overall objectives of the educational system of which that school is a part and must determine how the school fits into the system and contributes to the objectives. He or she must be analytical about the organizational divisions within the school, and about how well they function in relationship to each other. He or she is concerned about the balance of educational experiences the students receive, as well as the quality of those experiences. The well-being of the employees, their productivity, and their continuous professional improvement are other major concerns. These administrative responsibilities extend beyond direct managerial functions. In a like manner, the school superintendent's responsibilities are broader and more far-reaching than those of the principal. The superintendent's job involves less attention to the direct and specific aspects of management and more emphasis on the broader aspects of educational leadership. The school's director of athletics shares many of these broad administrative responsibilities. (More about the responsibilities of the superintendent and principal appears in chapter 2.)

The above descriptions imply that an effective administrator must be proficient in three general skill areas: *conceptual skills, technical skills*, and *human-relations management skills*. (Technical skills and human-relations skills are addressed in subsequent chapters.) Conceptual skills extend beyond management skills. They incorporate a sense of: (1) what the position of administrator entails, (2) the mission and functions of the organization, (3) the nature of education in a free society, and (4) an awareness of educational trends and current issues. Table 1-1 lists some current trends and issues in education, physical education and athletics that were salient at the time this book went to press. Those who aspire to become administrators should familiarize themselves with these trends and issues through the media and by reading the professional literature.

Conceptual skills also imply that administrators are philosophically grounded, that their training and life experiences have introduced them to values and principles that underlie their professional obligations. Yet, it is apparent that many students majoring in physical education (kinesiology/human movement) do not enroll in phi-

Table 1–1
Current Trends and Issues in Education, Physical Education, and Athletics

Education
Increase in school violence, and accompanying security implications
Consolidation of schools and related transportation issues
Efforts to abolish recess in elementary schools in order to add more in-class time
Home schooling: ramifications for public schools
Alternative schools: magnet schools, charter schools, contract schools
School vouchers
Site-based management
Distance learning, online courses
Required state-administered achievement tests
Increase in ethnic diversity of student population with accompanying language issues
Undesirable commercial intrusions into schools
Inadequate support of inner-city schools

Physical education
Health-related physical fitness
Youth obesity and related health problems as a result of inactivity and dietary habits
Accommodating disabled students: Individual education programs (IEPs), mainstreaming, etc.
Deficiencies in state physical education requirements
Reorganization of campus recreation and intramural programs at college level
Lack of adequate intramural programs in public schools
Technology-enhanced instruction: PDAs, pedometers, heart monitors, etc.

Athletics
Gender equity
Elimination of nonrevenue sports at college level
Commercialization of intercollegiate athletics
Inappropriate emphasis on competition in lower grades
Eligibility issues in relation to grades, residence, and home schooling
Role incompatibility of student/athlete, especially at college level
Improved preseason physical exams
Exercise-induced trauma and dehydration (especially in football)
Drug testing
Certification of coaches
Shortage of women coaches

losophy courses during their undergraduate education. In an attempt to remedy this deficiency, the following sections introduce the basic tenets of philosophy that inform the administrative process. Although the following discussion is necessary, it may not be sufficient. The reader is encouraged to explore further in this area of study.

PHILOSOPHICAL FOUNDATIONS

The term *philosophy* derives its roots from classical language, denoting "love of wisdom." It implies the need to find the real truths and values and apply them correctly to our personal and professional lives. Those who are regarded as philosophers are usually viewed as deep thinkers who have unusual insights and express superior wisdom. Philosophers think more analytically than the ordinary person, concentrating on life's fundamental values and vital issues. Most of us don't consider ourselves philosophers, nor do we plan to become philosophers in a formal sense. However, all of us need to become more philosophical as students and as future teachers, coaches, and administrators.

Philosophy has two important themes: gaining a greater understanding of life and the world in which we live, and sharpening our ability to think clearly and logically. Both of these insights are of vital importance to an educational administrator.

Branches of Philosophy

In order to comprehend more clearly the nature and meaning of philosophy, it is helpful to identify the processes that make up the branches of this discipline. These include the following (Hartvigsen, 1992):

- *Metaphysics* attempts to answer the question, "what is reality?" It is the study of the ultimate nature of being; the science of fundamental causes and processes; the nature of mind, body, and self; and the nature of the universe. Theology and cosmology are considered branches of metaphysics.

- *Epistemology* addresses the question, "what is truth?" It is concerned with the sources, nature, and limits of knowledge and the related issues of authority, validity, and principles.

- *Logic* is the study of how we make valid inferences; what constitutes sound reasoning, valid arguments, and logical thought; and the criteria by which they are judged.

- *Ethics* attempts to answer the question, "what is the highest good?" It addresses issues of obligation and duty, the question of what is right and wrong, the criteria for determining good and bad, and the foundation on which moral character rests.

- *Aesthetics* is concerned with the nature of art and the quality of beauty. It addresses questions of objectivity and subjectivity in relation to personal taste and preference, and the ways we discriminate between beauty and ugliness. Its basis is appeal and quality of form or technique.

- *Political philosophy* is the study of the various forms of government and social structure, such as anarchism, aristocracy, communism, democracy, and socialism. Much of the behavior of both humans and institutions is political in terms of causes and results.

Within the framework listed above, philosophers utilize the following methods for arriving at beliefs: reason, observation, experience, intuition, and revelation (Ziegler, 1980). Philosophers disagree on the reliability of these several methods, however. Such disagreements, coupled with diverse perspectives on the fundamental questions of philosophy, have led to differing systems of beliefs or "schools" of philosophy. Familiarity with these schools of thought assists us in grounding our own personal and professional philosophies.

Philosophical Schools

The prelude to aligning with a particular school of philosophy is known as *eclecticism*. Eclecticism is the selection of elements from different systems of philosophy, often without regard to possible contradictions among them. Most people start out as eclectics. They have personal beliefs or ideas about life, just as they have favorite remedies for aches and pains. However, their beliefs are fuzzy and poorly defined because they are not knowledgeable or haven't really analyzed matters thoroughly. Others express definite beliefs based on superficial thought, emotional reaction, or incorrect information. For these reasons, this stage of philosophical development often is referred to as *naïve* eclecticism.

Others graduate into a position of well-defined beliefs based on thorough knowledge and careful analysis. The result of such systematic reflection often leads to a coherent pattern of beliefs shared by others, which may center around a teacher or a school, and acquires a label or rubric to characterize it. The systematic study and elucidation of a particular perspective and set of beliefs is a "school" of philosophy. Some prominent formal schools of philosophy are idealism, realism, pragmatism, and existentialism. All of these have had a substantial influence on the nature of education.

Idealism is traceable to the ancient Greek philosopher Plato. It places special value on ideas and ideals as products of the mind, in contrast to a world perceived through the senses. It envisions Truth with a capital "T" as a perennial ideal. In education, it is aligned to a traditional curriculum with the teacher as the center of learning. Idealists feel that schools should be concerned with developing character as well as teaching knowledge.

Realism originated with Aristotle, one of Plato's pupils. It places great emphasis on the observation of nature, on the scientific method, and on systems of classification. Realism sees the world as authentic outside of the human mind. Knowledge then derives from experience. In education, realists are open to alternative approaches to learning such as laboratory classes. They believe that schools should teach *how* to learn as well as *what* to learn.

Pragmatism is a modern realistic philosophy often identified with the American philosopher John Dewey. It sees the world as constantly changing. Humans must interact with the world to understand it. Truth is derived from what works in practice. Pragmatists believe that schools should allow students to explore the world through play experiences beyond the classroom. They emphasize group learning and the problem-solving model.

Existentialism is less a systematic philosophy than a permeating influence on other philosophies. Existentialism is based on the belief that our very existence precedes systems of ideas, truths, or formal methods for determining reality. Consequently, reality is highly subjective and problematic. Existential educators believe strongly in student-centered learning with a maximum tolerance for individual differences and self-expression. They are suspicious of schools that attempt to standardize and regiment learning.

When educators are not completely comfortable with any one school of philosophy, they adopt the position of *syncretism*. Syncretism combines selected aspects of the various philosophical systems while attempting to resolve conflicts among them. This is eclecticism on a higher level—what might be called *sophisticated* eclecticism. The syncretist is like the political independent who agrees with some (but not all) positions of the major political parties, so aligns with neither.

Our philosophic beliefs shape our basic approach to the professional endeavors of teaching, coaching, and administration. Practical applications of the fundamental beliefs and methods of philosophy may not be readily apparent to the reader at this point. They will appear more relevant in the following discussions of applied philosophy, values, goals, principles, and theories, as each touches on issues addressed by the philosophical schools.

Philosophy of Education

We perform many practices in education without ever questioning *why*—the "why" often being rooted in philosophy. By developing an educational philosophy, we can better understand and explain why schools are the way they are, and why we as teachers, coaches, and administrators do what we do. Educational philosophy determines our viewpoint about the ultimate goals of education in a free society, the nature of schools, the nature of the instructional process, and the content of the school curriculum. Philosophy of education carries over into political philosophy, which addresses the importance of an educated citizenry in a free society.

Educators have been criticized for not being philosophically grounded. We are accused of going from one learning fad to another, joining whatever parade is currently passing by, without reflecting on where we are going. A school administrator should be able to defend his or her approach to education in terms of beliefs and values. Teachers and coaches should be able to justify what they are doing in the classroom and on the athletics fields. All educators should be more reflective. Plato was speaking to all of us when he admonished, "The unexamined life is not worth living."

Philosophy of Physical Education

Physical education should be concerned with the fundamental relationship between mind and body, and with the body's role in learning. We know now what Plato didn't know: that humans are integrated beings; the physical, cognitive, and emotional are inseparable. This understanding leads us to ask some basic questions: What does learning through the medium of human movement add to the educational experience? What values can be taught through games, dance, and physical competition? What does it mean to be physically educated, and what is the place of physical education in the school curriculum? These fundamental issues rely on philosophical foundations in their responses.

A broader philosophical view encompasses the meaning of the pursuit of physical excellence (the *citius, altius, fortius* of the Olympic motto), the breaking of records, the transcendent experience of victory, and the existential angst of defeat. Humans define themselves by their physical achievements. We are the species that can climb the highest mountains, swim the wave-chopped channels, and soar like birds on artificial wings. We are sensitive to the aesthetics of human movement: the beauty of a pirouette and the precision of a high dive. We are curious about the way in which athletes experience their bodies, about the relationship of the body with technology and the external world. In this sense, we all are philosophers whether or not we realize it (Weiss, 1971).

The schools of philosophy listed above are difficult to grasp in the context of what actually goes on in the physical education classroom on a daily basis. We find idealists, realists, pragmatists, and even a few existentialists in schools. Following are simple examples of these different philosophical approaches applied to a teaching task, to assist the reader in appreciating their relevance. The lesson for the day is hitting a ball with a bat.

The *idealist* instructor believes that there is a correct batting stance and a perfect swing. She has practiced and perfected the technique of batting, has a picture of it in her mind, and demonstrates it for her students, who are then directed to imitate her example. The teacher drills the students and corrects the imperfections in their efforts until they know how to execute correctly. She bases her students' grades on correct form.

The *realist* bases his approach to teaching batting on scientific studies that suggest that some methods are better than others, depending on individual differences of batters. He explains his approach in scientific terms, employing the principles of center of gravity, forward momentum, and bat speed, and then works with individual students to develop a stance and a swing that fits their needs. He tests the results and bases their grades on performance data converted into standard scores.

The *pragmatist* distributes balls and a variety of bats to her class, divides them into small groups, and instructs them to experiment on different stances and swings. She presents hitting a pitched ball with the bat as a problem to be solved. She tells the students to be creative, as there may be an effective approach to hitting that no one has yet tried. She reminds her students that the batting technique they should adopt is the one that works the best for them. Cooperating with others in the learning process is a major part of the students' grades.

The *existentialist* tells his students to forget momentarily that there is a team sport called baseball. He directs them to experience individually the feel of swinging a bat as a way of expressing themselves. He encourages them to create their own unique approach and style to batting. Whether or not they hit the ball consistently is less important than what they gain from the experience. There are lessons to learn from failure as well as success. He finishes by suggesting that batting is not for everyone; some students may prefer taking up another activity. This learning experience is difficult to translate into a grade.

BUILDING A PERSONAL PHILOSOPHY

Educational administrators can have a significant influence on schools, students, and the community. They should behave morally and ethically in carrying out their professional responsibilities. They must believe, represent, and foster desirable attitudes and practices. This implies that perhaps the most important characteristic of an administrator is a personal philosophy, as this influences all else. In striving to be an effective administrator, Socrates' admonition to "know thyself" is eminently applicable.

The question is not whether each of us has a philosophy, but rather what our philosophy includes. With a poorly developed philosophy, a person lacks clear understanding and direction. Certainly, life cannot be highly satisfying or successful without a life view that leads to a purposeful existence.

It is often helpful for a person to put into written form what he or she truly believes, because putting thoughts into writing contributes clarity to one's ideas and concepts. The English philosopher Francis Bacon reminds us that writing makes an *exact* person. The process of writing forces decisions about one's philosophical beliefs that otherwise might never be made. Prospective administrators are challenged to attempt this exercise. Building (or perfecting) a personal philosophy involves putting one's thoughts into a logical order beginning with the most basic items, including:

1. Where am I now, and where am I going?
2. What should I strive toward?
3. What are my goals, and how can I accomplish them?
4. Do I have responsibilities to others and to society? If so, what are they?
5. What are my beliefs (creed)? Are my daily actions consistent with them?

Answers to questions such as these will go a long way toward perfecting the fundamental aspects of your philosophy. The nature and strength of your responses to these questions should influence your daily actions and approach to life's challenges. Before beginning the process, some practical advice is in order:

1. Don't start wondering, "What do I believe, in general?" Be more precise and proceed with an orderly design. "What do I believe about . . . ?" is easier to answer.
2. Don't think that a written account of your philosophy has to be fancy or filled with "big words." It is possible to be complex and still express yourself in simple terms.
3. Don't believe that you have to agree with the "great minds," or the half-great minds, or your friends, or even your instructors. Think for yourself.
4. Don't write anything that is not sincere or truthful. In philosophy, loss of integrity is loss of everything.
5. Don't be ashamed of beliefs that may be linked to the idealistic, the beautiful, the artistic, or any other area that symbolizes a keen sensitivity and awareness of life, thought, and experience above the common and the ordinary.
6. Don't just skim the surface. Dig deep into your basic beliefs and concepts, and don't stop digging just because you hit a barrier. Barriers are made to be broken.
7. Finally, don't make hard work of the process. Let the words come easily. Begin by being open and expansive. Later, you can refine and condense your statements.

Values

The task of ferreting out what you believe with reference to your own specialized discipline—say, physical education—is not easy. This relates to even larger questions, such as the purposes of education in general, and those of life itself. For any particular individual, the purposes of life and education depend on a personal *system of values*.

If you wish to refine and perfect your system of values, you must ask the question, "Why?" Why have you selected a particular idea or held to a certain premise? If you have the fiber to persist in this line of questioning, you begin not only to think crisply but also to really understand. Moreover, because there are often good reasons for selecting any one of several alternatives, you begin to understand yourself better and become more tolerant, more understanding, and more respectful of others and their views.

A person's values shape life by influencing one's thoughts and actions in both personal and professional relationships. Indeed, the kind of person that an individual is as an employee, neighbor, friend, teacher, or administrator is strongly influenced by his or her values (or lack of values).

Some of our values change as we develop and expand our understanding. More experience, further education, and broader exposure may alter our values. In turn, our values affect our experiences and in fact determine which experiences we have by the choices we make.

It is said that values cannot be taught; but certainly teaching values is possible in the sense that values may be influenced by one's experiences, and one of the processes of education is to provide carefully selected experiences. However, it is also true that people do not consistently adopt values simply by being instructed to do so. The development of values results from a complex interaction between a person and his or her experiences.

It is, therefore, important for teachers, administrators, and others who influence educational programs to make value-based decisions in educational settings. Values influence such things as (1) one's objectives in education, (2) the selection of learning experiences and how they are arranged, (3) the relative emphasis on different subject matter, (4) the provision of facilities and other resources in support of the various areas of education, (5) teaching methods, (6) the administrative organization and procedures, and (7) the standards by which employees are to conduct themselves.

The development of values is a major function of education. As adults especially concerned with the development of young people, it is imperative that we understand the importance of providing positive models. Furthermore, as parents, teachers, and administrators, we have a responsibility to provide an abundance of developmental experiences for children and youth (as well as adults) that contribute to sound values.

Each of us has established a value system, however perfect or imperfect, which guides and directs us in all of our activities and influences our behavior. It forms the basis for our aspirations, dedications, and actions. To a large extent, our system of values dictates the meaning of life, forming the foundation of what we are and what we hope to become. It determines our most marvelous intentions and structures our actions.

However, we must recognize that our system of values is never complete or final. It is always more or less in a state of development and flux, due to new experiences and changing circumstances. In view of this, each of us must consistently try to keep our values on track, developing in a manner that will be for our good and the good of others whose lives we influence. For this reason, it is especially important for each of us to remain thoughtful and open-minded about new ideas and changing circumstances.

All that has been said about personal philosophy and values should be applied toward increasing one's authenticity. The authentic person is one who has a high degree of internal consistency in terms of an integrated philosophy that displays unity and coherence. Psychologist Abraham Maslow has said that authenticity is the reduction of phoniness and fuzziness to the zero point. Authentic people know where they stand, don't put on an act, and refuse to hide behind a facade. They project a sense of integrity and stability.

PHILOSOPHY APPLIED TO ADMINISTRATION

In education, the philosophies of those in leadership positions have significant influence on the nature of programs, their underlying purposes, established goals, and methods employed. The decisions leaders make are, in a sense, expressions of their philosophic beliefs about education and life. Therefore, it is of crucial importance for educational leaders to cultivate personal and professional philosophies that are conducive to the objectives and desired outcomes of the educational system.

Ideas about education in general, and physical education and athletics in particular, have developed from analysis of social and individual needs and from experiences with the total environment, including social, economic, religious, and political developments. Ideas originate as untested, unrefined, and unvalidated notions. These notions, along with opinions and intuitions, form general concepts that help formulate theories, principles, and laws, which in turn help to form the basis of our programs and our procedures. Unproved and untested ideas become verified through reasoning, constructive and critical thinking, analysis, evaluation, interpretation, and the testing of hypotheses. All of these are valuable methods that apply to the administrative process.

The philosophical methods described above help us to determine and apply useful aims, goals, and objectives to the educational mission. The philosophical elements of administration discussed in the following sections follow a deductive approach (i.e., from general to specific) beginning with a mission statement that addresses the purposes of the educational endeavor. Figure 1-1 provides an overview of the steps in this process.

Mission

Part of the responsibility of administrators is to articulate and implement the strategic intent and purposes of their organization. Open any college catalog, and you will find at the front a statement of its mission, which establishes what the school is statutorily charged with doing and what it is expected to achieve. An educational organization's mission statement identifies the purposes for which it was created, its boundaries, its activities, and its future direction. The school's mission represents the foundation for educational programs that will enhance the development of students. Indeed, the primary task in stating the mission is to identify how the school can affect the lives of stu-

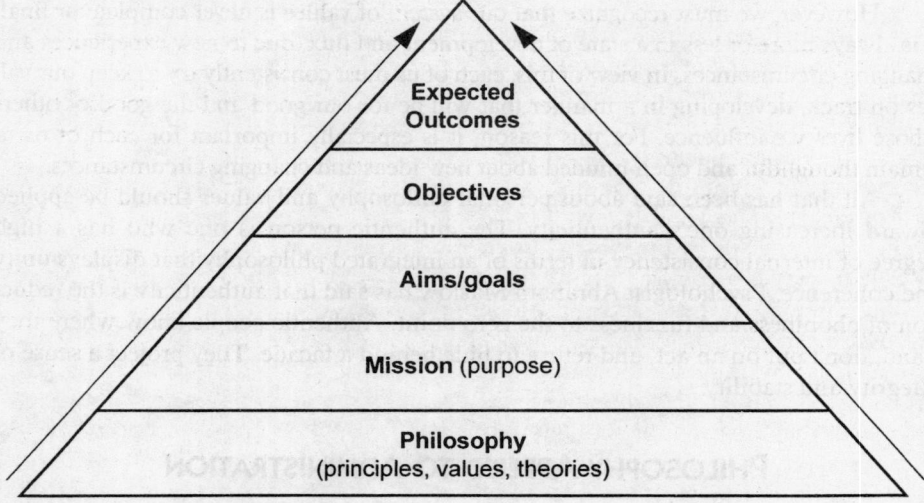

Figure 1–1. Hierarchy of philosophical steps for administrators

dents in order to bring about positive change. To complete this task, the mission must be translated into aims, goals, objectives, and outcomes (Castetter & Young, 2000).

Following is a mission statement for the professional preparation component of a department of health, physical education and recreation (Utah State University, 2000):

> The mission of The Department of Health, Physical Education and Recreation within the broader mission of Utah State University and The College of Education is: 1) To provide high quality education and training, leading to teacher certification in Health and Physical Education, and 2) To provide professional, managerial, and analytical skill development for potential HPER educators, leaders, and directors of community based agencies, institutions, businesses, and programs.

Aims

Aims are derivative of mission and roughly synonymous with *goals*, but they may be employed as a transition between these two concepts. An aim may be defined as a broad statement of purpose or a desired result. It provides a direction toward which to point, like a bull's-eye on an archery target. For administrators, the aim sets a remote level of achievement toward which efforts are directed. In the school setting, an aim is usually thought of as an inclusive goal stated in a manner that gives general direction to the broad spectrum of the educational program. Given that a statement of aim is similar to a mission statement, it often substitutes for the latter. Following is an example of a statement of aim for a high school department of physical education (Litcham High School, 2002):

> The Physical Education Department's primary aim is to equip all students with the knowledge, skills and enthusiasm to maintain a healthy lifestyle into adulthood— regardless of their physical abilities. All the activities covered are designed to develop skills, teach students to work as part of a team, or as individuals, and to reach their full potential. A further primary aim is to give an understanding of all roles involved, from playing a sport to officiating, coaching and spectating, and a sympathy for all others performing around them.

Goals

Goals are broad, remote statements that inform more specific *objectives*, which in turn shape *expected outcomes* (or results). Such a structure of intent gives definite purpose and direction to the program and to those involved with it. Having clearly stated goals that are both challenging and realistic is of utmost importance. Administration without them is similar to going on a trip with no plan in mind and, therefore, no determined route to follow. Socrates expressed it well when he said, "For a ship without a destination, no wind is favorable." Organizational success clearly depends on having coherent goals and mapping out an exact plan to reach them.

Goals may be expressed from different points of view. For example, goals are set by administrators, teachers, students, and those who sponsor the educational program.

There can be goals for the educational program as a whole and goals for specific areas such as physical education and athletics. In addition, each individual has some personal goals that relate to the situation and should be considered in the total plan.

Goals that are carefully thought through and properly stated imply a strong element of stability. They should not be thought of, however, as static and unchanging. As circumstances change, it becomes necessary to revise goals. One of the challenges of educational leaders is to maintain the right balance between keeping goals current with changing circumstances and maintaining sufficient stability and continuity.

Goals should be stated in terminology that is understood by those involved. They should be clear, concise, and meaningful. Goals should express concepts to which people relate, which motivate them and cause them to aspire to achieve. They should be measurable, and the time frame for their accomplishment should be stated. Each goal should include the implied *action* and expected *results*. It is interesting that sometimes the process and the goal are one and the same. This is largely true of teachers who carry on the process of effective teaching—a worthy goal as well as a process.

Developing worthy goals is a complex undertaking; but when the goals are formulated properly at each level of administration and in connection with each aspect of the program, the process of education is greatly simplified (Swanson, 1979).

Objectives

Objectives are more precise statements identifying what is to be accomplished, by whom, and by when. Educational objectives are specific goals for which one strives in the attainment of knowledge, skills, and appreciations. The achievement of specific objectives fulfills the aim or goal, which is more general and remote. Recalling the archery metaphor, the specific objectives to accomplish the aim of hitting the bull's-eye are: notching the arrow, drawing the bow, sighting the target, and releasing the arrow. Thus, objectives are steps toward achieving the aim.

Following are some examples of objectives relating to the school physical education program:

1. Students completing physical education will acquire a range of motor skills that promote minimal success in a variety of individual and team sports.
2. Students completing physical education will demonstrate a desire to engage in physical recreation during leisure time.
3. Student completing physical education will appreciate dance performance.
4. Students completing physical education will attain a level of body fat indicative of health-related fitness.

Over the past several decades, a trend has emerged to translate general objectives into performance or behavior objectives, which state expected learning outcomes in readily observable terms.

Expected Outcomes

Expected outcomes (often used interchangeably with *performance objectives*) are statements of anticipated results. They serve as short-term ends on which to focus, and they provide the evidence of attained objectives. An expected outcome in a beginning swimming class might be for each student to "swim 100 yards nonstop by the end of the semester." For a basketball team, an expected outcome might be to achieve "shooting 40% or better from the field during the season." A track athlete might have an expected outcome of "running the mile in 4 minutes 30 seconds by midseason." For an administrator, some expected outcomes might be (1) to transfer student records to electronic files by the end of the school year, (2) to assure that every professor in the school adheres to the prescribed schedule and guidelines for conducting final examinations, and (3) to get teachers to post their schedules indicating the hours available to meet with students and adhere to the posted schedule.

Outcomes imply a minimally acceptable level of performance and an agreed-upon basis for determining the performance achievement. These requirements presuppose an application of standards and criteria.

Standards

The term standard refers to a defined level of performance against which other performances or achievements are compared. Standards may be quantitative or qualitative. For example, it might be determined that ninth-grade boys should be able to run the 100-meter dash in 14.2 seconds in order to receive a rating of good (In other words, 14.2 seconds is the standard for a good rating). Some universities hold to the rule that a student must meet a standard of a 3.0 GPA (grade point average) in order to qualify for graduate school. Each teacher determines certain standards or levels of performance a student must meet in order to earn a particular grade. Standards evolve from criteria, and they serve as guides toward assessment and achievement. They help to evaluate the success of students, teachers, and curriculum content. In educational programs, reasonable and useful standards should be applied for the benefit of the students and the programs.

Criteria

Criteria form the basis for judgment, and they contribute to establishing standards. As with standards, criteria may be either quantitative or qualitative. The criteria for winning a high-dive contest are largely qualitative. The contestants must meet predetermined aesthetic standards applied to diving form. However, these criteria are routinely quantified with a rating system. The quantitative criteria for being admitted into a selective college include high school grades and test scores. These numbers are meant to represent quality of learning. Thus, a defined set of criteria and standards form the basis for making various evaluations.

The high jump can be used to clarify the difference between the concepts of standard and criterion. The criterion for judging a successful jump is clearing the crossbar. The standard is represented by how high the bar is set. If the bar is set at five feet and the jumper clears it, the standard has been met based on the criterion employed. Validity and reliability are essential characteristics of criteria. Without these characteristics, criteria would be misleading and useless. (For more information about standards, criteria, validity and reliability, refer to chapter 11.)

The Role of Principles and Theories

Principles

A principle may be defined as a fundamental truth, a uniform result, or a universal guide. Principles reflect values and the state of knowledge. They form the basis for defining methods and procedures; they guide conduct and behaviors. Principles can serve as cornerstones for administrative function or inform a particular body of knowledge. Principles are used in a variety of ways in education. For example, some principles give direction to the overall educational effort, such as the *seven cardinal principles* stated by the U.S. Commission of Education in 1918 (see chapter 13). For many years, these principles served as a guide for the public education programs in the United States. They are still referenced, as they are still basically sound. Certain physiological and mechanical principles (or laws) apply in athletics conditioning and performance. Examples are the *law of use* as it applies to the human organism (with proper use the organism improves, while in the absence of use it deteriorates), and the *principles of leverage* applied to movement of body parts and the application of external objects.

Principles of administration have been formulated to guide administrators in making appropriate and effective decisions (discussed further in chapter 2). Following are representative principles:

1. The assignment of responsibility should be accompanied by the delegation of authority to correspond with the responsibility.

2. Division of labor creates more productivity.

3. The span of control of an administrator should not exceed the ability to supervise employees adequately.

4. An employee should receive orders from one superior only.

5. Each division of an organization has a clearly defined sphere of duties, purposes, and required skills.

The profession of education also formulates principles of ethical conduct. For example, teachers and coaches should avoid intimate relationships with individuals over whom they act in an evaluative or supervisory capacity.

Theories

A theory is a general proposition useful in explaining phenomena in a particular field of study—like Isaac Newton's theory of gravity in the field of physics. There are theories of education and administration as well. Theories are particularly useful in the formulation of administrative practices. A theory shapes policies and procedure after sufficient evidence has been accumulated to support it. Administrative theory precedes and informs administrative practice. In other cases, theories become discarded because attempts to prove their validity and usefulness are unsuccessful.

Administrative theories are too numerous to discuss comprehensively. An example of a theory is presented below in brief form to help the reader understand the place of theories in the administrative process. Theories may be value laden and reflect philosophical beliefs. They can be viewed as a starting point for the development of models of behavior. The following theory offers two opposing views of human nature.

Douglas McGregor's theory of management has endured over time. It instructs the policies and practices of contemporary human-relations management in the workplace. Influenced by the humanistic psychology of Abraham Maslow, McGregor argued that managers must shift their view of human nature from a perspective which he called Theory X to one labeled Theory Y.

Theory X views subordinates by nature as:

*   disliking work,
*   lacking ambition,
*   irresponsible,
*   resistant to change, and
*   preferring to be led rather than to lead.

Theory Y, in contrast, views subordinates as naturally:

*   willing to work,
*   willing to accept responsibility,
*   capable of self-direction and self-control, and
*   capable of creativity.

McGregor was one of the first theorists to argue for a situational approach to management. He described the central principle of his theory as creating conditions such that the members of an organization can direct their personal goals toward the success of the organization (Starling, 1998).

Theories that endure over time mature and evolve. McGregor's theory has been built upon and synthesized by later theories, notably William Ouchi's model of Japanese-style collaborative decision making known as Theory Z.

Now that the reader has acquired a sense of what is implied by the general process of administration, we can move to a discussion of the structure within which administrative actions are carried out: the organization. As we move through subsequent chapters, the reader will be introduced to specific aspects of management, and finally to the object of these efforts: the programs in physical education, athletics, and intramural recreation.

References and Recommended Readings

Castetter, William, & I. Phillip Young. 2000. *The Human Resource Function in Educational Administration*. Columbus, OH: Merrill.

Hartvigsen, M. F. 1992. Unpublished papers on establishing a personal philosophy. College of Physical Education. Provo, UT: Brigham Young University.

Horine, Larry. 1999. *Administration of Physical Education and Sport Programs*. 4th ed. New York: McGraw-Hill.

Litcham High School. Physical Education Department. 2002. "Sports Education." Norfolk, England (UK): Authors. Online: <http://atschool.eduweb.co.uk/litchs/Sport.html>.

O'Connor, Tom. 2001. "A Brief Overview of Administration and What Administrators Do." *Distance Education in Criminal Justice*. Online article: <http://faculty.ncwc.edu/toconnor/417/417lect01.htm>.

Starling, Grover. 1998. *Managing the Public Sector*. Philadelphia: Harcourt Brace, pp. 373–75.

Swanson, J. 1979. "Developing and Implementing Objectives in Physical Education." *Journal of Physical Education and Recreation*, 50(3): 68.

Utah State University. Department of HPER. 2000. *Mission Statement*. Logan, UT: Authors. Online: <http://www.coe.usu.edu/hper/Mission.html>.

Weiss, Paul. 1971. *Sport: A Philosophic Inquiry*. Carbondale, IL: Southern Illinois University Press.

Ziegler, Earle. 1980. "Philosophical Perspective. *Journal of Physical Education and Recreation*, 51(11): 40.

Online Sources

Philosophy of Education Society
http://www.ed.uiuc.edu/EPS/PES-yearbook/
The Society maintains a search engine that allows directed access to scholarly papers on aims and objectives of education, personal philosophy, liberal education in a free society, and other pertinent issues.

2 Understanding Organizations

Throughout history, organizations have been created to help people accomplish collectively what they could not achieve individually. The bureaucratic structures of civil administration are traceable to the Han Dynasty in China, beginning in the second century B.C.E. The need to organize is apparent in the Old Testament account of Moses, who was confronted by numerous obstacles in leading his people out of Egypt. Contributing to the problems was his lack of organization; he had not established a hierarchy of authority and responsibilities. Moses' father-in-law, Jethro, recognized this lack of organization and offered the following advice to Moses:

> Hearken now unto my voice, I will give thee counsel . . . thou shalt provide out of the people able men . . . and place such over them (the people), to be rulers of thousands and rulers of hundreds, rulers of fifties and rulers of tens. . . . (Exodus 18:17–21)

Essentially Jethro was saying, "We must get organized."

Contemporary leaders are still struggling with organizational problems. Lack of organization contributes to friction and poor relationships, diminishes confidence, causes poor utilization of organizational resources, and ultimately generates chaos. Conversely, good organization creates order out of chaos, thus facilitating achievement of the organization's goals. In addition, good organization builds confidence and improves morale.

An organization can be broadly defined as "an arena where human beings come together to perform complex tasks so as to fulfill common goals" (Narayanan & Nath, 1993). In order to accomplish this mission, organizations must: (1) establish a structure of roles through determination of the activities, (2) group these activities, (3) assign each group of activities to a manager, (4) delegate the authority to carry out the activities, and (5) provide for coordination of authority and information horizontally and vertically within the organizational structure (Koontz & O'Donnell, 1986).

Organizations are not difficult to understand. Indeed, all of us have had personal experiences with them. As sociologist Amitai Etzioni (1964) pointed out, we are born in organizations, educated in organizations, work in organizations, and play

21

in organizations. This observation suggests the diversity in types of organizations—from hospitals to soccer leagues. Yet, all organizations share common features. Some elements of organization can be explained with diagrams (organization charts), whereas others are better explained by defining responsibilities associated with positions and accountability within the organization. In order to help the reader better understand organizations and organizational theory, this chapter is separated into two major sections: an explanation of organizational structures (with a brief discussion of organizational context), and descriptions of selected administrative positions in educational organizations.

ORGANIZATIONAL STRUCTURE

Organizational structure specifically refers to such elements as the number of departments, the span of control, and the extent to which the organization is centralized or decentralized. Principles of organizational structure have developed over time based on theory and practice. Hitt (1984) has provided ten useful principles to enhance the structure and functions of an organization:

1. The key activities necessary in order to accomplish the organization's goals must be clearly defined.
2. These activities should be grouped on some logical basis.
3. The responsibilities of each division, department, unit, and job should be clearly defined.
4. Authority should be delegated as far down in the organization as practicable.
5. Responsibilities always should be accompanied by an equal amount of authority.
6. The number of persons reporting to each manager should be of a reasonable size to accommodate an appropriate span of control.
7. The organization should be designed to provide the right mixture of stability and flexibility, thus making it a well-structured, fluid enterprise.
8. The organization should be designed for perpetuation and self-renewal.
9. The structure should be evaluated on the basis of its contribution to the organizational goals.
10. The leaders of the organization should be responsive and able to accommodate new needs and changing circumstances.

A general principle is that organizational structure follows function, as Drucker (1964) has noted. This means that the functional aspect of an enterprise should determine the structural characteristics. This is a reasonable assertion, because the only purpose of structure is to accommodate and enhance function. Thus, organization is a means to an end rather than an end itself. Sound structure is a prerequisite to a healthy organization, but it is not health per se. The test of a healthy enterprise is not the beauty, clarity, or perfection of its structure. It is the functionality of the organization.

Organization Charts

Most organizations utilize organization charts to represent and clarify how they are structured. Charts assist executives and managers in conceptualizing the organization as a whole, and they illustrate the levels in the hierarchy and the width of the span of control. They also can reveal structural faults such as duplication of effort. Charts have their limitations, however. Although they provide a picture of formal relationships, they do not depict the informal relationships that always exist in organizations (Jackson, 1987).

Organization charts employ certain conventions. *Rectangular boxes* indicate positions/people in an organization. The lines joining the boxes demonstrate the relationships between the positions. Line structures expand vertically through delegation of authority and horizontally through division of labor. In traditional chart rubrics, we refer to the verticals as *lines* and the horizontals as *staffs* (hence "line and staff chart"). The vertical lines indicate the *chain of command* (see figure 2-1) from the superior to the subordinates.

The principle of authority states that the chain of command should be clear to everyone, and that each subordinate reports to only one superior. (This latter provision is violated when a teacher/coach reports both to the athletics director and to the chair of the teaching department.)

When too many people report to one superior to be given adequate attention, subordinate manager positions are created, to which staff report in order to reduce the *span of control* (Jackson, 1987). Figure 2-2 illustrates the use of subordinate managers when head football coaches have too many assistant coaches reporting to them. In this situation they create the submanagerial positions of offensive and defensive coordinators. Recall that a horizontal staff on the chart indicates *division of labor* (also referred to as "job specialization"), which is defined as the degree to which overall tasks of the organization are broken down and divided into smaller component parts. Using the example in figure 2-2, distinct responsibilities under a defensive coordinator might be assigned to a backfield coach, a linebacker coach, and a defensive line coach. Division of labor is one of two opposing forces that operate within organizations. The other is *coordination*, referring to the process of linking the activities of the various divisions of the organization. These two opposing forces must be balanced so that together they contribute to the overall efficiency of the organization (Griffin, 1996). Coordination in the above example would imply—at the minimum—that the offensive coordinator knows what the defensive coordinator is doing, and vice versa. This requires adequate communication.

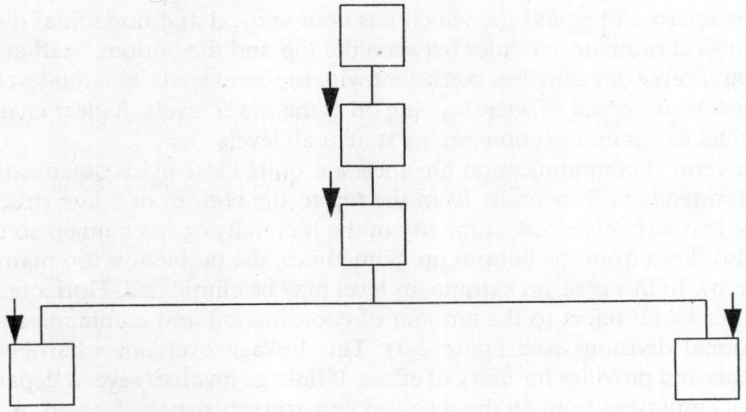

Figure 2–1.   Line and staff chain of command

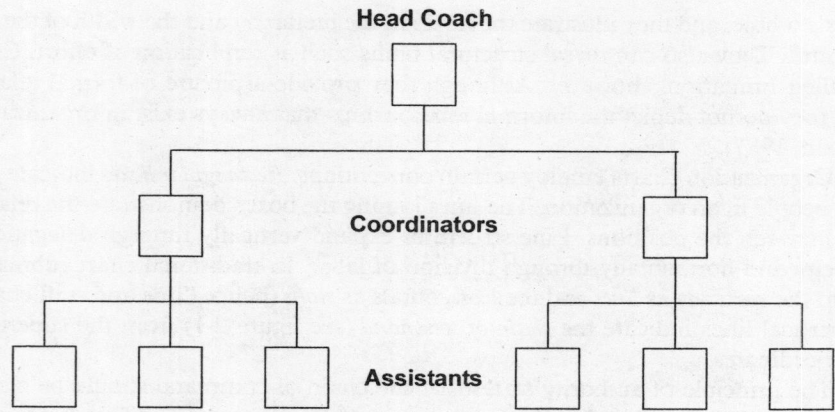

Figure 2–2. **Organization chart of a coaching staff with subordinate managers (coordinators)**

Coordination and Communication

Organizational structure as a command-and-control hierarchy is a legacy of the church and military bureaucracies of the preindustrial age. Traditional bureaucracies functioned like stovepipes—information flowed vertically from top to bottom. No longer can organizations operate on a "sell-down" model where management's vision is communicated to subordinates; organizations need a shared vision in order to function effectively. In the Information Age, administrators responsible for managing knowledge workers ("anyone who works for a living at the tasks of developing or using knowledge" [Drucker, 1964]) know that they cannot be managed in the traditional sense. Often they have specialized skills that are not understood by management. Moreover, with the advent of the Internet and "intranets" (LANs, or local access networks), more information flows horizontally within organizations. These developments have implications for communication and coordination within modern organizations (Tapscott, 1996).

The extent of coordination and communication among the elements in an organization is referred to as *linkage*, which has both vertical and horizontal dimensions. Vertical links coordinate activities between the top and the bottom. Staff at the lower levels should carry out activities consistent with top-level goals, and top-level administrators must be informed of what is going on at the lower levels. A clear chain of command facilitates communication among staff at all levels.

The vertical communication channels are quite clear in an organization chart. Information tends to flow easily from the top to the bottom of a line structure. It is important that administrators at the top of the hierarchy open channels so that information also flows from the bottom up. Sometimes, the problem is too many levels in the hierarchy. In this case, an extraneous level may be eliminated. Horizontal linkage, on the other hand, refers to the amount of coordination and communication across organizational divisions (see figure 2-3). This linkage overcomes barriers between departments and provides for unity of effort. If linkage involves several departments, a *task force* of employees from all the involved departments may be formed. A task force constitutes a temporary committee or team in which each member represents his or

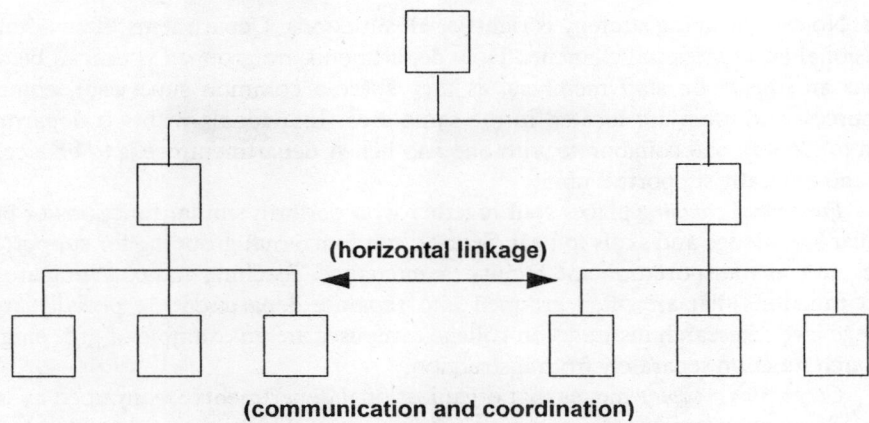

(horizontal linkage)

(communication and coordination)

Figure 2–3. Organization chart illustrating horizontal linkage

her department. This strategy often is used with short-term issues or problems. For example, task forces often are formed in schools to address accreditation of programs. The task force is terminated when the task is completed (Daft, 1998).

Responsibilities of individuals on a task force often transcend their job descriptions. Bridges (1994) has argued that "job positions" as we know them are an artifact of the Industrial Age. In the current postindustrial era, as schools move away from mass learning methods (the teaching paradigm) to more technology-based individualized instruction (the learning paradigm), teachers' work will become "de-jobbed." The de-jobbed instructor is more likely to move through temporary work situations, where work is defined by a particular project. In the future, more work will be shared and accomplished by teams.

Sharpe and Templin (1997) define the *team approach* as administrators and teachers: (1) sharing leadership roles, (2) being held mutually accountable, (3) providing collective work "products," and (4) collaborating in problem solving to meet the organization's objectives. They offer the example of a school where the principal designed teams so that every teacher in the school participated on a team. Each team consisted of four teachers from different subject-matter areas plus the principal, a liaison from the school district, and a teacher educator from a university. At regularly scheduled monthly meetings, the team decided on shared goals for their group, and on a timeline. In addition, all the teams in the school met together to determine which goals would be shared across teams. They addressed such issues as discipline challenges, teaching strategies, and curricular or policy disagreements. Increasingly, educational institutions are utilizing the team approach along with other "postbureaucratic" strategies within existing organizational hierarchies (discussed in the following section).

Organizational Design Alternatives

Organizational design indicates three things: needed work activities (e.g., teaching/coaching), reporting relationships (discussed above), and departmental groupings. Options for departmental groupings include *functional* grouping, *divisional* grouping, and *geographic* grouping. School districts and colleges utilize all three groupings. No one grouping strategy is right for all situations. Departments often combine divisional and functional elements. How departments are grouped is crucial because it has an impact on staff members, as they share a common supervisor, common resources, and often are located in the same area. Individuals within a department tend to identify and collaborate with one another. A department needs to be a coherent and mutually supportive unit.

Functional grouping places staff together who perform similar functions or bring similar knowledge and skills to bear. Schools use functional grouping for support services such as transportation and facility maintenance. Teaching and coaching are distinct functions that are often grouped into separate departments, especially at the college level. Research institutes on college campuses are an example of grouping the research function separately from instruction.

Geographic grouping means that organizational departments are grouped by location. Geographic grouping is common within school districts using site-based management at the elementary, middle, and high school levels. Colleges tend to segregate departments by location and establish branch campuses in strategic areas to serve their clientele more efficiently.

Divisional grouping means that individuals are organized according to what service they perform or product they produce. The academic department has been the traditional organizational unit in schools and colleges, organized around common and similar academic areas (e.g., Department of Kinesiology; Department of Health, Physical Education & Recreation) and generally provided with its own budget. The number of faculty in a department can range from as few as five or six individuals to three or four dozen.

The *matrix structure* (see figure 2-4) is utilized when an organization finds that none of the above groupings work adequately. This alternative structure features a strong form of horizontal linkage. In doing so, it creates a dual hierarchy that seems to violate a basic principle of administration; however, this approach will work when basic conditions are met. The vertical and horizontal lines of authority are given equal recognition and some shared responsibilities, while other functions remain distinct. This structure requires administrators who are comfortable with sharing authority and responsibilities. Figure 2-4 is a matrix structure utilized within a college (or "school") of HPER, which offers graduate and undergraduate professional preparation programs as well as service courses within the general education requirement.

The dean of the college serves as the top leader. The vertical hierarchy is represented by the chairs of the three departments: Physical Education, Health Education, and Recreation Administration. The horizontal hierarchy is represented by the program directors of the graduate program, the undergraduate professional preparation program, and the general education courses (e.g., activity classes, freshman health classes) that serve the general education requirement. The department heads' responsibilities pertain to functional expertise such as regulations and teaching standards. The program directors are responsible for coordinating programs and have authority over subordinates for such activities as class scheduling, course content, and exams. For this structure to work, the principal leaders involved should have good conflict-resolution skills and be willing to spend much time in meetings (Daft, 1998).

Postbureaucratic Structures

Bureaucracies are characterized by complex hierarchical structures, standardized procedures and regulations, strictly defined job positions, and a resistance to change.

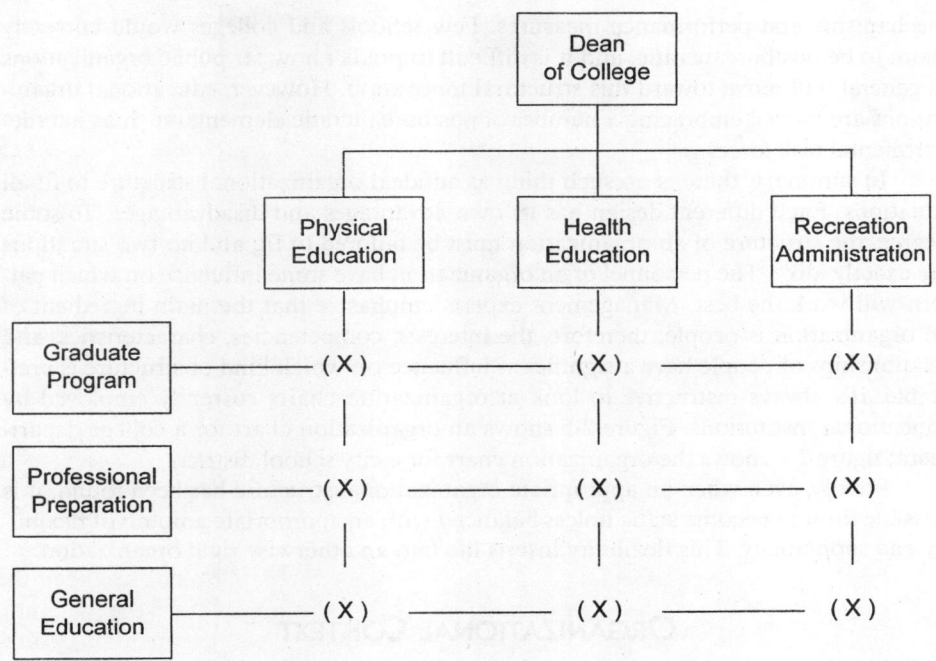

Figure 2–4. A basic organization chart for a college of HPER, utilizing a matrix structure

When organizations take on bureaucratic characteristics, they often are subject to criticism for delays, red tape, pettiness, ritualistic attachment to rules, concentration of power, and lack of responsiveness to clients. Individuals who work in such organizations are referred to pejoratively as "bureaucrats" (Chandler & Plano, 1988). Bureaucracies are not without merit, however. Complex organizational structures are necessary to carry out many of the functions and services we take for granted. The U.S. mail could not be delivered efficiently without a government bureaucracy. The quadrennial staging of the Olympic Games could not occur without a complex sports bureaucracy, the International Olympic Committee.

However, many organizations are moving away from traditional bureaucratic structures. They are experimenting with management systems that rely more on employee work groups to carry out innovations and solve problems. This new organizational model has been referred to as *postbureaucratic*. Many of the changes have come out of human-relations theory. The perception is that "good people get trapped in bad systems." Accordingly, the system is altered to create employee-powered work environments. Organizational structure is "flatter" (less hierarchical) and depends less on strict job descriptions and classifications. Regulations give way to general principles and guidelines. Managers have greater flexibility in deploying people. Mission, vision, and values are more important than following specific job tasks. Employees are encouraged to innovate and are allowed the freedom to fail without being punished for taking reasonable risks (Leavit & Johnson, 1998).

Not all employees are comfortable with these changes in organizational structure. Postbureaucratic elements should be gradually introduced with built-in feedback mechanisms and performance measures. Few schools and colleges would currently claim to be postbureaucratic, and it is difficult to predict how far public organizations in general will move toward this structural innovation. However, educational organizations are indeed embracing a number of postbureaucratic elements, such as interdepartmental task forces.

In summary, there is no such thing as an ideal organizational structure to fit all situations. Each different design has its own advantages and disadvantages. To some degree, the structure of an organization must be tailored to fit, and no two situations are exactly alike. The personnel of an organization have some influence on which pattern will work the best. Management experts emphasize that the main ingredient of an organization is people; therefore, the interests, competencies, characteristics, and relationships of people have a significant influence on which kind of structure is preferable. It's always instructive to look at organization charts currently employed by educational institutions. Figure 2-5 shows an organization chart for a college department; figure 2-6 shows the organization chart for a city school district.

Finally, even when an appropriate organizational structure has been found, it is possible for it to become static unless balanced with an appropriate amount of flexibility and spontaneity. This flexibility inserts life into an otherwise rigid organization.

ORGANIZATIONAL CONTEXT

Organizations have contextual dimensions as well as structural dimensions. *Contextual dimensions* refer to the organization as a whole, with its internal and external environments. At least five contextual dimensions are worth noting (Daft, 1998):

Size. Because organizations are social systems, the number of employees typically determines their size. Other measures in educational settings include the number of students or the quantity of assets. Schools range in size from a hundred students in a small elementary school to 50,000 students on a single college campus. Some school systems are so small that all the employees know each other. Others are so large that two individuals can work for the same school district or college and never cross paths.

Organizational technology. This refers to the techniques that change organizational input into output. In schools, input includes students, money, and instructional resources. The output is advancement of knowledge and educated human beings. Input is changed into output through techniques employed in various settings, whether a "smart" classroom, a library, a computer lab, or a practice field.

External environment. This includes all the elements outside the boundary of the organization. Schools must adjust to the economic environment, the political environment, and local demographics. Some schools are caught up in budget crunches, others in struggles over racial integration. Others are affected by violence and poverty in the communities they serve. Schools constitute *open systems* that must interact with and respond to their environment in order to carry out their mission.

Goals and strategy. Goals and strategy set organizations apart from their competitors. A strategy is a plan of action taken in order to achieve goals. Schools must distinguish between appropriate and inappropriate goals, as they are called upon to address a variety of social problems. Likewise, schools must adopt a strategy that is both appropriate and efficient. Clearly, public schools have goals and strategies distinct from other educational institutions such as online universities or private academies.

Department of Exercise Science, Sport, Physical Education, and Recreation

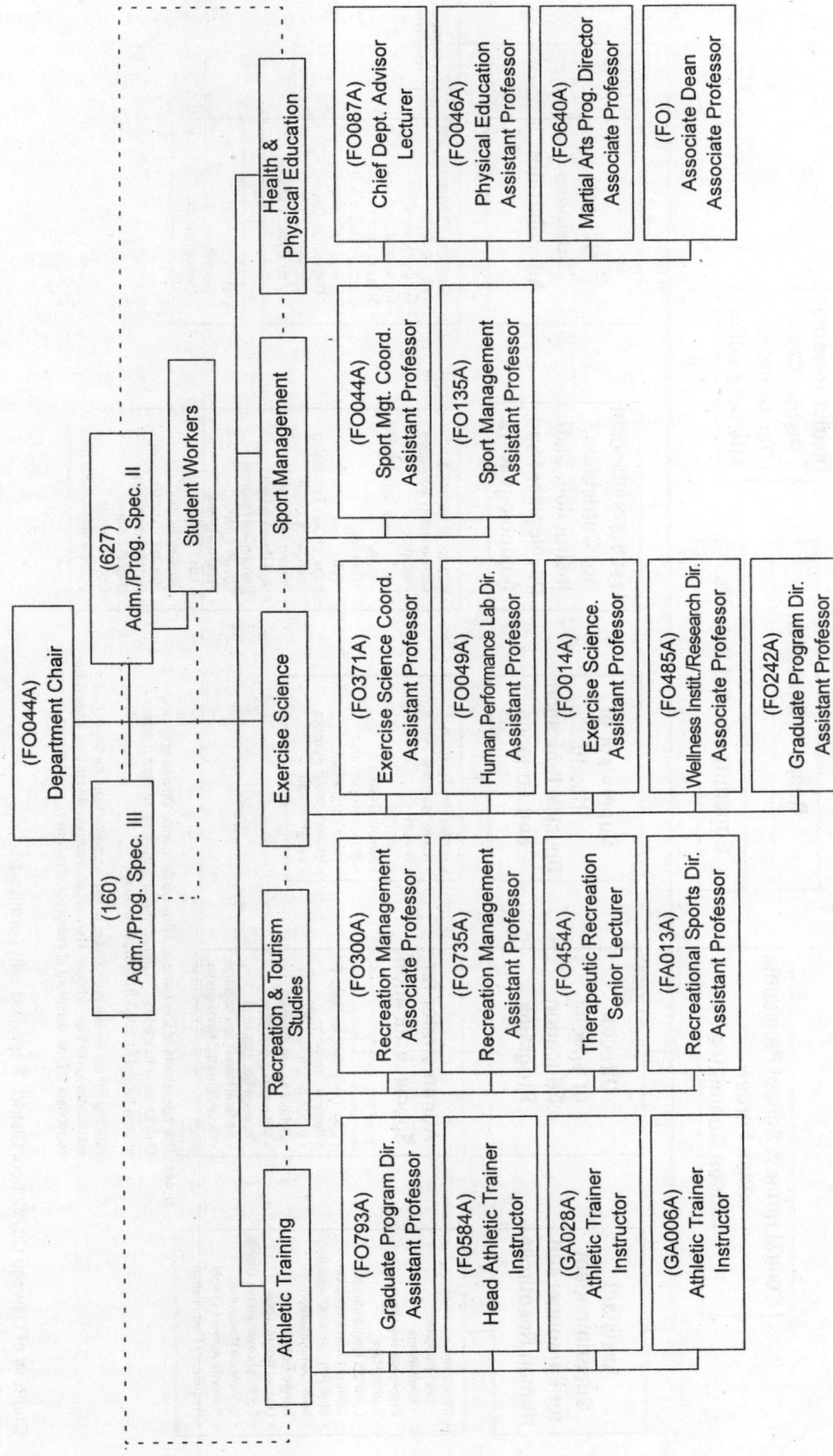

Courtesy of Department of ESPER, Old Dominion University. Reprinted with permission.

Figure 2–5. Organization chart for a college department

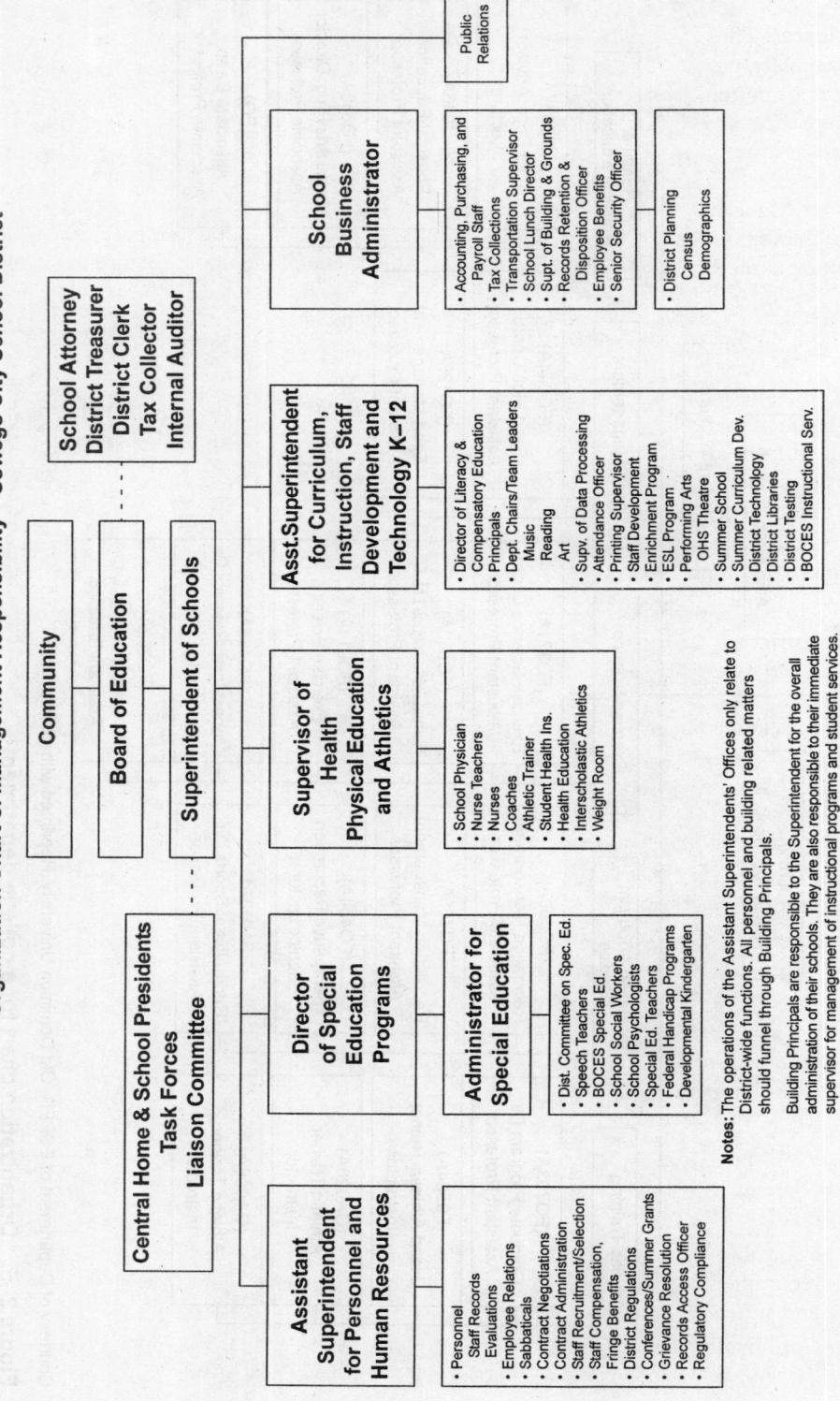

Organization Chart of Management Responsibility—Oswego City School District

Community

Board of Education

Superintendent of Schools

School Attorney
District Treasurer
District Clerk
Tax Collector
Internal Auditor

Central Home & School Presidents
Task Forces
Liaison Committee

Assistant Superintendent for Personnel and Human Resources

· Personnel
  Staff Records
  Evaluations
· Employee Relations
· Sabbaticals
· Contract Negotiations
· Contract Administration
· Staff Recruitment/Selection
· Staff Compensation,
  Fringe Benefits
· District Regulations
· Grievance Resolution
· Conferences/Summer Grants
· Records Access Officer
· Regulatory Compliance

Director of Special Education Programs

Administrator for Special Education

· Dist. Committee on Spec. Ed.
· Speech Teachers
· BOCES Special Ed.
· School Social Workers
· School Psychologists
· Special Ed. Teachers
· Federal Handicap Programs
· Developmental Kindergarten

Supervisor of Health Physical Education and Athletics

· School Physician
· Nurse Teachers
· Nurses
· Coaches
· Athletic Trainer
· Student Health Ins.
· Health Education
· Interscholastic Athletics
· Weight Room

Asst. Superintendent for Curriculum, Instruction, Staff Development and Technology K–12

· Director of Literacy &
  Compensatory Education
· Principals
· Dept. Chairs/Team Leaders
  Music
  Reading
  Art
· Supv. of Data Processing
· Attendance Officer
· Printing Supervisor
· Staff Development
· Enrichment Program
· ESL Program
· Performing Arts
  OHS Theatre
· Summer School
· Summer Curriculum Dev.
· District Technology
· District Libraries
· District Testing
· BOCES Instructional Serv.

School Business Administrator

· Accounting, Purchasing, and
  Payroll Staff
· Tax Collections
· Transportation Supervisor
· School Lunch Director
· Supt. of Building & Grounds
· Records Retention &
  Disposition Officer
· Employee Benefits
· Senior Security Officer

· District Planning
  Census
  Demographics

Public Relations

Notes: The operations of the Assistant Superintendents' Offices only relate to District-wide functions. All personnel and building related matters should funnel through Building Principals.

Building Principals are responsible to the Superintendent for the overall administration of their schools. They are also responsible to their immediate supervisor for management of instructional programs and student services.

Courtesy of Oswego City School District. Reprinted with permission.

Figure 2–6. Organization chart for a city school district

Organizational culture. This refers to underlying values, beliefs, and norms shared by employees. These translate into ethical behavior, commitment, and efficiency, and they ultimately provide the "glue" that holds the organization together. Often the indicators of culture are unwritten but are discernable in anecdotes, slogans, dress, and ceremonies. We would expect a military academy and a magnet school for the arts to have distinct organizational cultures, although their organizational structures might have similar features.

Contextual dimensions provide a basis for analysis that cannot be seen by the casual observer. Although they are not readily apparent on an organization chart, they require careful consideration when designing and administering an organization.

ADMINISTRATIVE POSITIONS IN ORGANIZATIONS

The most crucial features of an organization are its people, their assigned duties, and their functional relationships. Much insight into organizational relationships can be obtained from knowing the basic responsibilities associated with each administrative position. Following are descriptions of selected administrative positions in schools and colleges. (See chapters 6 and 15 for further discussion of job descriptions.)

Superintendent

Typically, local school systems are known as districts, and all of the schools within a district are managed under the administrative leadership of a superintendent. In general, the superintendent's responsibilities are to give direction to the district educational system in accordance with state laws, regulations, and guidelines, and with the directions and counsel provided by the district school board. The board is the local governing authority to which the superintendent is responsible. The superintendent, as the chief administrator for the school district, manages the educational system in accordance with the actions and wishes of the school board.

A typical school superintendent performs the following specific functions:

1. Prepares the agenda for meetings of the school board and meets with the school board.

2. Finalizes the hiring, fosters professional development, and provides for the evaluation of all school district employees.

3. Provides leadership to the board of education and the professional staff in defining, clarifying, and modifying educational goals and programs consistent with sound educational practices, democratic ideals, and community expectations.

4. Plans for, manages, and evaluates negotiations with organized employee groups in the district.

5. Maintains and strives to improve communications within the school district and between the school district and the community.

6. Plans and supervises the budget development process to ensure that budget recommendations made to the board reflect prudent planning, efficiency, and sufficient budgetary support of district programs and priorities.

7. Continually assesses the general climate and conditions in the schools through informal visits in schools and in classrooms. This includes formal

and informal conversations with the administrators, teachers, and students; regular administrative staff meetings; meetings with the officers of the district education association; and open-agenda meetings with the school faculties.

8. Investigates concerns raised by district residents directly or through members of the board of education.

9. Carries out other responsibilities required by law, contractual agreements, and board policies.

A school superintendent should be a well-prepared educational administrator and not a political appointee. The professional requirements of the position vary considerably among school districts. Some districts are small and have meager educational resources, while other districts have large and complex school systems. It has become common for the superintendent to hold a doctorate degree as well as being a highly regarded and reputable educator.

The professional staff of the superintendent's office often includes associate or assistant superintendents, with one in charge of each of the levels of education—elementary, middle school, and senior high school. The staff also includes supervisors or specialists in curricular and extracurricular areas such as physical education, music, art, and athletics (see figure 2-6). A business manager and legal counsel are also important staff positions. The kind of staff and its size vary with the complexity of the school district. Some very small districts amount to little more than the superintendent and a secretary, whereas others consist of a large and highly specialized staff.

School Principal

The principal is the chief administrator of a particular school and the one directly responsible to the school district. In general, the principal's responsibilities are to provide effective administrative leadership for all aspects of the school program. Typically, this involves the following:

1. Participation in the selection and hiring of teachers and other personnel in the school.

2. The assignment of teacher loads and the utilization of facilities in connection with the teachers' assignments.

3. Supervision of curriculum development and management of the curriculum for effective results.

4. Administrative leadership of noncurricular activities and programs.

5. Definition of the role of student government in the school and its administrative supervision.

6. Supervision of the school's teaching process.

7. Budgetary planning and management in connection with all aspects of the school.

8. Management of the community-relations program within the broad framework of the district public-relations effort.

9. Professional development and enrichment experiences of school personnel.

10. Administrative leadership toward desirable student values, behavior, and discipline.

11. Supervision of the school's planning function, both short-term and long-term, with respect to goals, faculty, facilities, and programs.

The requirements of a principal's job also vary with the size of the school. The job of principal of a school of 200 would be much less demanding than that of a school of 2,000, even though the two jobs would include the same elements. The principal is responsible to the superintendent or, in some cases, to an associate or assistant superintendent, as designated.

College Dean

Typically, a university is divided into several colleges (or schools), with each encompassing several departments related in subject matter. The dean is the chief administrator of a college or school. The heads of the several departments report to the dean, and the dean, in turn, reports to a provost or vice president. Following are the basic administrative responsibilities required of a dean:

1. Coordinate and facilitate personnel matters such as:
   a) selecting and appointing new faculty members;
   b) determining faculty and staff salaries;
   c) dealing with rank advancements, leaves of absence, terminations, and retirements; helping determine faculty assignments and evaluating faculty loads and effectiveness;
   d) receiving and evaluating student and faculty suggestions and complaints;
   e) supervising and working effectively with the heads of the several departments.
2. Oversee the preparation and management of college and departmental budgets.
3. Review and act on proposals for program and curriculum changes.
4. Review and act on department heads' recommendations relative to classified personnel, part-time faculty, and student assistants.
5. Appoint and supervise college committees as needed.
6. Serve as the liaison between the college and other colleges and schools and between the college and the central administration.
7. Serve as the focal point of relationships between the college and nonuniversity organizations and agencies.
8. Supervise nondepartmental functions of the college such as centers, laboratories, and institutes.

Associate or Assistant Dean

A dean may allocate a variety of responsibilities to associate or assistant deans. Among the more logical responsibilities to delegate are:

1. Reviewing and evaluating faculty assignments and loads.
2. Supervising assigned college committees.
3. Supervising service aspects of the college closely related to academic programs.
4. Managing the college budget (or portions of it).

5. Coordinating facility scheduling, managing equipment, and ordering textbooks.
6. Interviewing and selecting student assistants and part-time faculty.
7. Supervising nondepartmental units and programs.

Department Chair

Department chairs exist at both public school and college levels. They are key members of the administration because they manage programs at the grassroots level. They are the "front line" administrators who work directly with the faculty and students. The basic responsibilities of a college department chair include:

1. Coordinating student advisement and supervising progress toward completion of student requirements.
2. Determining the need for faculty and formulating recommendations for such to be considered by those in higher administrative positions.
3. Reviewing faculty salaries and participating in recommending salary adjustments.
4. Preparing class schedules and arranging for facilities and equipment in support of classes.
5. Coordinating faculty assignments, generally supervising faculty involvement, and evaluating faculty productivity and performance.
6. Providing recommendations relative to faculty advancement, leaves of absence, terminations, and retirement.
7. Organizing and supervising departmental committees.
8. Supervising departmental staff members in addition to faculty (clerical personnel, student assistants, and others).
9. Preparing and managing the departmental budget.
10. Coordinating curriculum and program planning and modifications.

Athletics Director

Many of the responsibilities of an athletics director are similar or equivalent to those described for department heads, with the exception that directing athletics does not involve matters relating to class instruction, such as curriculum, scheduling classes, and ordering textbooks. It does involve the following responsibilities not included in the description of department chairs:

1. Scheduling athletics contests.
2. Conducting athletics contests.
3. Promoting and selling tickets.
4. Implementing rules of athletic and academic eligibility.
5. Providing and managing athletics areas and facilities, and acquiring and caring for athletics equipment.
6. Providing sports medicine needs.
7. Dealing with the sports media.
8. Coordinating the total public-relations effort with regard to athletics.
9. Negotiating contracts with media and corporate sponsors.

The complexity of an athletics director's position varies greatly from that of a high school or small college program to that of a major university program. In the latter case, there is a much greater amount of public interest (and pressure) and, therefore, much more public-relations involvement, large crowds to accommodate and control, high-pressure game situations that must be properly managed, and extensive financial involvement. The basic elements of an athletics director's position, however, are essentially the same as described, whether the program is large and complex or small and relatively simple.

References and Recommended Readings

Bridges, William. 1994. "The End of the Job." *Fortune*, 130(6): 62–74.

Chandler, Ralph, & Jack Plano. 1988. *The Public Administration Dictionary*. 2nd ed. Santa Barbara, CA: ABC-CLIO.

Daft, Richard. 1998. *Organization Theory and Design*. Cincinnati: South-Western Publishing.

Drucker, Peter. 1964. *Managing the Results*. New York: Harper & Row.

Etzioni, Amitai. 1964. *Modern Organizations*. Englewood Cliffs, NJ: Prentice-Hall.

Griffin, Ricky. 1996. *Management*. 5th ed. Boston: Houghton Mifflin.

Hitt, William. 1984. *Management in Action*. Champaign, IL: Sagamore Publishing.

Jackson, John. 1987. "Organizational Structure." *Journal of Sport Management*, 1: 74–81.

Koontz, Harold, & Cyril O'Donnell. 1986. *Management: A Systems and Contingency Analysis of Management Functions*. 2nd ed. New York: McGraw-Hill.

Leavit, William, & Gail Johnson. 1998. "Employee Discipline and the Post-bureaucratic Public Organization: A Challenge in the Change Process." *Review of Public Personnel Administration*, 18(2): 73–81.

Narayanan, V. K., & Raghu Nath. 1993. *Organization Theory: A Strategic Approach*. Boston: Irwin.

Sharpe, Tom, & Tom Templin. 1997. "Implementing Collaborative Teams: A Strategy for School-Based Professionals." *Journal of Physical Education, Recreation & Dance*, 68(6): 50–55.

Tapscott, Donald. 1996. *The Digital Economy: Promise and Peril in the Age of Networked Intelligence*. New York: McGraw-Hill.

Online Sources

ERIC Clearinghouse on Educational Management
http://eric.uoregon.edu/trends_issues/ organization/
The site includes abstracts, discussion, links, and resources on school organization.

3   Administrative Leadership

Leadership is a highly sought-after and highly valued commodity. Organizations seek individuals who have "leadership ability" to fill administrative positions. Leadership is also one of the most studied topics in administration and management. Fundamental questions are posed by students of leadership: What is a leader? What behaviors distinguish leadership? Why do certain leadership approaches function well in some situations and not in others? These questions and others will be explored in this chapter. In the process, we shall examine leadership models, forms of leadership, leadership patterns and types, and the characteristics of successful leaders.

Dictionary definitions of leadership place emphasis on guiding, directing, and presiding, but the leader must also be an enabler and an influencer. Being an enabler means that instead of just being "the boss," the leader also is a resource person to the organization. The leader's functions include facilitating, as well as directing and guiding. Instead of merely issuing orders, the leader must maintain a healthy flow of communication up, down, and across the organization. A leader also is an influencer. The current textbooks define leadership as a process of influencing a group of individuals to achieve a common goal (Northouse, 1997). British Field Marshall Bernard Montgomery described leadership as "the capacity and will to rally people to a common purpose." This requires a leader to be inspirational as well as influential. Leadership influences the actions of individuals and, as a result, influences organizations. The key word is *influence*. A person who is well liked and admired but does not influence people and events is ineffective as a leader. In short, leadership is getting people to do what the leader wants them to do.

In educational organizations, leadership is a tool for achieving shared objectives, and its success must be judged in terms of what is accomplished. Exerting significant influence in an organization is a complex process. Influence is exercised through an intricate system of authority, responsibility, persuasion, and communication. It is often assumed that those in administrative positions make the greatest leadership contributions, but this is not necessarily true. Some people actually exert more positive influence from nonadministrative positions. This is not meant to minimize the importance of administrators. The point is that nonadministrators can often be just as important in terms of overall contributions to organizational leadership. This circumstance illustrates the distinction made between assigned leadership and emergent leadership. *Assigned leadership* is based on occupying a position within an organization, such as school principal or director of athletics. When others perceive an individual in an organization as an influential member of the group—regardless of job title—this person is exhibiting *emergent leadership* (Northouse, 1997). Coaches seek out emergent leaders among their athletes, as they are the potential team captains.

Clearly, leadership is multifaceted. An effective leader must be able to persuade others to achieve objectives, have the human qualities that bind a group together and motivate it, and demonstrate the art of coordinating and stimulating groups and individuals to achieve desired ends. Leadership may also be described as the capacity to help others achieve their own goals.

In addition to the influence of direct leadership, employees are also governed by standardized policies and procedures, organizational loyalty, and other factors built into the environment. This fact emphasizes the importance of creating conditions conducive to motivation and to effective performance by the staff members. The point is that leadership can be either direct or indirect. *Direct leadership* involves immediate contact between the leader and those being led. *Indirect leadership* is purposeful influence on people without direct contact with them.

An example of direct leadership is a department chairperson's supervisory influence on how well other employees perform their responsibilities. From such a position, the leader is expected to direct the program and account for its accomplishments. A teacher is a direct leader in the classroom, regularly influencing students' lives along a set pattern toward educational goals. Examples of indirect leadership are an educator who writes materials that influence those who read them, or one who communicates indirectly to a large number of people through delegates. Another form of indirect leadership is the *role model*. Both colleagues and students are influenced by models with whom they relate. A person who has achieved unusual professional standing is often identified as a model by others who aspire to the same kind of success. Outstanding teachers and coaches serve as worthy models for students and for others in the profession. The effective administrator serves as a model for those who aspire to administrative positions.

In describing the importance of administrative leadership, management specialist Peter Drucker (1993) stated:

> . . . the spirit of an organization is created from the top. If an organization is great in spirit, it is because the spirit of its top people is great. If it decays, it does so because the top rots; as the proverb has it, "trees die from the top down." No one should ever be appointed to a senior position unless . . . willing to have his or her personal and professional character serve as a model for the organization.

STYLES OF LEADERSHIP

The basic leadership styles are *autocratic, participative,* and *laissez faire* (the latter two are democratic in nature). The main distinctions between autocratic and democratic leadership appear in the contrasts between totalitarian and democratic governments. Less pure forms of these contrasting styles can be found in various organizations, and among individual leaders within the same organization. In a general sense, autocratic leaders think of others as working *for* them and subject to their commands, whereas democratic leaders think of others as working *with* them.

The Autocratic Leader

Strict *autocratic leadership* often stems from an authoritarian personality characterized as harboring protective and paranoid tendencies, combined with a lack of security, and frequently bolstered by arbitrary displays of authority. Autocrats tend to be remote from other members of the group and demand to have attention and activities centered on them. The apparent stability of the administrative unit is somewhat artificial, maintained largely under threat. Typically, any deviation or disloyalty is met with chastisement in such a rigid, authoritative structure. Negative expression directed at the organization or leader is discouraged, although some "constructive criticism" may be tolerated.

The autocratic leader projects a "boss" image and insists on monopolizing decision-making power. Often seen as "heavy-handed" and "hard-boiled," this leader issues orders and expects subordinates to be unfaltering in their subservience. Typically, such leaders are hard workers and hard drivers. Many autocratic leaders threaten and employ negative sanctions to enforce their authority. They may show little concern for the needs and problems of others. Subordinates often respond to such leadership by being tense, subtly resentful of the "system," and quarrelsome among themselves. This leadership style seems particularly inappropriate for schools, given that the majority of the staff is college educated and could provide valuable input into decision making.

Forms of authoritarianism exist that depend on more benign techniques toward gaining support by providing incentives to subordinates. A common type of benevolent authoritarianism is *paternalism*. Traditionally adopted by men in power (hence its name), it places the leader in the guise of "father figure." The paternal leader implores his subordinates to let him make all the important decisions and, in return, he will take care of their needs. When leaders employ the metaphor of "family" to characterize an organization, this is often a tip-off to a paternalistic perspective. Paternalism has an unsavory political history, as it was used to keep women and minorities "in their place." One of the main drawbacks of paternalism in an organization is that it preempts mutual adult-level interactions. Although some subordinates respond positively to benevolent forms of authoritarianism, they still do not share in important decision making. (Leadership textbooks employ the term *maternalism* when this style is adopted by a woman administrator.)

Authoritarian leaders of either the strict or the modified pattern usually insist on exact compliance with rules and regulations, and they place heavy emphasis on mechanical considerations of operational efficiency. In fact, adherence to rules frequently becomes so important to the authoritarian that the achievement of organizational goals is subordinated to procedural perfection. An old adage states that failed organizations tend to be overmanaged and underled.

The Participative Leader

Participative leadership emphasizes the concept of planning *with* people, not *for* them. It represents an intermediate position between autocratic and laissez-faire styles. The essence of this democratic style is: "Those who are affected by decisions have a voice in making decisions." Although staff members participate in and contribute to decision making, the leader still makes the final decision after considering input from others. The leader shares authority and decision making with subordinates but retains the final authority and the associated responsibility. Democratic leaders view followers as partners in work and attempt to promote a team spirit.

Leaders issue directions and expect adherence to rules under this system, but the system of incentives emphasizes positive rewards over negative sanctions. Goal achievement is stressed at least as fully as adherence to rules and procedures. Democratic leadership is less efficient than autocratic leadership, as decision making can be time consuming and cumbersome. The advantage of the democratic method is that it adds stability, since it does not depend too much on the judgment of one person. Checks and balances exist that are not present in autocratic leadership.

William Hitt (1985), an advocate of participative management, listed the following ten characteristics of this style of leadership:

1. Confidence and trust prevail throughout the organization.
2. Staff members feel free to discuss job problems with their supervisor.
3. There is group participation in goal setting.
4. Personnel at all levels feel responsibility for the achievement of organizational goals.
5. Information flows down, up, and across units.
6. There is substantial cooperative teamwork throughout the organization.
7. Staff members are actively involved in decisions that influence their work.
8. Responsibilities for review and control are widespread throughout the organization.
9. Control data are used for self-guidance and coordinated problem solving, rather than criticism and chastisement.
10. Formal and informal processes merge.

Highly participative leadership can have some drawbacks. While minimizing heavy-handed administration may reduce tension among subordinates, it is also true that too much emphasis on the participative technique may adversely affect productivity. In its extreme form, it can become almost leaderless, unstructured, and chaotic. The preferred degree of democratic leadership is one in which the leader exercises enough command to move things along at the desired pace and keep the organization responsive and alive, while showing adequate sensitivity toward other members of the organization and permitting sufficient involvement and influence.

The Laissez-Faire Leader

This pattern of leadership allows extensive freedom to act without interference from the leader. Its essence is captured in advice given by U.S. statesman Adlai Stevenson that a leader should "hire good people and then leave them alone." The *laissez-faire leader* is available to serve as a resource person and to assist in accomplishing individual or group goals. Such leaders will generally provide materials, information, guidance, and support when needed, but little uninvited direction. They encourage coworkers to set their own objectives, make their own decisions, and proceed on their own. Laissez-faire leaders often appear to be bystanders, observers, or subtle mediators without getting deeply involved in the activities of the group. Lao Tsu, a contemporary of Confucius, taught that "a leader is best when people barely know that he exists. . . . When the work is done, his aim fulfilled, they will all say, 'We did this ourselves.'"

Management experts and behavioral scientists generally agree that the probability for disorganization, instability, low output, and chaos is just too great to advocate the laissez-faire style. However, in selective circumstances such an approach can work. Possible examples might be a group of competent and dedicated researchers, a mature departmental faculty at a university, a law firm staff, or a medical staff in a clinic.

In terms of administrative leadership styles, it is important to consider the following points:

1. The personal characteristics of the administrator will influence the choice of a particular leadership style.
2. The leadership style will determine how the person carries out the administrative functions.
3. How the administrator carries out the functions will have a profound influence on the overall effectiveness of the organization.

Leadership Continuum

A continuum graph is a helpful method of viewing the complete range of leadership styles from one extreme to the other. Different scholars have presented this concept in a variety of forms. Figure 3-1 presents a leadership continuum illustrating the range from the extreme autocratic to the extreme democratic approach. At one end is the autocratic administrator who makes all the decisions, which the other employees are expected to accept. This leader not only sets the goals for the group and makes the final decisions but also discourages subordinate participation. At the other end is the extreme democratic administrator, who not only permits but also actually encourages extensive participation by subordinates.

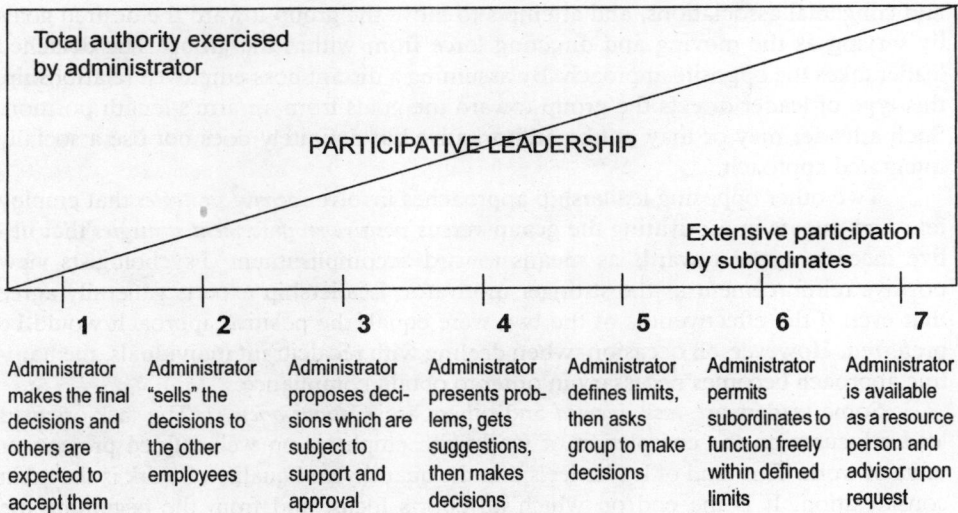

Adapted from Tannenbaum and Schmidt, 1973.

Figure 3–1. The leadership continuum

Contrasting Patterns of Leadership

Within the three leadership styles just discussed, specific patterns emerge that have distinguishable characteristics. For example, a leader might be basically either autocratic or democratic and still exhibit certain characteristics included in the patterns described here. The contrasting patterns represent extremes, and most leaders function somewhere between. The study of these patterns can help administrators better understand the strengths and weaknesses of their own approach and help them to visualize the alternative approaches that might be considered.

An interesting distinction exists between failure-avoidance and success-seeking leadership. The *failure-avoidance* strategy is based on the idea that one succeeds by simply avoiding failure (like the football coach who opts for a tie rather than risking a strategy to win). In organizations, such leaders emphasize adherence to the status quo. Change tends to be discouraged, because one measure of success is the smoothness of the operation and the absence of problems. The failure-avoidance process often hinders accomplishment of significant results. There is a tendency to accept relatively low productivity as long as it meets minimal standards.

The *success-seeking* style is just the opposite. Such a leader views success as relative and favors the highest possible level of achievement. Goal accomplishment is the principal task; and this type of leader seeks innovative problem solving, accepts reasonable risk, and promotes dynamic change as a sign of organizational vitality and success. A leader who stresses this style does not expect the operation always to run smoothly. Adherence to defined procedures and regulations is relatively unimportant if the expected results are to be achieved. This leader is not afraid to fail in trying something innovative.

Another interesting contrast is between the *integrated innovator* and the *detached director*. The integrator's approach is socially oriented. Such a leader wants to be "one of the group" who exercises leadership as a fully involved member, strives for close and congenial associations, and attempts to move the group toward the desired goals by serving as the moving and directing force from within the group. The detached leader takes the opposite approach. By assuming a distant boss-employee relationship, this type of leader directs the group toward the goals from an arm's-length position. Such a leader may or may not be authoritative but definitely does not use a socially integrated approach.

Two other opposing leadership approaches involve *coercive strategies* that employ fear and threats in motivating the group versus *positive reinforcement strategies* that utilize incentives and rewards as means toward accomplishment. Psychologists view positive reinforcement as the stronger motivator. Leadership experts generally agree that even if the effectiveness of the two were equal, the positive approach would be preferred. However, on occasion, when dealing with recalcitrant individuals, the hardline approach becomes necessary in order to obtain compliance.

Some leaders are *task oriented* and others are *relations oriented*. The task-oriented leader focuses on the completion of work with emphasis on well-defined projects or units of work. This kind of leader feels that the quantity and quality of work is the main consideration. It is the end on which all efforts focus, and from the beginning the emphasis is on the final result rather than the methods and working relationships involved. Conversely, the relations-oriented leader gives primary consideration to people. This approach is based on the idea that concern for individuals and relationships is a worthy end in itself and, in addition, will produce the best overall results in productivity.

Another interesting difference in leadership styles is *functional* versus *formal*. Functional leadership depends greatly on highly workable relationships between the leader and others in the organization, which earn cooperation and support for the leader. By this method, the members of the organization reinforce the functional leader's authority. In contrast, the formal leader claims the authority that comes with the position and utilizes that authority in a structured manner to enforce rules and policies. This formal authority enables the leader to have things done a particular way because "the boss says so."

LEADERSHIP TYPES

The process of reducing complex human behaviors into "leadership types" results in a substantial loss of individual detail. No person exactly fits any ideal type, yet a description of types provides guidelines by which one can analyze and understand one's own and others' leadership behaviors. Although leaders are in some degree prisoners of their personalities and experiences, few are so constrained that new alternatives are entirely closed to them. The particular types discussed here are climbers, conservers, zealots, innovators, reformers, advocates, diplomats, experts, philosophic leaders, symbolic leaders, charismatic leaders, and managers. The literature on leadership identifies and explains the basic characteristics of these types (Banovetz, 1971).

Climbers, as the name implies, are self-propelled in their quest for improved status. They are energetic in seeking out avenues of personal progress and self-aggrandizement. They tend toward empire building. When a frontal assault is blocked, the climber will likely regroup and then try to move in a different direction to increase the sphere of influence and recognition. Climbers tend to be perceived as abrasive because of their aggressive approach. Many climbers, however, are effective leaders because they are "doers" who efficiently attack the situation.

Conservers tend to maintain the status quo. They resist and even resent change, since it has an unsettling effect on the structured procedures and relationships. Conservers are more interested in security and stability than in power and progress. They want to trim their management unit to a simplified, clear-cut package and then hold it intact. They prefer organizations that rely extensively on formal regulations and protocol. Conservers often quote the old adage, "If it ain't broke, don't fix it." In the current era of rapid change, this approach has become less and less functional.

Zealots have visions of greatness for both themselves and their programs, and they manifest an evangelistic zeal for the improvement of the organization—as they see it. They are generally aggressive, determined, and hard-driving. Unlike climbers, however, zealots primarily have the interest of the organization at heart. In contrast to conservers, they are impatient, wanting to improve and innovate. Their mantra is, "If it ain't broke, break it." Their point is that today's organizations must keep abreast of change by breaking old, outdated patterns.

The aggressiveness of zealots sometimes irritates people who do not share their visions. Opposition, however, means relatively little to zealots, who are noted more for aggressiveness than for human relations. They are excellent instruments for stirring up an apathetic organization or getting a new one off the launching pad. They make better task force leaders than overall administrators because they place their "sacred objectives" first, often to the neglect of broader and more far-reaching goals.

Innovators share characteristics with both zealots and reformers. They have the unique ability to envision or develop better methods, techniques, and procedures.

Innovators are energetic doers who can bring about positive changes. Their approach is to search out or create better methods and apply them. General Eric Shinseki, a radical innovator, took over as Chief of Staff of the U.S. Armed Services in 1999. When his efforts to transform the military met with resistance, he admonished, "If you don't like change, you're going to like irrelevance even less." Innovative leadership is an important characteristic among both teachers and school administrators. An adequate amount of it can make the difference between the educational program being alive and interesting or dormant and drab.

Reformers are often associated with political, social, and religious unrest, but they also appear in organizations. They want people to turn against the present system and its established methods. Reformers search out and create new systems in order to instigate a reformation. Reform efforts are not limited to governments and churches. They occur in less obvious settings, including the educational system. A. S. Neill, who founded the radically democratic Summerhill School in Great Britain, is an example of a reformer. When reformers become agitators their efforts may be unpalatable, but they often produce constructive results. There are situations when it seems that no other approach can bring about the desired changes.

Advocates are concerned with improvement of the organization, especially the part they represent. They are tigers at fighting for their personnel and programs. Unlike zealots, who are basically loners, advocates are usually responsive to the ideas and influences of their superiors, peers, and subordinates. Unlike climbers, who are ever self-serving, advocates will at times promote programs that do not benefit them personally but have long-term favorable implications for the organization. Whereas zealots tend to take on all adversities, advocates usually engage in conflict only if supported by their colleagues. Externally, they defend their group or department in a partisan manner. Internally, they tend to be fair and impartial.

Diplomats are found in practically every organization. They have a strong inclination to intercede in conflict and transcend office politics, and they try to stand above parochial interests. They seek to reconcile conflicts and disagreements by pointing to the overall objectives or mission of the organization, or by emphasizing the fair and equitable side of the situation. Diplomats are sometimes located in the rank and file of the employees, and they perceive themselves as spokespersons for the group, usually in carrying messages to or negotiating with the management. Diplomats who are in administrative positions perceive themselves as spokespersons for the organization.

Expert leaders are recognized for their specialized knowledge. People look to them for expert opinion and professional guidance. What they say as administrators or teachers takes on unusual significance. Recognition as an expert increases a person's influence in almost every respect. Each profession has its own group of experts who exert a strong influence within the profession and beyond it. Each organization also has individuals who are more or less recognized as experts and whose influence is felt accordingly. Expert leaders often reside in high-tech departments in organizations, such as computer services.

Philosophic leaders place emphasis on ideas and ideals. Leaders of this type can have significant influence on both goals and program results. They focus on the broad consequences of actions, are aware of and act on principles, and are sensitive to ethical consequences. Such leaders may appear remote and sometimes may be understood by their followers. This type tends to be most effective in organizations like universities where the workforce is highly educated.

Symbolic leaders rarely are found in educational organizations. Sometimes dynamic teachers or coaches can be symbolic leaders to their students, and occasionally a school board will have a member who is more symbolic than functional. Symbolic leaders derive their strength from tradition and association with their status in the organization or the community. The symbolic leader fills essentially the same purpose regardless of individual characteristics or ability, assuming that the individual is respected and accepted by others. Symbolic leadership usually provides a stabilizing influence. An example might be a nationally known, retired executive who is placed on a board of directors.

Symbolic leaders often have a great deal of charisma. In terms of leadership, charisma is a kind of "personal magic." The *charismatic leader* captures the imagination of the followers, inspiring their loyalty and enthusiasm. Such persons lead and influence others by means of their personal appeal and direct motivation. We have all known coaches who have led by the force of their personalities.

Managers are caretakers. They place their focus on making sure things are operating efficiently and seeing that resources are available and being utilized fully. This type of leader takes on a coordinating and supporting role. Managers are concerned with implementing procedures and providing direction toward the desired goals. The manger is not particularly brilliant in conceptual skills and motivation but acts more as a facilitator.

THEORETICAL APPROACHES TO LEADERSHIP

Leadership is a complex process that has been studied empirically for at least a century. The early studies focused on prominent political, social, and military leaders. Later research was directed at studying leadership in business and public organizations. Leadership was recognized as a multidimensional concept, and a range of theoretical approaches emerged from the research. Leadership theories can be both descriptive and prescriptive.

Currently, at least three major perspectives for explaining leadership are evident. Leadership has been described as: (1) dependent on a set of traits or characteristics that an individual possesses, (2) a set of behaviors or actions that an individual performs, and (3) an interactive process that depends on the particular situation. The following discussion will briefly introduce the reader to these three perspectives.

The Trait Approach

The *trait approach* focuses on the leader rather than on the followers or the situation. The underlying assumption is that certain people have identifiable leadership traits, and that procedures should be implemented to select the right person (with these traits) to lead an organization. Leadership trait questionnaires have been developed to assist in recruiting the right person. Traditionally, these traits are thought to be somewhat fixed psychological structures that are difficult to develop through training (evoking comments such as "He is a born leader," or "She is a natural leader"). This view restricts leadership to those individuals who have these inborn characteristics.

Several studies have confirmed the relationship between certain traits and leadership success. Literally dozens of leadership traits have been postulated over the years, and the complete list of traits became unwieldy. Scholars have analyzed the many studies of leadership and culled the following five traits that seem paramount in importance (Northouse, 1997).

Intelligence means having strong verbal and perceptual abilities, and a capacity to reason. Experience indicates that the leader's intellectual ability shouldn't vary too much from that of followers, as this will cause problems with communication and cohesion within the organization.

Self-confidence means feeling certain about one's skills and competencies. It implies self-esteem and self-assurance, and the belief that one can make a difference. Self-confidence allows leaders to feel that they are doing the right thing and that their efforts to influence others are appropriate.

Determination refers to the desire to get the job done and includes characteristics such as initiative, persistence, dominance, and drive. Determined leaders are proactive (rather than reactive), are able to assert themselves, and persevere in the face of obstacles.

Integrity is the quality of honesty and trustworthiness. Leaders with integrity adhere to their principles and take responsibility for their actions. This quality inspires confidence in followers, who see the leader as believable and worthy of trust.

Sociability refers to the inclination to seek out social relationships, to be friendly, outgoing, courteous, and tactful. Sociable leaders create cooperative relationships with followers, while being sensitive to their needs and concerned with their well-being.

The Style (Behavioral) Approach

The *style* or *behavioral approach* emphasizes the style of leaders. It defines leadership as a series of actions toward employees. As a result of many studies on leadership style, researchers settled on two general kinds of behaviors as most important: task behaviors and relationship behaviors (sometimes referred to respectively as *initiating structure* and *consideration*). *Task behaviors* concern goal accomplishment. They assist employees in achieving objectives. *Relationship behaviors* help employees feel comfortable with themselves and with others in the organization. Some leaders appear strong in both behaviors; other leaders are strong in one and weak in the other, or weak in both. Certain situations may require varying levels of each type of behavior. Successful leaders are able to discern when to emphasize structured task orientation and when to emphasize relational behaviors. Good leaders provide structure for their subordinates, and they nurture them (Northouse, 1997).

Unlike the trait approach, this approach to leadership incorporates the idea that individuals can learn these leadership behaviors and improve their effectiveness, although the inclination to emphasize one type of behavior over the other may be in part the consequence of personality variables. (See the section on Directing in chapter 4 for further discussion of these two types of leader behaviors.)

The Situational Approach

The *situational approach* to leadership extends and refines the behavioral approach. The basic premises are that different situations require different types of leadership, and leaders must be flexible in their leadership style over time and with different groups of people. The situational approach recognizes that employees act differently when doing various tasks and may even act differently during consecutive stages of the same task. Both a directive and a supportive dimension of leadership behavior have to be applied depending on the given situation. Situational leaders must assess the commitment and competence of their subordinates, changing the degree to which they are directive or supportive based on this determination.

Directive behaviors imply establishing goals and methods of evaluation, giving directions, setting time lines, and defining roles. In short, this means deciding what is to be done, how it is to be done, and who is responsible for doing it. This function, for the most part, constitutes one-way communication. *Supportive behaviors* involve two-way communication. Examples of supportive behaviors include sharing information, asking for input, problem solving, praising, and listening. Studies have shown that new teachers are more satisfied with principals who are highly structured and directive, while the performance of more experienced and mature teachers seems unrelated to the leadership style of their principal (Northouse, 1997). This finding illustrates the situational nature of effective leadership.

Every leadership situation has four basic elements: (1) the leader, with abilities, personality traits, responsibilities, and authority; (2) the followers, with their abilities, personalities, and responsibilities; (3) the situation, with its special features and dynamics; and (4) the purpose or task with which individuals, programs, or the organization are concerned. Recognition of these elements suggests that individuals who emerge as leaders in one setting might not emerge as leaders to the same degree in other circumstances. Different situations and circumstances call for different leadership qualities.

Selected Theories of Leadership

Various theories of leadership have been developed that are more formal than the above three approaches. There is not room in this chapter to cover all of them. The following sample introduces the reader to four theories that are current in the literature.

Exchange Theory (Leader-Follower Interaction)

This theory centers on the interactions between leaders and individual followers—relationships that sociologists call *dyads* (i.e., two persons in interaction). In vertical dyads (leader/follower), negotiations take place that cause subordinates to do things beyond the scope of their formal job responsibilities, and the leader in turn does more for them. This situation often leads to in-groups and out-groups in organizational units, if such interactions are selective. Out-groups can be dysfunctional for the entire unit. Successful leaders attempt to establish positive negotiations with all subordinates to prevent out-groups from developing. Effective leader-follower exchanges lead to greater organizational commitment, better job attitudes, greater participation, and less employee turnover. Leaders are encouraged to offer subordinates opportunities to take on new roles and responsibilities (Northouse, 1997).

Transactional leadership is a theoretical model based on exchange. Transactional leaders exchange things of value with subordinates that will advance both their agendas. This strategy utilizes *contingent reward* (rewards received by followers contingent upon certain behaviors). Subordinates realize that it is in their own best interests to do what the leader wants.

Transformational Leadership

This theory evolved from transactional leadership but goes beyond the transactional emphasis on satisfying the needs and self-interests of subordinates. Instead, transformational leaders articulate a vision of the organization, provide a model consistent with that vision, foster acceptance of group goals, and provide support to individuals. Transactional leaders attempt to influence the values and aspirations of subordinates in order to motivate them to transcend their own interests for the sake of the organization. When this is done, subordinates tend to achieve higher goals, perform work beyond the minimal levels specified, and report higher job satisfaction. Job satisfaction, in turn, leads to lower turnover, less absenteeism, and higher productivity. While developed in the business sector, this leadership style has been applied successfully in school settings including athletics departments (Aminuddin, 1998).

Organizational Leadership (Multiple Leaders)

Most of the previously discussed models of leadership have emphasized the concept of a single leader and many followers rather than leadership in a pluralistic sense. The organizational leadership model asserts that leadership is not found in one individual's traits or behaviors but is a characteristic of the entire organization. The assumption is that leader roles overlap and complement each other, that leadership shifts from time to time and from person to person. This model emphasizes shared leadership. It disputes the assumption that leadership is a possession of one individual; instead, it asserts that leadership may be exercised by a team. Organizational theorists believe that "invisible" leadership among lower-level staff members occurs normally within organizations (Méndez-Morse, 2001).

Gender-based Theory (Women as Leaders)

Women continue to be underrepresented in administrative leadership roles, even in professions like teaching where women have outnumbered men. Women often face a "glass ceiling," an invisible barrier that prevents them from rising higher in an organization. Stereotypes about women largely have hindered their ability to move up to higher leadership positions. Feminists have pointed out that perceived differences in abilities and interests between men and women aren't innate but rather are the result of socialization and lack of equal opportunity. Studies have shown that girls who competed with boys (e.g., in sports) during their youth are more likely to move into nontraditional roles as adult women (Coats & Overman, 1992).

Women can learn the skills perceived as male strengths—such as focusing on tasks more than relationships—and, at the same time, enhance their feminine leadership traits like empathy and nurturing (just as men can acquire feminine leadership skills to build on their masculine skills). Combining traditional masculine and feminine strengths leads to an *androgynous* (gender-neutral) *leadership* model, which may offer the best of both worlds. Leadership consultants point out that twenty-first century organizations are moving away from highly directive, top-down management toward a self-directed, team-based approach. The more participatory networking style of women managers makes them good candidates for leadership positions in this emerging work environment.

Women also have been more sensitive to creating diversity in the workplace. As organizations compete in the global economy, the diversity of skills and perspectives that comes with race, gender, and culture becomes a competitive advantage (Smith, 1997).

COMMON LEADERSHIP MISTAKES

Most textbooks focus on what educational leaders *should* do rather than on what they *should not* do. Arguably, knowing what not to do may be as important as knowing what to do, given that the negative fallout of one mistake may be far-reaching and offset the effects of doing the right thing most of the time. This approach to leadership is practical in that the behaviors leaders should avoid are probably fewer than those they should exhibit.

Many of the mistakes school administrators make fall into the category of poor human-relations skills. Projecting a lack of trust or an uncaring attitude are the two behaviors most frequently associated with this deficiency. Leaders perceived as distant don't call teachers or coaches by their names, fail to delegate, and fail to compliment their staff. Generally, leaders who display these shortcomings have overemphasized "task orientation" at the expense of "people orientation."

Other types of mistakes educational leaders commonly make include: lacking vision, lacking knowledge about instruction and the curriculum, lacking ethics or character, being too controlling, forgetting what it is like to be a teacher or coach, being inconsistent, showing favoritism, failing to hold staff accountable, failing to follow through, making snap judgments, and utilizing poor communication skills.

Regarding poor communication skills, the example most frequently given is failure to listen. Doing paperwork in the presence of visitors and not maintaining eye contact are behaviors suggesting such a failure. Another major problem with interpersonal communication is the failure to provide feedback. On the receiving end, some administrators reprimand teachers in front of their colleagues instead of doing it privately. Just as it can be detrimental to reprimand students in front of the whole class, it is also unprofessional to reprimand teachers in front of their peers.

The premise of this leadership focus is that the overall climate of a school is affected by the number of mistakes an administrator makes, and that administrators make mistakes that could be avoided if they were aware of them. Leadership experts caution administrators against "sweeping mistakes under the rug." They emphasize the importance of admitting one's mistakes and moving on. Although acknowledging a poor decision is tough, the sooner it is done the better. Failing to admit a mistake is a mistake in itself, and too many mistakes can cause administrators to lose their jobs (Bulach, Pickett, & Boothe, 1998).

ACHIEVING SUCCESS IN LEADERSHIP

Most students have held leadership positions on athletic teams, in school government, in their church, or in the community. Following graduation, many will take on additional leadership responsibilities in their chosen profession. Generally, success in leadership is achieved by discovering one's best talents, skills, and abilities and applying them where they count the most. Once a person accepts a particular position or job, delimiting influences assert themselves. Thus, the challenge is to achieve success within the framework of the circumstances. Even the bleakest of circumstances might be overcome. An old maxim counsels, "If life gives you lemons, make lemonade." There's a lot to be said for maintaining a positive attitude.

Motivation

The literature on self-motivation has moved beyond the doctrines of positive thinking promulgated by early advocates like Dale Carnegie and Norman Vincent Peale. Contemporary models of motivation are more likely to be couched in cognitive psychology. Cognitive therapists agree with proponents of positive thinking that changing one's beliefs changes the results. Beliefs show up in the inner dialogue that we have with ourselves, in our "voice of conscience" that talks to us about life. We are advised to listen to our inner voice and to reshape beliefs about what is happening. For instance, after experiencing an embarrassing failure, a leader shouldn't develop a defeatist attitude. Instead, the inner voice should acknowledge, "I made a mistake, but I can get through it and learn from it." Don't fall into a negative pattern of worry or self-criticism, which can only make matters worse. Pay attention to the inner voice but be sure that it is realistic. Realistic thinking will lead to realistic consequences and a sense of control (Johnston, 2000).

Motivation means providing incentives that sustain dedication to a cause. Three of the dominant motivators are: (1) the need to feel important and recognized, or the need for identity; (2) the need for refreshing change or new stimulation; and (3) the need for a feeling of security. Those in leadership positions should not allow their personal needs to take precedence over the needs of the organization they serve, however. Individuals respond to both internal and external motivations. These include:

1. Recognition and praise from peers, parents, and members of the public (external).

2. Ego reinforcement, which comes from a feeling of conquest or fulfillment (internal).

3. Material gains such as money or material proof of achievement in the form of medals, certificates, trophies, and the like (external).

4. Emotional fulfillment derived from the achievement of a goal (internal).

External sources of motivation—the cheering of a crowd, slaps on the back, handshakes, certificates, medallions, and monetary rewards—are helpful; but these often have only temporary effects. Once their influence has faded, the person once again depends almost totally on internal motivation. One's inner thoughts, ideals, and aspirations are the sources from which the more permanent motivational energy originates and flows.

Thus, it is an error to depend too much on others for incentive. Creating and sustaining the necessary level of motivation is largely a "do-it-yourself" project. One of the greatest needs for sustained motivation is steady progress toward self-actualization. People must feel that they are growing, developing, and progressing in the areas of their particular interest. Musicians must make music, painters must paint, and leaders must lead in order to feel fulfilled.

Personal Traits to Emphasize for Success

Current models of leadership seem interested in what leaders do to achieve success. They define leader effectiveness in terms of leader behaviors. Notwithstanding the recent de-emphasis on trait theory, we can identify certain personal traits that promote success in assuming a leadership role. Desirable traits can be strengthened within certain limits, and this affords the potential for self-improvement. In addition to those listed in the previous section on trait theory, the following traits can contribute to achieving personal goals. It would be well to develop these traits in yourself if you aspire to become a leader.

Specialized knowledge. In the Information Age, administrators have found that more knowledge and skills are required than they have the ability to acquire. In this situation, the best strategy is to engage people as consultants or assistants who can help provide the specialized knowledge.

Imagination. Creative and ingenious ideas are trademarks of effective administrators. As a person conceptualizes goals, it is important to be imaginative about how to achieve them. The mind should be constantly stretched toward new ways of thinking.

Active mind. Our minds need to be developed continually through study and novel experiences in order to maximize the capacity of our memories to store valuable information. This process results in an expanded base from which new ideas and better conclusions can be drawn.

Desire. Desire is the starting point for action. Be both definite and realistic about the level of achievement you desire. Determine clearly what you are willing to sacrifice or the price you are willing to pay in terms of dedication and effort to accomplish your desire.

Enthusiasm. Enthusiasm can change would-be drudgery into a pleasant and challenging pursuit. It can make a difference as to whether you feel stimulated to continue, recover from setbacks, motivate others to assist with your plan, or sustain the necessary effort to ensure success.

Initiative. Success comes to the person who is willing to act. While others may have more ability, they may be irresolute in acting on their ideas. Initiative is necessary to begin the tasks at hand.

Persistence. Many people have good ideas and formulate sound plans, but when the going gets tough, they falter. Lack of persistence deprives them of a full measure of success. Leaders must have the faith to endure.

Open-mindedness. Open-mindedness is the willingness to try to understand different points of view, even though you may not agree with them. Prejudice, partisanship and other such habits close the mind to new ideas (what the philosopher Karl Jaspers called a "prison" of set opinions and convictions). A leader who is unable or unwilling to be open-minded runs the risk of contributing to distortion of facts, diversion of truth, and perpetuation of misinformation.

Sixth sense. This is sometimes referred to as "seat-of-the pants" administration. It cannot be taught; it comes from the accumulation of experiences. It is less a product of the conscious mind and the other five senses than of the subconscious. Sixth sense suggests intuitiveness, the pursuit of a certain course because it "feels" right—the equivalent of a poker player's "knowing when to hold them and when to fold them."

These traits alone cannot guarantee success, either singularly or collectively. They remain secondary to a leader's orientation to the job and behavior toward colleagues. The list does suggest a plan of self-improvement for students of administration who wish to hold leadership positions in the future. In the final analysis, personal traits and behaviors are reciprocal. Cognitive psychologists inform us that beliefs translate into outward behaviors, while behavioral psychologists promote behavior modification as a way to reshape attitude and character.

References and Recommended Readings

Aminuddin, Yusof. 1998. "The Relationship between Transformational Leadership of Athletic Directors and Coaches' Job Satisfaction." *The Physical Educator*, 55(4): 170–75.

Banovetz, James. 1971. "Leadership Styles and Strategies." *Managing the Modern City*. Washington, DC: International City Management Association, pp. 109–21.

Bulach, Clete, Winston Pickett, & Diana Boothe. 1998. "Mistakes Educational Leaders Make." *Eric Digest*, #122. Online article: <http://www.ed.gov/databases/ERIC_digests/ed422604.html>.

Coats, P. Boyne, & Steven Overman. 1992. "Childhood Play Experiences of Women in Traditional and Nontraditional Professions." *Sex Roles: A Research Journal,* 26(7/8): 261–71.

Drucker, Peter. 1993. *Management: Tasks, Responsibilities, Practices.* New York: Harper.

Hitt, William. 1985. *Management in Action.* Columbus, OH: Batelle.

Johnston, Dan. 2000. "Improving your Thinking and Mood." *Awakenings: Lessons for Living.* Online article: <http://www.lessons4living.com/depression5.htm>.

Méndez-Morse, Sylvia. 2001. "Leadership Characteristics that Facilitate School Change." Southwest Educational Development Laboratory. Online article: <http://www.sedl.org/change/leadership/welcome.html>.

Northouse, Peter (Ed.). 1997. *Leadership: Theory and Practice.* Thousand Oaks, CA: Sage.

Smith, Dayle. 1997. "Women and Leadership." In P. Northouse (Ed.), *Leadership: Theory and Practice.* Thousand Oaks, CA: Sage.

Tannenbaum, Robert, & Warren H. Schmidt. 1973. "How to Choose a Leadership Pattern." *Harvard Business Review* 51(3): 162–80.

Yukl, Gary. 2001. *Leadership in Organizations.* 5th ed. Upper Saddle River, NJ: Prentice-Hall.

Online Sources

ERIC Clearinghouse of Educational Management
http://eric.uoregon.edu/hot_topics/ index.html#school_leadership
Click on "School Leadership." The site provides discussion and a list of research abstracts.

4

What an Administrator Does

In a general sense, an administrator provides direction to the organization and assures that its resources are fully utilized. To employ a metaphor, the administrator has the responsibility to "shine light on the track and put steam in the boiler." One person alone cannot accomplish all these responsibilities; therefore, above all else an administrator is someone who gets things done through other people.

The broad term *administrator* is incorporated into various titles, including superintendent, president, dean, principal, director, and manager. In a school setting, the athletics director, the intramural director, and the physical education department chair all have major administrative responsibilities. However, the dichotomy in education between those who teach and those who administer is artificial. Most teachers and coaches have administrative duties, and some administrators share in the teaching. All are educators who should be building and planning together for a common purpose.

MODELS OF ADMINISTRATION

It's important to understand that what an administrator does takes place within a specific context or setting. Even though educational administration differs from other administrative settings, we often look to the fields of public and business administration for models of how to manage educational organizations. At any point in time, management culture tends to be dominated by one school of thought or model. Recent management approaches have included Management by Objectives (MBO) and Total Quality Management (TQM). Each approach has its proponents. Management consultants and writers of textbooks tend to adopt the theory currently in vogue. Management models become popular, peak in influence, and are superseded by the next new idea. They seem to come and go approximately every decade. (For a discussion of management models, including MBO and TQM, see Starling, 1998.)

There is no good reason why administrators must follow the latest school of management thought. On the other hand, just because an idea is new does not mean that it should be dismissed. There are reasons why one particular approach is better than another, depending on the situation. Moreover, an administrator who has experience in one approach may have difficulty adapting to another. Usually administrators tend to use the approach with which they are most familiar. Administrators can be much more effective if they are able to select a management approach that is most appropriate to the desired needs or goals of their organization. This adaptability or eclecticism can prove especially useful in today's rapidly changing environment (Arveson, 1998).

Regardless of the management model adopted, administrators must perform some basic functions for an organization to succeed. The performance of these functions has changed with emerging technology, but the functions themselves have stood the test of time.

ADMINISTRATIVE FUNCTIONS

Broadly speaking, the administrator engages in those functions required to realize the organization's goals. The idea of a set of standard administrative functions carries back to Luther H. Gulick in his "Notes on the Theory of Organization" (1937). His acronym, POSDCoRB, stood for planning, organizing, staffing, directing, coordinating, reporting, and budgeting. Gulick's original list has been refined and condensed several times. For purposes of the present discussion, the functions an administrator performs are grouped into the following five categories: planning, organizing, staffing, directing, and controlling.

Planning

Long ago, Confucius spoke to the wisdom of planning when he said, "A man who does not think and plan long ahead will find trouble right by his door" (Yutang, 1938). Planning is not bound by history or culture. The great structures and projects throughout time have required extensive planning. Obvious examples are the construction of the pyramids and the Roman aqueducts, construction of the Panama Canal, and the landing on the moon. Certainly, the range of things for which people plan in modern times has widened. As social life gets more complex, so does the variety of phenomena that demand forethought (Clarke, 1999). While planning is necessary, it is not sufficient in and of itself. Plans alone cannot make an enterprise successful. Action is required. The role of planning is to focus action on goals. Without adequate planning, we are left with a series of random actions that end in chaos.

Functionally, planning is deciding in advance what to do, how to do it, when to do it, and who should do it. It bridges the gap from where we are to where we want to go. Planning makes it possible for useful things to occur that would not otherwise happen. The planning function involves establishing goals and arranging them in logical order from immediate to long-range. A plan provides the strategies to follow in order to achieve the set goals. It is like a recipe or a road map—a spelled-out guide of where to go and how to get there. A plan also provides standing information for employees. For example, a budget is a financial plan that allows employees conducting programs and activities to stay within their resource allotment. Thus, planning has great utility in reducing uncertainty for organizations.

Planning—especially high-technology planning—implies expertise. Administrators claim mastery and thoughtfulness about certain issues in order to make plans. Claims of expertise are always claims that somebody should be left out of the decision loop. At the same time, planning is an inherently interactive process. It should involve getting people together and hearing their ideas, so that the planning is done *with* them and not just *for* them. Thus, planning is unavoidably political.

Planning has both symbolic and functional value. Leaders of organizations use plans as tools to convince followers to "buy into" the organization's goals. Plans are much more than blueprints for future action; they are also rhetorical devices used to persuade. Just as memoranda and speeches are directed at others, so are plans. Plans are a public declaration that the organization has deliberated carefully about its purposes and goals and has developed the requisite wisdom and power to carry them out (Clarke, 1999).

Planning should accomplish two broad objectives: it should enable the administrator to foresee and control situations more effectively, and it should help an administrator shape the future of the organization. Planning to accomplish these objectives requires a level of certainty about what is going on. The less certain the situation, the more difficult it is to develop an effective plan. Thus, flexibility is crucial in planning because it deals with the future, which is not entirely predictable. Sometimes it becomes necessary to revise the plan before the final goals can be reached. (It would be foolish to continue working with an obsolete plan.) Consequently, periodic review of plans and timely updating should be part of the process. Time is another important element in planning. Planning involves setting goals to achieve within a certain time frame. Intermediate targets should be defined to serve as indicators that progress toward the goals is taking place according to schedule.

Most organizations do annual planning as a matter of course. Administrators should supplement short-range plans with strategic planning. *Strategic planning* is a procedure that has been widely used in the corporate sector and has been successfully adopted by school districts. In the strategic planning process, an organization determines where it wants to be in the future and what active steps are necessary to arrive at that point. A strategic plan is global in its outlook. It defines strategies, goals, and the actions necessary to achieve the organization's vision over a three- to five-year period. School districts that implement strategic planning usually form a planning team to carry out this task. Examples of school districts' strategic plans can be accessed on the Internet (for example, the strategic plan for Alachua County Schools in Florida at http://www.sbac.edu/~wpops/strategic/).

Models have been developed to assist administrators in carrying out the planning process. Following is the *logical incremental planning model*, which contains six steps with recommended strategies (Starling, 1998).

1. *Need Sensing*
   a) Shop widely for new ideas and important signals.
   b) Sense need for change in vague or undefined terms.

2. *Building Awareness*
   a) Create study groups.
   b) Create new options.
   c) Improve information.

3. *Broadening Support*
   a) Force discussions; probe positions; explore options.
   b) Identify opponents, but do not threaten them.

    c) Encourage ideas ("trial balloons") with which you agree.

    d) Change perceptions.

4. *Pockets of Commitment*

    a) Launch exploratory projects.

    b) Keep options open, and control premature momentum.

    c) Develop a committee to educate/neutralize opponents and build momentum.

    d) Watch for a crisis or event that allows attention to focus on a particular goal.

5. *Building Commitment*

    a) Champion someone who genuinely identifies with the goal.

    b) Design budgets, programs, recruiting, and reward systems to reflect the goal.

    c) Reassign supporters and persistent opponents as necessary.

6. *Continuing Dynamics*

    a) Adapt original goals to unknown realities.

    b) Push forward each administrative unit separately, depending on resistance met.

    c) Ensure that the new consensus doesn't become inflexible.

At the completion of the planning process, a visible product emerges: the plan. With the plan as a blueprint, the administrator can now move on to the next function.

Organizing

As noted in chapter 2, organizing involves identifying the responsibilities to be performed, grouping the responsibilities into departments or divisions, and specifying organizational relationships. The purpose is to achieve a coordinated effort among all the elements in the organization. The primary focus is on definition of responsibilities and relationships. Among key questions are: (1) How many different tasks make up a particular job? (2) Which particular tasks fit together logically into one position? (3) Which positions fit together logically to form a division or department? (4) How do the divisions and departments fit together to form the complete organization?

Span of Control

One important question in organizing is how many employees should report directly to an administrator. Obviously, there is a limit to the number of persons an individual can supervise effectively. Traditional administrative theory advocates a narrow span of control. Most federal agencies keep the number of principal subordinates to less than twenty. Common sense suggests that the optimum span of control is contingent on the characteristics of a particular situation and the nature of the work performed, not on some magic number. The amount of time the administrator has for supervising employees is one important consideration. If a series of duties is required that takes time away from the supervisory function, then the span of control may have to be reduced.

Other important factors in determining span of control include (1) the abilities of administrator and subordinates, (2) the degree of coordination the particular kind of work requires, (3) the frequency of new problems that arise within the administrator's realm, and (4) the potential difficulty in solving those problems. The effective span of control, then, varies according to these factors. A common fault among administrators is to try to cover too broad a span. When this happens, the administrator can improve effectiveness by narrowing the span, thus relinquishing some of the control to others in the organization.

Delegation of Authority and Responsibility

Authority can be defined as the right to invoke compliance by subordinates on the basis of one's formal position in an organization. Authority is distinct from *power*, the ability to coerce compliant behavior of subordinates through the use of rewards or sanctions. Authority represents institutionalized power (Chandler & Plano, 1988). From an administrative point of view, this includes the right to control people and situations and to allocate the organization's resources toward accomplishment of its objectives.

In an educational or business organization, ultimate authority resides with the governing board that represents the owners or the taxpayers. The board grants authority to the chief executive, who in turn delegates authority to others. When authority is delegated, an obligation is placed on the recipient to exercise it properly. Delegating authority does not relieve the administrator of overall accountability. The administrator is still responsible for accomplishing the objectives of the organization, accounting for the use of its resources, and to some degree the actions of subordinates.

As an organization grows, it becomes increasingly difficult for one person to manage everything. Therefore, it is necessary to rely on subordinates. Ultimately, the administrator has to choose between losing command by trying to control too much directly (centralizing) or retaining orderly control by delegating authority and responsibilities (decentralizing). If a head coach tries to coordinate all phases of the game, he or she will do a mediocre job at best. A solution is to delegate some responsibilities to a qualified assistant. Through delegation, the coach retains general control of the overall program while relinquishing direct control over certain aspects.

Centralization and decentralization can be considered the extreme ends on a continuum with numerous intermediate points. Effective decentralization depends on: (1) the ability of the top administrator to delegate authority (and responsibility) to subordinates, and (2) the ability of the subordinates to accept and handle it well. Researchers have found that, in many cases, decision making would be vastly improved if the top administrator made fewer decisions. Often, others have more time to consider the influencing circumstances adequately, and they are sometimes better informed on the specifics that should affect decisions (Arnold, 1978). Failing to delegate effectively compromises one's ability to function in a high administrative position.

Delegation involves:

1. Selecting competent personnel who can accept and handle delegated duties.

2. Defining the delegated responsibilities clearly.

3. Allowing an adequate amount of freedom for the subordinates to handle the assignments their own way within the framework of correct procedures and sound practices.

4. Accepting accountability for the results.

Delegation is distinct from simply assigning work to someone else. Delegation means that you pass on to a subordinate a portion of your *responsibilities*, along with the *authority* to carry them out, and the *accountability* for how well they are carried out. Some responsibilities are delegated with guidelines to define the framework within which they are to be conducted. In other cases, responsibilities are delegated with

total freedom as to how to handle them. In analyzing whether you delegate sufficiently and effectively, you should ask yourself the following questions:

1. Are you unavailable because you are usually too busy? If so, are there valid
   reasons for this, or do you spend too much time doing things that someone
   else ought to do?

2. What would happen if you were suddenly taken out of circulation for a few
   days? Some administrators take pride in the idea that the place cannot get
   along without them. Such indispensability, however (even if true), indicates
   an organizational weakness.

3. Is the operation slowed down and productivity diminished because others
   must wait for you to fulfill all your responsibilities? Are others waiting for
   decisions that you should have made last week?

4. Are communication channels too rigid, requiring subordinates to channel
   information through you and not permitting them to cut across departmental lines, provided they keep you properly informed?

5. What does an analysis of your homework show? Does your briefcase bulge
   each night and on weekends with correspondence and reports?

6. Are there certain administrative matters that someone else could handle
   instead of you, even if not quite as well—or perhaps even better?

7. Are there responsibilities that someone other than you could do at less
   expense, in view of the dollar value of your time?

8. Could a secretary or administrative assistant do many of the things you do,
   and would the additional freedom of your schedule more than compensate
   for the cost of the extra staff person through improved management and better results?

Why Some Administrators Fail to Delegate

Although administrators usually understand the theory of delegation, many are
unwilling to apply it. Failure to delegate is usually inherent in the individual rather
than the situation. Here are some interesting reasons gleaned from the literature that
explain why administrators resist delegating (Drucker, 1999).

1. *Fear of being found out.* Some administrators are not as well informed as they
   like others to believe. There is a tendency to defend against this by keeping
   others detached and uninvolved.

2. *Overdeveloped sense of perfection.* Such administrators have the attitude that in
   order to have it done right, they have to do it themselves. They try to do it
   all, and the result is an organization that gets "gummed up" at the top level.

3. *A reluctance to admit that others know more about certain aspects of the operation.* If
   competent personnel are given ample opportunity for development, some
   will know more about certain responsibilities than the chief administrator.
   Some administrators find this hard to accept.

4. *Fear of not getting credit.* Here, a fundamental insecurity manifests itself—the
   desire to get personal credit. This form of pettiness is found in high-ranking
   administrators as well as in lower-level supervisors.

5. *Fear of subordinates progressing too fast.* This can seem like a threat to an administrator who is afraid of "slipping."

6. *Fear of poor relationships.* Some administrators have such a strong desire to be well liked that they avoid demanding firmly that subordinates carry their share of the load.

The appropriate delegation of authority and responsibilities ties in with staffing and staff development. Delegating to subordinates improves their knowledge and abilities, and thus their level of competency. When recruiting internally to fill a position in the organization, better qualified candidates will be available if administrators have shared duties and responsibilities widely.

Staffing

An organization is composed of people carrying out specific responsibilities that lead toward set goals. Effective staffing means filling the job positions with the right people at the right time. Human-resource managers can help in staffing by recruiting and screening people, but the ultimate responsibility for effective staffing rests with administrators. The discussion of the staffing function will be brief. (Staffing and staff development are discussed at length in chapter 6.)

Hitt (1985) provides the following principles that inform the administrative functions of staffing and staff development:

1. A clearly stated organizational mission should be the beacon that guides the staffing function.
2. A human resource plan should be developed to serve as a road map in the staffing and staff development function.
3. Selecting the right people at the right time is essential to effective staffing; mistakes in this regard are expensive.
4. The organization should provide an effective staff training program to meet the need for new knowledge and skills.
5. Long-term career development should not merely be "left to chance"; it should be systematically planned.
6. The primary purpose and final result of performance appraisal should be the improvement of job performance.
7. Each element of staffing and the staff development program should be evaluated periodically in terms of cost/benefit and contribution to organizational goals.

Directing

Directing is leading employees in a manner that achieves the comprehensive goals of the organization. This involves proper allocation of the organization's resources and providing an effective support system. The premise underlying the function of directing is that the organization's goals are accomplished through the work of other people. Thus, a good director must have exceptional interpersonal skills and be able to motivate subordinates. (Refer to the section on leadership in this chapter.)

Approaches to Directing Subordinates

Current leadership models offer approaches for directing subordinates. According to the *transactional leadership* model, leaders motivate subordinates by appealing to their self-interest. Motivation strategy focuses on meeting employees' needs. The transactional leader should be alert to the different kinds of needs that individuals feel and should strive to fulfill these needs to the greatest extent possible on the job. The reader can refer to psychological theories of human needs, such as the one developed by Abraham Maslow (1998). According to the transactional model, it is helpful for an administrator to know as much as possible about the traits of the individual employees being supervised. With this kind of information and insight, the administrator is better prepared to work effectively with and for the employees in helping to fulfill their needs while working toward accomplishing the organization's goals.

Empirical studies on the effectiveness of transactional leadership yielded mixed results, however. In the mid-1970s, James McGregor Burns introduced the concept of *transformational leadership*. Transformational leaders went beyond satisfying individual needs (or tried to shift the focus to higher psychological needs such as self-actualization). Generally, they focused on a common purpose, addressing intrinsic rewards and developing commitment with and in subordinates. According to this model, leaders and followers raise one another to higher levels of motivation. Burns saw transformational leaders as individuals who appeal to higher ideals, such as justice and equality (Méndez-Morse, 2001).

In a dynamic environment, the transformational leader must be able to impose change upon the organization. To do this, an administrator must successfully carry out the following activities (Daft, 1998):

1. *Create a new vision*. The leader must help the organization break free of old patterns and structures, processes, and activities that are no longer useful, and must spread his or her strategic vision throughout the organization.

2. *Mobilize commitment*. The leader must garner widespread acceptance of the new vision and must mobilize commitment among all the units in the organization.

3. *Institutionalize change*. The leader must see to it that the new values and practices are permanently adopted. This means that major resources must be devoted to transforming the organization. In short, the new vision must be "institutionalized."

People Emphasis versus Production Emphasis

One of the crucial issues in directing employees is the correct balance of emphasis on people versus emphasis on production. Some would argue that such a choice is unnecessary because one need not exclude the other—the emphasis can be placed on both. However, different administrators seem oriented more in one direction than in the other, and circumstances sometimes influence the balance of emphasis. By its very nature, education is a people-oriented enterprise. However, education also has a product (e.g., the educated student), which is at the focus of the group effort. Clearly, both emphases must be taken into account.

Studies of organizational climate in schools have led to operational definitions of production emphasis and people emphasis, employing the respective labels "initiating structure" and "consideration." *Initiating structure* reflects the extent to which the leader attempts to organize work, work relationships, and goals. A leader high in initiating structure emphasizes schedules and specific work assignments, establishes channels of communication, and sees to it that the followers are working up to capacity. *Consideration* reflects the extent to which the leader maintains job relationships characterized by mutual trust, respect for subordinates, and regard for their feelings. A leader high in consideration listens to staff members and is approachable. Several studies have found that teachers perceive a higher commitment to organizational goals under a leadership characterized by high consideration; regardless of the level of initiating structure (John & Taylor, 1998).

The specific relationship between these two areas of emphasis has been analyzed in some detail. Leaders exhibiting consideration and initiating structure behaviors can be grouped into four quadrants.

In figure 4-1, a *Quadrant I leader* is low on consideration and high on initiating structure. This leader is production-oriented and interested in getting the work done, often forgetting in the process that he or she is dealing with human beings. The *Quadrant II leader* demonstrates both consideration and initiating structure behaviors. Such a leader is efficient and effective in managing both people and tasks. The *Quadrant III leader* is high on consideration but low on initiating structure. This leader maintains a friendly relationship with the subordinates and is concerned about subordinate welfare but is ineffective in getting things done. The *Quadrant IV leader* is low on both consideration and initiating structure. This leader's management approach results in a lack of group cohesion and task ineffectiveness.

The message seems clear: to be an effective leader/director, one must demonstrate a high level of people skills balanced with an optimal level of production emphasis. One strategy that good managers use to accomplish this is to make their subordinates feel personally responsible for their work, and encourage them to report regularly on their achievements.

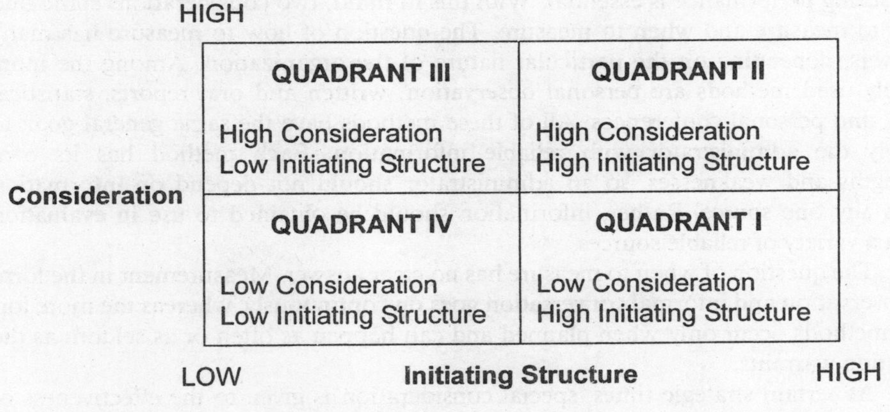

HIGH

QUADRANT III High Consideration Low Initiating Structure	**QUADRANT II** High Consideration High Initiating Structure
QUADRANT IV Low Consideration Low Initiating Structure	**QUADRANT I** Low Consideration High Initiating Structure

Consideration

LOW   **Initiating Structure**   HIGH

Source: Adapted from Masih & Taylor, 1998.

Figure 4–1. Leadership behavior quadrants for initiating structure (production emphasis) and consideration (people emphasis)

Controlling

Recall the adage, "A chain is no stronger than its weakest link." Administrators don't always know in advance which link of the operation is the weakest or when it will break. Therefore, controls are needed to ensure quality in all areas and to detect potential or actual deviations from the organization's plan. The control function's purpose is to ensure high-quality performance and satisfactory results while maintaining an orderly and problem-free environment. It involves setting and achieving performance standards and applying the organization's policies and procedures. (Policies are discussed in chapter 5.) The control function is essentially a monitoring process for the purpose of identifying and addressing weaknesses. It might also be viewed as an ongoing process of evaluation that involves defining standards, measuring actual results against the standards, and taking appropriate actions to prevent problems and overcome weaknesses. It requires keeping a stern hand on the rudder during turbulent times.

Information Management

Schools, like other organizations, have been influenced by the information revolution. To effectively control today's organizations, administrators must understand the technology of gathering and utilizing information to implement controls. An *information management system* is a mechanism used for collecting, organizing, and distributing data throughout the organization. information systems have five basic components: input, processing, storage, control, and output (reports). Control is a device for determining that the information is of sufficient quality and is timely and complete enough to have relevance for its utilization. The controlling function also must-have the capability to change the output (Starling, 1998). Computers play an indispensable role in facilitating information management systems.

Measuring Performance

If the organization is to know whether its work is proceeding according to plan, measuring performance is essential. With this in mind, two considerations stand out: how to measure and when to measure. The question of how to measure has many answers, depending on the particular nature of the organization. Among the more widely used methods are personal observation, written and oral reports, statistical data, and personal conferences. All of these methods have the same general goal: to supply the administrator with reliable information. Each method has its own strengths and weaknesses, so an administrator should not depend on information from any one source. Rather, information should be obtained to use in evaluation from a variety of reliable sources.

The question of when to measure has no clear answer. Measurement in the form of observation and informal conversation goes on continuously, whereas the more formal methods occur only when planned and can happen as often or as seldom as the situation warrants.

At certain strategic times, special consideration is given to the effectiveness of the organization and its personnel. For example, at the end of the school year, certain factors are critically evaluated to determine what changes would be beneficial. Another strategic time is when a vacancy in a position occurs. It is then appropriate to evaluate the effectiveness with which the job has been performed, the strengths and weaknesses of the person who vacated the position, whether the position should continue, whether the job description should be altered, and exactly what qualifications should be required of the one who will assume the position. Major financial and narrative reports also provide strategic opportunities for evaluation and control. In education, measurement focuses on the quality of the programs, student progress and achievement, and teacher effectiveness. (Financial controls are discussed in chapter 7; assessment of programs and staff is covered in chapter 11.)

Taking Corrective Action

Measurement and evaluation can focus attention on deviations from accepted procedures and standards, and this should naturally lead to the next step—appropriate corrective action. In some cases, only minor adjustments are needed. Occasionally, however, the results of evaluation indicate a need for major changes. Thus, in one case corrective action might involve a change of procedure, a minor shift of resources, or a realignment of personnel so that the organization can regain its direction; another case might call for a major reformulation of plans or a major shift of emphasis.

Behavior Aspects of Control

The objective of control is to get people to work more effectively toward the organizational goals, and especially to identify and eliminate weaknesses in the total process. The administrator, in exercising the control function, must consider how those who are affected by it will react. Controls are often unpopular among those being controlled. The administrator must be prepared to deal with negative responses. A negative reaction can often be avoided or lessened if the manager is careful about interpreting and implementing control measures. Important in this respect are: (1) the clear interpretation of organizational goals, policies, and procedures, and (2) the education of employees about the control measures necessary to achieve them. Employees must be able to see how their work contributes to the goals of the organization and appreciate that measuring individual performance is legitimate and necessary. Employees will better understand the reasons for adjustments when the results of evaluation clearly show a need for corrective action.

The administrator should encourage employees to participate in setting performance standards that are both challenging and attainable. Schools often require teachers to set their performance goals in writing at the beginning of the school year. They are then reviewed at year's end. The value of this exercise cannot be overemphasized because it is almost a certain guarantee of personal commitment. Low standards fail to motivate employees toward achievement, whereas standards that seem so high as to appear remote will fail to stir enthusiasm and are likely to result in disinterest and discouragement. If standards are to induce positive results, they must be challenging, realistic, obtainable, and mutually agreed upon.

MAKING DECISIONS

Decision making is inherent to all of the previously discussed administrative functions. Making a decision means selecting one course of action from various alternatives. Decisions involve judgments based on a combination of information and reasoning. They often involve breaking big tasks into smaller parts, thereby simplifying the problem and the process. Teaching an individual to make good judgments is difficult—some would say impossible. One can learn how to organize the steps in decision making; how to avoid certain mistakes and pitfalls; and how to gather, organize, and utilize facts. Yet, even within the framework of all this, some administrators are much better at making sound decisions than others.

Decisions can be categorized into problem-solving and goal-oriented decisions. *Problem-solving decisions* are generally made to facilitate routine day-to-day operations. *Goal-oriented decisions* weigh the anticipated results in light of the organization's purpose and mission. Both kinds of decisions are necessary for the effective functioning of the organization. Some administrative decisions have little significance, whereas others have great effect. It is necessary to clearly recognize the importance of the decision and its potential consequences. Decisions are often labeled *right* or *wrong*, however, these are extreme labels. Terms that are more meaningful are *strong* or *weak*, *appropriate* or *inappropriate*, *sound* or *unsound*, *good* or *poor*. Decisions have a better chance of being sound if they are made in an organized manner. Five basic steps are involved in decision making (Starling, 1998):

1. Identifying the problem (or opportunity).
2. Gathering facts.
3. Making the decision (choosing among alternatives).
4. Implementing the decision.
5. Evaluating the decision.

The following discussion takes the reader through the first three steps. (The last two steps loop back to the administrative functions of directing and controlling.)

The first step in decision making is for the administrator to identify the problem accurately. Sometimes, decision makers misconstrue the real problem and then apply the wrong solution. Part of this task is to determine whether the problem is unique or generic. Most problems that administrators face are generic—part of a pattern stemming from underlying causes. The underlying causes are seldom obvious, however, so administrators tend to see each problem as a unique and isolated event. The result of this tendency is to treat the symptoms of the problem rather than its root causes. Consequently, administrators end up making more decisions than are necessary. Administrators should avoid cosmetic solutions to problems.

Once the problem is defined, the next step is to frame the response (i.e., to determine what are the upper and lower limits of the decision). Upper limits, which refer to how far an administrator can go, include: (1) legal limits, (2) limits of available resources, (3) limits of available time, (4) limits due to previous commitments, and (5) limits based on available information. Lower limits refer to what *at least* must occur to solve the problem. The final decision must fall within these parameters (Starling, 1998).

Actually, most decisions probably begin with an opinion. Administrators tend to rely too much on personal opinions and on past experience—both mistakes and successes—in making decisions. Personal experience is not a reliable guide in decision making. Administrators must try to visualize the broad picture and recognize that their personal experience makes up a small and inadequate sample. When gathering facts, administrators should cast their net widely, looking at others' experiences, consulting experts, examining the norms, and collecting statistical data.

Analytical Decision Making

Effective administrators do not rely totally on facts in decision making. They employ a particular style or model of thinking. One such model is *analytical thinking*, which attempts to break decisions down into components and to define specific problems by isolating them, thus making them manageable. Analytical thinking can incorporate *cost-benefit analysis*—an attempt to assess the costs and benefits of a decision. Real benefits are those derived by the clients. In education, the primary clients are the students—among others in the community. Administrators attempt to look at direct benefits (e.g., improved learning) as well as indirect benefits (e.g., better public relations), sometimes called by-products. Benefits are either *tangible* (which means measurable) or *intangible*. In education, it is more difficult to measure tangible benefits than in the private sector, which uses profits as a measure. Costs also can be direct or indirect. For example, when a college decides to utilize a grant to fund a new program, this decision often requires support services that must be covered by the operating budget for the term of the grant and beyond. These indirect costs must be taken into account (Starling, 1998).

Analytical thinking also employs *decision rules*. One type of decision is a yes–no decision. The administrator's decision to implement a program is based on whether the benefits of implementation exceed the costs. A second type of decision involves a choice between two mutually exclusive projects, each of which would be analyzed for its cost-effectiveness before deciding which one to implement. A third type of decision involves the administrator determining at what level a solution will be implemented, based on budget constraints. In addition, there are *opportunity costs* attributable to doing one thing rather than another, which causes the organization to forego other opportunities. For example, by deciding to build a new building, the school loses the opportunity to utilize money in a renovation fund to modernize an existing building.

Administrators may employ *decision trees* in these situations. A decision tree utilizes decision forks to indicate consequences and to help avoid pitfalls. The primary decision (D1) is divided into alternative *x* and alternative *y*. Either alternative leads to subsequent decision forks (D*n*), which help to ascertain the end results (Starling, 1998). Figure 4-2 illustrates a simplified decision tree for the above example.

Decision making often employs the systems approach. This technique requires the organization to identify key decision makers, who then look at the problem as a system with interdependent components. The systems approach often relies on modeling. Models allow the administrator to convert real-world situations into simplified and controlled representations, which are easier to conceptualize. Computer simulation allows the replication of actual situations with a high degree of accuracy. For example, complex inventory systems may be worked out in model form before implementation. One of the advantages of simulation derives from the counterintuitive nature of systems. In other words, things often don't react in ways we think they should. The systems approach reveals the unreliability of human intuition—the "seat of the pants" approach—to making decisions (Starling, 1998).

Obtaining Expert Opinion

Administrators must recognize their own limitations and seek assistance from experts who better understand the particular problem being addressed. In this regard, the considered judgment of one expert is often more useful than the opinions of

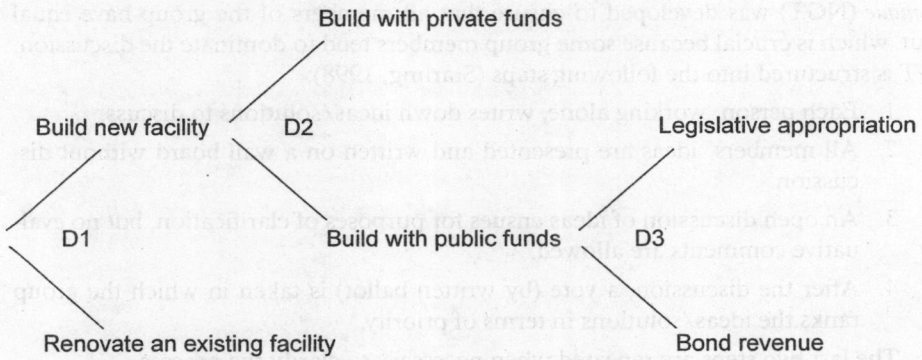

Figure 4–2. Decision tree for a building project

numerous nonexperts—the latter often consisting of little more than pooled igno-rance. This approach assumes that experts can be readily identified for the area in which the decision is being made.

Conversing with experts is not a means of letting others do your thinking, but rather a means of letting them help you think more clearly. One advantage of consult-ing an expert is that the interchange of ideas has a stimulating effect on the thought processes of the decision makers. The exchange often produces ideas that would not have occurred to either person in the absence of the interaction.

Sometimes the cost of consulting an expert from outside the organization seems prohibitive. However, administrators should appreciate that an expert can bring special training and new insights into a situation. Not using an expert might turn out to be more "costly" in time delays or wrong decisions due to improper analysis of the situation. Expert opinions can play a particularly important role in decisions involving technical knowledge (e.g., computer networks, natatorium management, financial accounting).

Group Decision Making

Recall that one principle of participatory governance is that individuals affected by decisions have a voice in making those decisions. Administrators should seek ways to more fully integrate staff consultation into decision-making processes. Staff partici-pation in decision making is associated with improving employee morale and esteem. In addition, participation in decision making causes the staff to be more supportive and cooperative in the implementation of decisions. A fundamental reason exists for relying on the group's expertise in decision making. Since groups bring together a broader perspective for decision making, group decisions are often more innovative than individual expert decisions. Although participatory decision models require a number of preconditions, these preconditions are usually met in the school setting.

Administrators should be cognizant of disadvantages as well as the advantages of group decision making. Group decisions are more complicated and time consum-ing, which is why groups should not be employed to make routine administrative decisions. Sometimes group dynamics lead to unsatisfactory compromise solutions, or groups may stifle dissenting opinions. Groups should be formed to work on specific problems in areas where they have expertise. Finally, the administrator needs to con-sider the potential group members' current workloads and their expectations about involvement, as well as their conflict-resolution skills.

Several strategies are available for group decision making. *The nominal group technique* (NGT) was developed to ensure that all members of the group have equal input, which is crucial because some group members tend to dominate the discussion. NGT is structured into the following steps (Starling, 1998):

1. Each person, working alone, writes down ideas/solutions to discuss.

2. All members' ideas are presented and written on a wall board without dis-cussion.

3. An open discussion of ideas ensues for purposes of clarification, but no eval-uative comments are allowed.

4. After the discussion, a vote (by written ballot) is taken in which the group ranks the ideas/solutions in terms of priority.

The last two steps are repeated when necessary to clarify the process.

A valuable technique implied in NGT is *brainstorming* to optimize creative think-ing and generate solutions by suspending criticism. Members freely generate ideas subject to certain rules, while judgment and evaluation are withheld until the idea-generating period has been completed. The greater the number and more radical the ideas, the better the chance to arrive at an innovative solution (Starling, 1998).

Faculty members historically have had the broadest role and greatest influence in decisions regarding curriculum and personnel (especially tenure and promotion). Within the last several decades, faculty participation also has become relatively well-accepted in institutional planning and the selection and evaluation of administrators. Traditionally, faculty involvement in budgeting has been limited by both administrative resistance and faculty ambivalence. However, in many educational settings faculty members are now more involved in budget planning, and they are gaining a better understanding of the technical basis and political dynamics of the budgetary process. There is also evidence that governing boards are becoming more sophisticated about the importance of obtaining meaningful input from the faculty in the decision-making process.

Cautions in Decision Making

Be aware of biases in human decision making. Clearly, individuals have limitations to how rational they can be. Sometimes, individuals choose the first alternative that satisfies the minimal decision criteria. This is known as *satisficing*. Busy administrators often lack the time to make complex decisions; so they must satisfice. Other biases in thinking have been identified (Starling, 1998):

1. Seeing only one dimension of the problem.
2. Giving too much weight to available or recent information.
3. Being overconfident. (We often mistake confidence for competence.)
4. Ignoring the laws of randomness. (Trends are extrapolated from a single event.)
5. Reluctance to analyze important decisions retroactively.

Timing is also important in decision making. Generally, administrators should avoid "quick fixes." In military combat quick decisions may save lives, and there are other settings where such an approach is necessary. However, in the educational setting, *timeliness* is a better standard. Likewise, an administrator should avoid *procrastination*. Most of us have a tendency to avoid making difficult decisions. It is enticing to wait for more information, to think it over longer, or to hope a development will occur that would make the decision easier, but too much waiting can be detrimental. Although there are occasions when making a decision should be delayed, effective administrators are able to make decisions expeditiously. Once a decision is made, it is important to stick to it, unless ample evidence surfaces to justify reversing it. Successful administrators make decisions in a timely manner and change them slowly, whereas making decisions slowly and changing them precipitously is a characteristic of unsuccessful administrators. Definite goals make it easier to reach sound decisions.

Sometimes decisions result from default. Administrators should recognize that "deciding not to decide" is a decision in itself. The absence of a conscious or forthright decision can lead to desirable or undesirable consequences and may result in a course of action quite different than would be the case if a decision were made.

A related problem is the practice of taking half measures. A *half measure* is a form of temporary compromise that only postpones a more permanent solution. Occasionally, administrators settle on half measures in order to stave off unpleasant action, or to settle on a compromise that will pacify those who resist forthright action. It is true that a half measure is occasionally appropriate. Sometimes it can furnish a temporary solution and buy time that is truly needed to work out the best long-range solution, but a half measure is detrimental when used as a hedge against the full measure that is really needed.

To summarize, educational administrators are faced with myriad decisions ranging from the routine (purchasing supplies) to the exceptional (program review). Decision dilemmas are often complex and carry significant implications for the decision maker, the staff, and the organization. The administrator must first decide who should make the decision. If the decision is indeed best made by the administrator, he or she should accumulate pertinent information in a timely manner; prioritize it based on the perspective of sources, the context, and omissions; consider environmental constraints (politics, finance, and local culture); and attempt to apply the principle of "least harm . . . greatest good." By using this approach to decision making, the administrator will improve the quality of decisions and develop a sense of trust and predictability among colleagues (East, 1997).

TIME MANAGEMENT AND STRESS MANAGEMENT

Time management is vital to administrative success. Efficiency is the art of managing time and causing it to serve you fully and well. However, the efficient use of time is an elusive goal. Studies show that administrators are frustrated by having more work to do and more people to see than they can satisfactorily handle in the time allowed. Although many administrators work longer hours than others do, and have above-average energy and powers of concentration, they often feel the stress of unfinished tasks and unresolved problems. Underlying this predicament is the concern that unfinished administrative work means delayed decisions and disrupted productivity. This situation must be addressed, for in order to manage others well, managers must first manage themselves. Unfortunately, the busier a person becomes, the less time there is to analyze procedures and to plan more effectively. Consequently, some administrators are always scrambling to keep from being overwhelmed by the mounting job tasks.

Time should be allocated in accordance with the priorities of one's responsibilities. A study of time use conducted by the American Management Association (1986) drew three conclusions about administrators: (1) They thought they were spending more time on certain responsibilities than was actually being spent. (2) They had failed to delegate enough of their responsibilities to subordinates. (3) They were spending too much time on relatively unimportant tasks. It became clear that administrators' use of time was not in accordance with real priorities. Periodically, every administrator ought to go through this kind of analysis.

Keep a detailed record for a week of exactly how your time is spent. This can be a real eye opener! Most people have analyzed their financial expenditures by keeping a record of everything spent and have improved their personal budgeting by this method. It is equally possible to improve the management of time by the same method. Figure 4-3 provides a sample form for recording and analyzing how an individual spends time.

Strategies for Managing Time

Control Your Schedule.

Parkinson's first law states that "work will expand to fill the time available." Most administrators never have enough time. When time is dear and job pressures mount, the conscientious person is inclined to press harder and harder to get things done. This process can lead to a vicious circle resulting in reduced time for planning. The consequent state of frustration and fatigue make working time less and less efficient. If you find yourself in this circumstance, take time to analyze your situation in a calm and collected manner. Arrive at a sensible plan for improving time management;

NAME _____ DATE _____

Hour	Description of Time Use
8:00	
9:00	
10:00	
11:00	
12:00	
1:00	
2:00	
3:00	
4:00	
5:00	
Evening	

Summary and Analysis of How Time was Used (Hours spent)

Meetings _____

Telephone _____

Paper work (routine) _____

Special problems _____

Reports _____

Teaching _____

Research _____

Luncheons and social _____

Other _____

Figure 4–3. Example of a time analysis form

then apply the solution with confidence and composure. When you are under time pressure, it is better to work smarter than harder.

Prioritize Your Attention.

According to Parkinson's second law, "we tend to devote time and effort to tasks in inverse relation to their importance." Indeed, administrators often spend a disproportionate amount of time on relatively unimportant matters. There is a tendency to give prompt attention to items that are simple and uncomplicated, routine and repetitive, because with such matters success is easy and immediate. In such an "activity trap," administrators become so involved in activities that they lose sight of the more important aspects of the job. Realistically, a busy administrator cannot afford too much time dealing with $100 matters if $1,000 problems are being neglected. It is true that routine items such as key distribution, custodial services, and paper supply must receive attention by someone, but these matters should be kept in proper perspective. They must not infringe too much on the time needed to address the main objectives of the organization.

Learn to Discriminate.

Although the term *discrimination* usually carries a negative connotation, it also has a positive dimension. A person in an administrative position must be highly discriminating relative to (1) which people (and how many) are permitted to occupy his or her time in conferences, interviews, and telephone conversations; (2) which particular reports, surveys, and other written material require the most time; and (3) which parts of the job can be handled directly as opposed to those that can be delegated. Unfortunately, many administrators are unable or unwilling to discriminate clearly between more important and less important matters.

Avoid Long Conversations.

Some people who aspire to be administrators are unsuccessful because they lack the ability to keep routine conversations short and to the point. Most conversations involving the administrator must be thorough yet relatively brief. When the important business has been covered, the conversation should end. A desirable knack in conversation is to bring up the important points and lead the discussion to a conclusion without wasting time, but without causing the conversation to seem rushed. Furthermore, it is important for the administrator to keep a proper balance in conversations and prod others to do the same. One-way conversations tend to be uninteresting and often unproductive. Conversations should result in an adequate exchange among those involved. The administrator is usually the one expected to give direction to the conversation and to keep it on track. The effective use of the other person's time as well as your own is an important consideration.

Manage Paperwork Efficiently.

A certain amount of paperwork is associated with administration, but it can easily become counterproductive. For example, it is not necessary to make and keep on file a copy of every routine written communication. Neither is it desirable for two or more officers in the same organization to file the same kind of information. Furthermore, there is no need to put every kind of communication in written form. In short, an organization easily can overburden itself with too much paper handling and filing. Many highly successful administrators "travel light" in this regard. The key is the ability to discriminate between what is necessary and what is not in terms of producing, handling, and filing written material. Technology has altered the decision to a choice of whether information should be stored in a computer file or maintained as a "hard copy."

Another aspect of paperwork is expediency in its handling. The ideal is to handle each paper only once, when possible. When you look at a particular written correspondence for the first time, decide what to do with it. Sometimes the decision is easy—discard it. In other cases, it should go to the secretary for filing. Perhaps a telephoned response should be made immediately, or perhaps a quickly-dictated reply would be better. Some papers should be passed on immediately to others in the organization; others may require delayed attention. Specific information about how to cope with this problem appears in chapter 5.

Avoid Excessive Travel.

Travel falls into two general categories: daily travel within the vicinity of your work, and out-of-town trips. Some people spend hours per day traveling to and from work and running errands. Keep in mind that time spent in this manner is time away from administrative duties. Not much can be done about commuting time unless you are willing to change the location of your residence. Perhaps reducing unnecessary errands has more time-saving potential. Are you spending time making deliveries that could be handled by a secretary or other office employees? Procrastination can cause too many last-minute hand deliveries that could be accomplished by mail in a more timely fashion.

Excessive out-of-town travel also can cause problems for administrators. Some who enjoy professional trips find various excuses to go when the need is suspect. If there is a legitimate reason to travel, perhaps another staff member could accomplish the objective of the trip as well or better than the chief administrator could. A few administrators seem to believe that there is a direct relationship between time spent on professional trips and job effectiveness, but this is not consistently true. An administrator should not resist a legitimate need to travel, as one's presence may be required at certain events, and one can easily lose touch with people with whom contact should be maintained. However, it is equally true that for some administrators less travel and fewer days away from the office would improve overall effectiveness. If you must travel, taking a laptop computer on trips will allow you to work while traveling.

Taking Advantage of Clerical Help

Is the office secretary underutilized? If you have a secretary who spends a majority of the workday doing strictly clerical chores, you may not be receiving the assistance you need. Consider some areas in which you can use more assistance. A secretary can save you precious time each day by planning a schedule, making appointments, acting as a barrier to interruptions by dealing with visitors, and enforcing short meetings by entering when a visitor's allotted time is up. The secretary who knows the details of your interactions with others can handle many phone calls and draft replies to routine correspondence. Delegating some of your regular tasks to a secretary can free up your time.

A veteran secretary who knows a great deal about the office can help you make decisions faster and more effectively. Someone who is familiar with the work flow and knows where bottlenecks develop or communication breaks down can suggest ways to keep up with the growing workload. This may mean devising and supervising more effective methods of filing, record keeping, and other paperwork. Efficient clerical assistance will allow you to concentrate more on important matters and spend less time dealing with crises that can grow out of existing workloads.

Other Time-Saving Ideas

Categorizing time usage is an effective strategy. List the things you have to do in order of importance. Work on the most difficult items first, at a time of day when you have a high level of energy and concentration. Similar activities should be grouped together (e.g., making phone calls, opening mail). Plan blocks of time when you can work uninterrupted. Some administrators find it profitable to come to the office early. Successful administrators who have good work habits have learned how to make themselves work without trifling or taking detours.

Some administrators advocate the *open-door policy*, which means that employees can come in without an appointment. While such a policy can contribute to congenial relationships and a friendly atmosphere, in most cases it is not the best way to achieve accessibility. Open doors often increase the wrong kind of communication—trivial and unproductive socializing. Interruptions are multiplied, and the administrator is frequently distracted from important matters. For administrators with demanding schedules, an unlimited open-door policy is usually a poor one. Controlled access through a congenial but discrete secretary is a better choice.

The *ten-minute head start* technique for effective time management is worth considering. Whenever you leave the office, time your schedule so that you can leave 10 minutes earlier than necessary, providing a time buffer to cover unforeseen delays or unexpected phone calls. This technique helps to guarantee promptness, which will have a positive influence on your image and the promptness of others, and it removes tension and frustration caused by last-minute rushing. Furthermore, it provides a little "gliding" time, which can be used constructively for thinking about the next important event.

Utilize current technology to save time. Consider the benefits of such devices as a laptop computer, a personal digital assistant (PDA), a cellular or cordless phone with enhancements (e.g., call screening), or a digital voice recorder. Such devices are not panaceas, however. Contrary to popular advertising, the latest pocket-sized, wireless techno-gadget alone won't automatically save time on the job. Often these gadgets simply redefine the task at hand. The trick is to embrace a technology only after it's been proven to be useful but not to wait so long that you miss out on important benefits. Such devices have both advantages and drawbacks (e.g,. a pager can be a boon to highly mobile administrators who must remain accessible, yet it can prove to be interruptive at times.) Also consider the learning curve; new technologies are increasingly sophisticated and require time-consuming effort to master (Bergeron, 1998).

E-mail is both a blessing and a curse. Used properly, e-mail can save time. Most organizations have established intranets (LANs) by which employees communicate with each other. This facilitates the distribution of announcements, meeting agendas, memoranda, and other important communications. However, these networks are often misused. It's far too easy to send a message with the click of a button. Consequently, e-mail consists of an indiscriminate mix of both the important and the trivial. Administrators can spend an inordinate amount of time sorting though incoming messages. One solution may be to establish policies for the use of e-mail. Some administrators have assistants or secretaries screen e-mail, but there is no simple solution to the problem. Voice mail can be subject to some of the same types of abuse.

Finally, restate your job description in terms of duties, responsibilities, authority, and accountability. Ask yourself what is expected and needed by way of continued professional preparation. This kind of analysis will help clarify whether your job is representative of the official description that appears in the operational manual. Perhaps your perspective of the job is different from that seen by others in the organization. If so, these inconsistencies should be resolved.

Once your job is seen in perspective, no matter how pressed you have been for time, it is likely that with reasonably good organization, adequate delegation, and good relationships your responsibilities will become more manageable. This is a logical assumption because, for most positions, a predecessor was able to handle the job and people in comparable positions are able to do so. The problem of insufficient time usually originates with the person rather than the position. If the job content is reasonable, then proper time management should serve as an effective solution. If job responsibilities really are too great, it will become clear by means of analysis, and this will help provide a basis for corrective action.

Controlling Stress and Tension

There are administrators with oversized desks and stuffed briefcases who are ulcer-ridden bundles of complexities, obsessed with the idea of success and burdened with endless responsibilities. This, of course, is a stereotype, but it depicts a prime candidate for job-related stress. Too many administrators feel immense strain in their jobs and suffer from tension that robs them of both enjoyment and efficiency. It is not possible to measure exactly the loss of efficiency caused by emotional stress. However, psychologists are able to point to a considerable body of evidence that efficiency suffers when one works under too much stress. Excessive stress not only cuts into productivity; in extreme cases, it also can disable an administrator. Even a modest reduction in emotional stress can often produce a sharp upswing in creativity, foresight, and judgment, and this can add significantly to the effectiveness of the organization.

Some people are more susceptible to stress than others are. A few individuals generate their own private emergencies and remain more or less constantly on guard against their versions of disaster. This is a fatiguing approach to life. Some stress-ridden administrators may be uncomfortable and perhaps unhappy, but they are not completely ineffective. However, learning to alleviate stress can make their lives and their jobs more productive and satisfying.

Clearly, too much stress can be destructive. But what about too little stress? What happens to efficiency when people are too secure, too comfortable, and too complacent? The answer is that efficiency tends to decrease. People have to be motivated if they are to achieve. Most people have to be at least a little bit apprehensive and keyed up to do their best work. Emotional stress is not, therefore, an altogether undesirable thing. A moderate level of stress may actually improve performance. Stress becomes harmful only when it exceeds a reasonable level of tolerance (becomes acute) or when it persists too long (becomes chronic).

Too much stress can throw judgment out of kilter so that it takes longer to make decisions, and it may even prevent a person from making a decision at all. It can narrow one's perspective to a highly personal viewpoint in which the objectives of the organization become secondary or obscure. It can exaggerate the magnitude of small problems at the expense of more important matters. It may inhibit an administrator's ability to solve problems or cause an administrator to obsess about a problem to the extent that it interferes with other items of importance. Stress can change a smooth, integrated thought process into a series of broken plays.

Sources of Job Stress

It is unlikely that a person can alleviate stress unless its sources are identified and controlled. Following are possible causes of stress among administrators (American Management Association, 1986):

1. *Inadequate command of the situation* can be both a cause and a result of stress. Stress can develop if a person is not holding things together, and problems develop faster than they are solved. Thus, the administrator becomes more of a problem solver than a forthright leader. The mode of operation is reactive rather than proactive.

2. *Pressures resulting from lack of delegation* might relate to the situation described in Item 1. If the administrator tries to give direct attention to too many responsibilities, it becomes impossible to cover them adequately, with the result that the job becomes overwhelming. When a person advances up the managerial ladder, there is a tendency to want to retain some responsibilities associated with one's former position. However, one cannot keep close tabs on every activity that occurs in the organization.

3. *Wavering confidence* or distrust in one's own ability can cause stress to increase rapidly in pressure situations. Less confidence, more stress, less success, even less confidence: it becomes a vicious cycle.

4. *Distrust of subordinates* can cause one to feel nervous about whether others are carrying out their responsibilities adequately.

5. *Insecurity* about whether others in high positions are displeased with one's performance may be a product of one's imagination or could be caused by lack of communication.

6. *Feelings of guilt about holding authority* cause some administrators to regret being separated from the rank and file, and they feel nervous about their suitability for a position of authority and the responsibilities they hold.

7. Other potential sources of stress, such as *problems at home, personal finances,* and *poor social relationships*, can carry over to the job and diminish an administrator's effectiveness.

Managing Stress

A variety of stress-management techniques are identified and described in some detail in most college health textbooks, including biofeedback, meditation, yoga, imagery, and a variety of relaxation techniques. A good diet, regular exercise, and sufficient sleep are prerequisites to managing stress. Stress management techniques also can be accessed online at *Mind Tools*. (See Online Sources at the end of the chapter for this site's URL.) Unfortunately, the person who suffers from stress often is unable to identify and control it. Once it develops, the individual may be unable to think clearly, analyze objectively, or respond to the situation. This is why organizations need to implement a stress management program for administrators and staff (see chapter 6).

References and Recommended Readings

American Management Association. 1986. *Workshop Notes in Time Management.* New York: Authors.

Arnold, J. D. 1978. *Make Up Your Mind: The Seven Building Blocks to Better Decisions.* New York: AMACOM.

Arveson, Paul. 1998. "Selecting a Management Approach." Rockville, MD: The Balanced Scorecard Institute. Online article: <http://www.balancedscorecard.org/basics/selecting.html>.

Bergeron, Bryan. 1998. "Taming Time with Technology." *Digital Doc.* Online article: <http://www.postgradmed.com/issues/1998/01_98/j98_dd.htm>.

Chandler, Ralph, & Jack Plano. 1988. *The Public Administration Dictionary.* 2nd ed. Santa Barbara, CA: ABC-CLIO.

Clarke, Lee B. 1999. *Mission Improbable.* Chicago: University of Chicago Press. Chapter one, "Some Functions of Planning." Online: <http://leeclarke.com/mipages/short_fd.pdf>.

Drucker, Peter. 1999. *Management* (paperback ed.). Woburn, MA: Butterworth-Heinemann.

East, Whitfield. 1997. "Decision-making strategies in educational organizations." *Journal of Physical Education, Recreation & Dance,* 68(4): 39–45.

Goslin, Lewis, & A. J. Rethans. 1986. *Basic Systems for Decision Making.* 3rd ed. Dubuque, IA: Kendall/Hunt.

Gulick, Luther H. 1937. "Notes on the Theory of Organization." In Luther Gulick and Lydal Urwick (Eds.), *Papers on the Science of Administration,* pp. 191–95. New York: Institute of Public Administration, Columbia University.

Hitt, William. 1985. *Management in Action.* Columbus, OH: Batelle.

Masih, John, & J. W. Taylor, V. 1998. "Leadership Style, School Climate, and the Institutional Commitment of Teachers." Graduate School, Adventist International Institute of Advanced Studies. Silang, Cavite, Philippines. Online paper: < www.aiias.edu/academics/sgs/info/v2n1/john_institutional_commitment.html>, 26 pp.

Maslow, Abraham. 1998. *Maslow on Management.* New York: Wiley & Sons.

Méndez-Morse, Sylvia. 2001. "Leadership Characteristics that Facilitate School Change." Southwest Educational Development Laboratory. Online article: <http://www.sedl.org/change/leadership/welcome.html>.

Starling, Grover. 1998. *Managing the Public Sector.* Fort Worth, TX: Harcourt Brace.

Yutang, Lin (Ed.). 1938. *The Wisdom of Confucius.* New York: Random House.

Online Sources

ERIC Clearinghouse of Education Management
http://eric.uoregon.edu/
This site includes trends and issues, hot topics, and a list of publications on various aspects of educational administration.

ManagementLearning.com
http://managementlearning.com/
The site offers a variety of topics, exercises, articles and reviews on management.

Mind Tools Bookstore. "How to Manage Stress," Mind Tools, Ltd.
http://www.psywww.com/mtsite/smpage.html
This site offers a discussion of how to master stress, including understanding stress, finding your best level of stress, stress management techniques, reducing long-term stress, and how to make your environment less stressful.

Work and Stress
http://www.nottingham.ac.uk/business/i-who/work_and_stress/
Work and Stress is a quarterly journal of stress, health, and performance.

Part II

Management Techniques

5

Management Skills and Procedures

We live in an era in which administration has become a matter of managing information as well as people. An important aspect of this management function is seeing that the right people get the right information at the right time. This chapter includes material about communicating effectively and about techniques for managing staff meetings, committees, private conferences, and an office—all of which facilitate the availability and exchange of information. The application of the guidelines set forth here can contribute significantly to improved effectiveness as an administrator.

COMMUNICATING EFFECTIVELY

Communication is the sharing of information. Verbal communication has two essential dimensions: speaking and listening. Barriers to adequate verbal communication include different languages, different mindsets due to background and experience, and inability to speak clearly or to listen. Written communication also has barriers. The ability to write and read effectively is fundamental. Communication problems are not new. More than 2,000 years ago, Plato recognized three reasons for miscommunication: (1) people want to be right all the time, (2) people refuse to stick to the subject, and (3) people do not know how to listen.

In all aspects of communication, it should be recognized that clear thinking is indicated by the proper selection of words. Words are the symbols that enhance our ability to think and express our thoughts. Administrators need to cultivate a superior vocabulary and a knack for employing the right words to make a point.

The process of communication is illustrated in the five steps: message conceived, message sent, message received, message understood, and message applied or implemented. The total purpose is not accomplished unless all of the steps are completed.

For a busy administrator, it is important that all aspects of communication, from highly official to unofficial, be performed effectively. Following are some important considerations that can improve one's communication ability.

1. What information should be communicated, to whom, and for what reasons?

2. How much explanation should be given?

3. Should the information be labeled confidential or restricted, or should no limits be placed on it?

4. Is it important for certain individuals to receive the information directly from the administrator?

5. Should certain persons receive it before others?

6. Through what sources should the information be passed, and how fast does it need to move?

7. What risk is there that the information will be misinterpreted, and what can be done to reduce the risk?

8. Will certain individuals create their own versions, thus distorting the information?

Some communication can better be accomplished *orally*, whereas other messages should be in *written form*—memos, letters, directives, or reports. Still other information is transmitted by mood, expressions, and gestures.

Oral Communication

Important prerequisites for effective oral communication are: speaking audibly so those listening do not miss words or phrases, using correct volume and tone to make the words appealing and interesting, and selecting words and statements that are easily understood. Placing the proper emphasis on words and phrases helps listeners to understand the true meaning. Keep wordiness, unfamiliar terms, and meaningless jargon to a minimum. As a general rule, business conversations should be concise and to the point. Oral presentations should be well thought out, carefully organized, and should follow a predetermined plan.

Many administrators effectively use silence (often combined with facial and body expressions) more or less unconsciously. However, conscious use of the technique of silence can add a subtle dimension to your communicative abilities. It can convey an attitude or opinion without making a commitment. Rapid talkers avoid the pause, and as a result they are more susceptible to repeating themselves, making misstatements, or overworking a point.

Since pauses can be as much a part of communication as words, administrators should strike a balance between the two in order to communicate effectively. A silence can have a cooling effect on an emotional situation. It can help to recapture the attention of those in a meeting or force a response from another person who wants to keep the conversation moving. Following a particular statement, it can give the statement special emphasis. During a negotiation, a period of silence or no reply places the pressure on the other person to keep the negotiation alive.

Written Communication

Good writing is much like good speaking. Writing, however, tends to be superior because one has more opportunity to organize and express thoughts clearly.

A written communication should have a theme that leads to certain conclusions. Some writing experts advocate stating the conclusions first and then developing the points that support those conclusions. Others believe that one should develop the theme and then state the conclusions. Whichever approach is taken, the theme must be developed logically and in as few words as possible, and the conclusions should be stated concisely and clearly.

Clarity is of utmost importance. One reason for vagueness and rambling is a lack of clarity in the writer's mind. When a memo or letter doesn't come off exactly the way you had planned, carefully analyze the exact purpose of the communication. With the purpose clearly in mind, take another look at its content and ask these questions: Did I accomplish the purpose in a simple, clear, and direct fashion? Does the communication contain any extraneous material? Is there any important information that needs to be added?

Organization is basic to clarity. The thoughts in a memo or letter should proceed in a logical sequence, one point flowing into the other from start to finish. The principle of logical organization applies to the overall content as well as to individual paragraphs, sentences, and even phrases. The ideas should run in a sequence that is easy for the reader to follow and absorb. To accomplish this, think your letter through step by step. Even small changes in content and sequence will often make a big improvement in its flow and clarity. These writing qualities aren't accidents—they are the end product of clear thinking and effort. Most well-written communications have gone through several drafts to maximize their organization and clarity.

Style is as important to writing as is organization. Keep your sentences and paragraphs "clean" and free of useless jargon. Use words and phrases that are clear and precise. Avoid words that might be unfamiliar to the reader or confuse the meaning of a statement. Be cautious about words that have generalized or relative meanings.

The tone of a memo or letter is influenced by the selection of words and phrases. Attention to tone can make a written message sound rude or polite, reserved or enthusiastic, personal or impersonal, complimentary or critical.

Usually *memos* are used for written communication within the organization. Memos normally contain these elements: name of receiver, name of sender, date, subject, body, and signature.

A *business letter* format is normally used to send communications outside your own organization. Only on rare occasions is the letter format used within the organi-

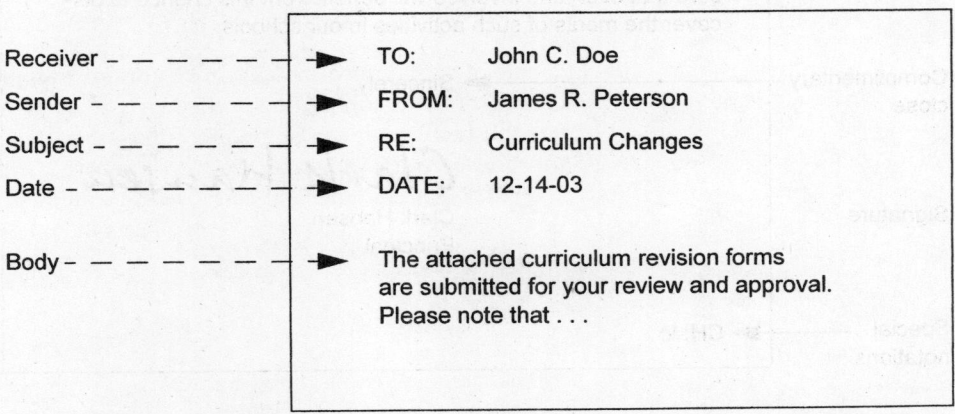

Figure 5–1. Components of a memorandum

zation. A business letter normally has the following elements: return address, date, inside address, salutation, body, complimentary close, signature, and special notations (if appropriate). In preparing the letter, give special consideration to the theme, conclusions, coherence, and tone.

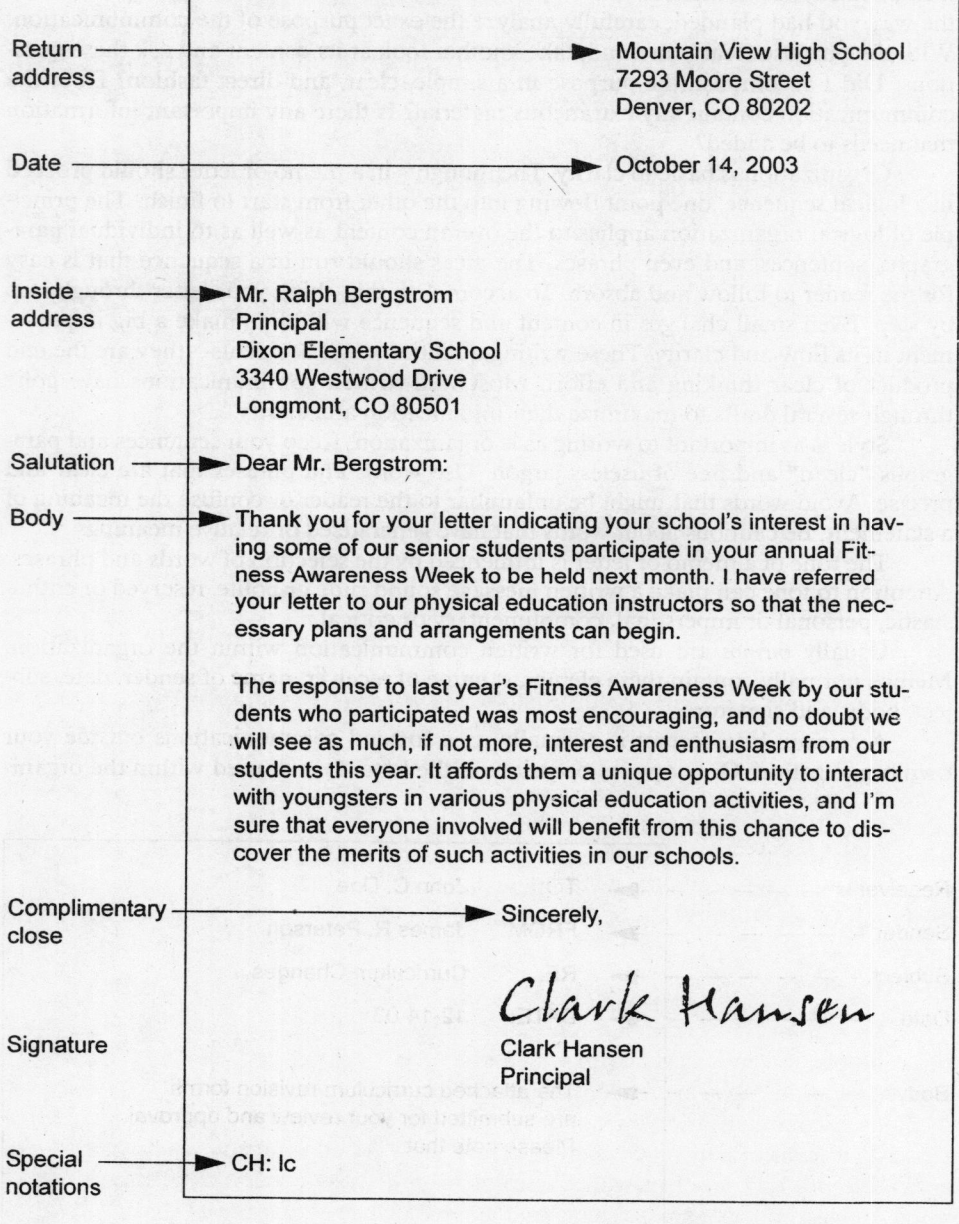

Return address	Mountain View High School 7293 Moore Street Denver, CO 80202
Date	October 14, 2003
Inside address	Mr. Ralph Bergstrom Principal Dixon Elementary School 3340 Westwood Drive Longmont, CO 80501
Salutation	Dear Mr. Bergstrom:
Body	Thank you for your letter indicating your school's interest in having some of our senior students participate in your annual Fitness Awareness Week to be held next month. I have referred your letter to our physical education instructors so that the necessary plans and arrangements can begin. The response to last year's Fitness Awareness Week by our students who participated was most encouraging, and no doubt we will see as much, if not more, interest and enthusiasm from our students this year. It affords them a unique opportunity to interact with youngsters in various physical education activities, and I'm sure that everyone involved will benefit from this chance to discover the merits of such activities in our schools.
Complimentary close	Sincerely,
Signature	Clark Hansen Clark Hansen Principal
Special notations	CH: lc

Figure 5–2. Components of a business letter

A *proposal* is a form of written communication based on a substantial amount of thought and study, because acceptance or rejection is often influenced by how well it is prepared. A proposal normally includes a cover letter explaining its purpose and nature, recommendations, and justifications.

Registered and *certified mail* are sometimes useful methods of communication. Both methods insure proof of delivery, as the recipient must sign a return receipt. Use either method when you want ensure that an important message or material reaches the intended destination. Certified mail is sufficient for most situations. Registered mail is slower and more expensive but is the safest way to send an item (which can also be insured against loss).

E-mail is appropriate for both internal and external communication under certain circumstances. It is much easier to compose and send, and it saves postal costs. Generally, e-mail is considered more informal than other forms of correspondence. For business correspondence, the sender should always employ upper-case letters according to grammatical conventions and review with Spellchecker before sending. E-mail messages have the added advantage of remaining as a record on the sender's computer. E-mail has almost completely replaced the use of telegrams (with the exception of money transfers).

A *written report* often consists of a combination of primary and secondary information (with primary information originating with the report's author and secondary information coming from other written sources). Usually, a report includes a letter of transmittal, an introduction, the main body (presentation of data), a summary, and conclusions. Sometimes the summary and conclusions are placed at the beginning and sometimes at the end. A report must be well thought out, so that it contains the essential information arranged in logical sequence and stated clearly and concisely. It is important that the information be current, accurate, and pertinent. The readability of a report and the impressions it leaves often have a substantial influence on its acceptability and the results it generates.

The written communication should not be overworked because it becomes cumbersome if done too extensively, and it is a less personal approach than a direct conversation. On the other hand, written communication can be beneficial because the opportunity exists to make sure that it says exactly what is intended, copies can be sent to other individuals, and it can be referred to in the future if need be. The reader also has the advantage of being able to study written communications to make sure the full meaning is understood. Written communication may be more efficient and convenient than oral communication when the accessibility of the recipients is an issue.

Mood

An administrator communicates either intentionally or unintentionally by mood. For example, the enthusiasm with which one greets a coworker can send a clear message. In addition, facial expressions, tone of voice, and the amount of attention given are methods of communication by mood. Some administrators purposely use expression of mood as a form of language, whereas others show mood unintentionally or inadvertently. The primary problem with this element of communication is the possibility of being labeled moody, inconsistent, and unpredictable. Administrators are generally more successful if they portray an even temperament and use a consistent approach.

Body language conveys attitude. Listening with legs crossed and arms akimbo expresses a lack of openness. Eye contact should be natural. Being on the same eye level as the person with whom you are conversing is less threatening than standing over them. Physical proximity should feel comfortable for both parties.

Emotional Responses

People in administrative positions sometimes feel a need to put somebody in place, set the record straight, discipline a subordinate, or react firmly to a colleague. It is easy to become emotional in situations of tension and stress, and there is a tendency to make sharp and sometimes abrasive statements. Remember that once something has been said it is on the record, and it will never be completely erased. Be sure that you say what ought to be said and what you would want to say when not influenced by emotion. Keep in mind that the administrator is expected to be mature, objective, and composed.

When you feel emotional about a situation and want to lash out, try to hold back until you have had ample time to think it over. If you are going to respond in writing, follow the motto "write now, send later." In other words, after writing the memo or letter, put it in your desk drawer for a day or more while your emotions settle; then read it. If you honestly feel that it should be sent, go ahead; but there is a good chance you will decide to rewrite it or not send it at all. Be very cautious about making hasty statements that you later may wish you could retract. Even though you cannot always control what happens, you can control your response to it.

Listening

Listening is the other side of communication. It is surprising what an administrator can learn about an organization and its people just by listening. In addition, there is an important psychological aspect to listening; being attentive to what others say is satisfying to them. Careful listening is especially useful in the case of grievances. A helpful technique is to repeat what the person stated, so that person knows you heard and understood what was said. Often, when a person has thoroughly stated a grievance in the presence of the administrator, the problem is more than half solved. The remainder of the solution takes some checking and follow-through, which is equally important.

There are many mind blocks to careful listening. Unfair prejudgments and stereotyping can prevent a person from effective listening. Anger or other emotions can also get in the way, as can sloppy habits in paying attention. Good listening is more than just letting sounds pass through your ears. It involves concentrating on the meaning of what is said and recording the important points in your mind (remembering).

Listening cannot always be casual and relaxed. Serious listening requires precise attention and serious concentration. It is not just hearing words; it is also implies understanding, interpreting, evaluating, and appraising intent and content. Often, effective listening involves "reading between the lines." Sometimes it includes interpreting what a less-than-effective communicator is trying to say. It also involves watching expressions and gestures, recognizing emphases on words and phrases, and mentally attaching the correct meanings to the statements. Some people hear but do not listen; others listen and hear but do not understand. Effective listening includes *listening, hearing,* and *understanding.*

Physical proximity may be a very real factor in determining the frequency of oral communication. Consequently, office layout is an important aspect of the communication system. How an office is arranged affects communication during conferences.

Sitting behind a large desk conveys a different attitude than inviting the visitor to sit next to you around a table.

Additional Ways to Communicate

Policy manuals (discussed later in the chapter) are used to communicate organizational policies and practices that have a relatively permanent application. Such manuals are often called administrative manuals. Two important uses of manuals are: to acquaint new organization members with the written policies and procedures, and to add consistency to the application of the organization's administrative methods. One inevitable result of the preparation and use of manuals is to increase the centralization of decision making, because a manual standardizes decisions that have previously been left to individuals. Many organizations currently maintain their policy manual online to make it more accessible to employees.

Telephones are effective instruments of communication if they are used efficiently. Have a secretary screen your calls, and set aside specified times in the day for returning calls. Grouping similar activities such as reviewing telephone messages and returning calls can help you use time more efficiently. When speaking on the phone, identify yourself immediately, be courteous, get to the point quickly but not in a hurried or abrupt fashion, and keep a pad and pencil handy to record information. Most administrators now employ a cell phone as well as a desk phone. Because a cell phone greatly increases one's accessibility, administrators should be selective in giving out their cell phone number.

Using the telephone can help lay the groundwork for a future meeting. This mode of communication allows for an exchange of ideas without the travel time of a face-to-face conference. A conference call among several people might accomplish as much as getting everyone together, and at a much lower cost in terms of both time and expense. It is also possible, with the use of a telephone speaker, to link large groups together for a free exchange of ideas so that everyone has access to the information. Voice-mail systems are another time-saving device that has become standard in offices.

Reading

Reading administrative material is another important aspect of communication. Concentrate on what you are reading by avoiding distractions. Underline or highlight important statements with a marker, or write notes in the margin or on a note pad. Remember that some material should be read thoroughly in an effort to get the full meaning, whereas other material can be skimmed for broad concepts and high points. For items that must be read thoroughly, find a place and schedule a block of time for concentrated reading without interruption.

STAFF MEETINGS

Administrators get involved with staff meetings of three different kinds: (1) meetings with other administrators, such as a department chair with area coordinators, or an athletics director with head coaches; (2) faculty meetings, including all or a portion of the teachers in the administrative unit; and (3) meetings with nonteaching personnel or students. Too often, staff meetings are viewed by those who attend as ineffective and inefficient. Thus, effective handling of staff meetings is an important administrative challenge. This involves several considerations: when and how often to call meetings, whom to include in each meeting, how long the meeting should last, and which items should appear on the agenda.

Unfortunately, a high percentage of meetings are less than completely successful due to lack of preparation and poor leadership. A successful meeting usually requires a strong leader who is dedicated to the participative process and knows how to implement it. It is the leader's job to see that the discussion moves toward the purposes of the meeting. The Standard Oil Corporation (subsequently AMOCO Oil and now British Petroleum), in its administrative manual, described a meeting chair as someone who. ". . . must plan, promote, lead, direct, inform, interpret, encourage, stimulate, referee, judge, moderate and conciliate."

Clear meeting objectives and time limits should be established and adhered to. If possible, the discussion leader should have a clock or a watch within view. With a time frame in mind and with a written agenda, the leader is obligated to move the meeting forward at an appropriate pace, and to accomplish the purposes of the meeting within a reasonable amount of time.

In the planning stages of a meeting, several important questions must be considered.

1. What are the purposes of the meeting? This should be answered clearly in the opening remarks. If members begin to digress, restate the purposes.

2. What is needed in the way of facts and figures? Should the leader present these or let someone else do it? Who is best prepared to do it?

3. Should certain information be copied in advance and distributed ahead of time or during the meeting?

4. Will there be any visitors? If so, are they fully oriented as to what is expected of them?

5. Would visual aids help? A picture can be worth a thousand words, but only if it is used effectively.

6. Would a demonstration be useful? Sometimes a well-planned demonstration can make a point that people otherwise will not grasp.

Be especially cautious about expounding in tedious detail, monopolizing the discussion, or permitting anyone else to do so. Former President Dwight Eisenhower claimed he once received the following advice from an aide: "Never neglect an opportunity to keep your mouth shut." Tactfully silencing others requires gentle diplomacy combined with firmness. It often seems that those who have the least to say take the longest to say it, but eventually even the most long-winded person has to take a breath. When he or she does, the leader can promptly take command of the discussion or pass it to someone else who will keep it moving.

An executive of a successful company offered this useful point (Tyson, 1980):

> Every person on our staff has better ideas than mine, but my job is to get the jury to agree. Most of the decisions made in our meetings are based on compromises, because many of the good ideas are not compatible with each other. However, occasionally, a diehard in the group, unwilling to bend, blocks the process of all the others. In such cases, I have to state a reminder that the purpose is *not to win a point but find a solution*.

One administrator who found that too much time was being wasted in meetings made a rule that everyone who wished to present a problem had to prepare and submit a memorandum answering these questions: What is the problem? What is the cause of the problem? What are the possible solutions? Which solution do you recommend? The number of meetings was drastically reduced as a result of this procedure, because the solution usually became apparent without the need for a meeting.

Meetings can be highly useful for training, building group morale, decision making, grievance settling, creative brainstorming, encouraging group action, and gathering or disseminating information. Unless a meeting is really necessary, however, it should not be held. When you call your next meeting, make it truly worth its time. Be sure your meetings do not support the claim of one meeting critic who said, "You can accomplish work or go to meetings, but you can't do both."

A Written Agenda

Meetings that can benefit from a formal structure should have a written agenda. For example, an agenda should always be considered for a faculty or staff meeting but often is unnecessary for committee meetings. A good rule to follow is, "If in doubt, prepare an agenda." When possible, distribute the agenda ahead of time (e.g., via e-mail) so that attendees can peruse it and bring it with them to the meeting.

An agenda offers several potential advantages. It forces decisions as to what should be included in the meeting; helps guarantee premeeting preparation; permits the arrangement of items into logical groupings and desired sequence; and provides an overview of what is going to be discussed, which gives each person a better opportunity to contribute to the discussion. Time is more effectively utilized with a written agenda, because it helps the discussion move systematically from one item to the next.

Rules of Order

Formal meetings, particularly those with a large assembly and/or an intricate business agenda, are facilitated by utilizing rules of order. These may be a version of Robert's Rules of Order (see Online Sources at end of the chapter) or a set of rules developed within a particular organization. If rules of order are used, a parliamentarian should be appointed to adjudicate disputes and interpret the rules. Rules of order do not serve well when several attendees are ignorant of the rules and others "in the know" use their knowledge to control the meeting or stifle dissent. Many school staff meetings proceed adequately without employing formal rules of order. Occasionally, rules of order are resorted to ad hoc, when the need arises. Groups such as faculty senates, which operate under a constitution or bylaws, commonly use rules of order.

COMMITTEES AND CONFERENCES

Committees and conferences can be among the greatest consumers of an administrator's time, yet guided group thinking is necessary for the proper conduct of any organization larger than a one-person shop. The potential fault lies not in the tool but rather in its use or abuse. The ability to conduct an effective conference or give productive leadership to a committee is a necessary administrative skill.

A *conference* is a meeting with one or more people for the purpose of analyzing a situation and receiving useful input or mapping out a plan to be enacted. A *committee meeting* between the administrator and a task force or standing committee is held for the purpose of receiving information from the committee or giving direction to it.

Some administrators attempt to apply conference and committee techniques where they have little chance of success. Conferences and committees are no substitute for executive action. Following are three useful guidelines:

1. It is advantageous to involve members of the organization in conferences and committees in order to capitalize on their ideas and expertise and to allow them meaningful input into management.

2. On the other hand, it is important to avoid obtaining judgments and opinions and then failing to give them fair consideration. When this happens, coworkers recognize that they are only being taken through the motions—that the conference or committee procedure is perfunctory and inauthentic.

3. Conferences and committee meetings may be highly formal or relatively informal. If the meeting is formal, an official record should be kept of the proceedings, decisions, and recommendations. In a less formal situation there may be no need for a written record, although sometimes a brief memorandum of the high points of the discussion serves well as a matter of clarification and record.

Leading the Discussion

In the conduct of a conference or a committee meeting, the leader serves as the catalyst who draws out information and gives guidance to formulate a plan of action based on the specialized knowledge of those present.

The right psychology must be employed by the group leader in order to handle potential difficulties. For example, if the group consists of a mixture of higher- and lower-ranking employees, the lower-ranking personnel may be reluctant to enter the discussion. One or more dominating personalities may intimidate others in the group. Perhaps one member talks too much but has relatively little to offer, or another leads the discussion astray.

It is important to have the members of the conference or committee say as much as they want without having the discussion become exhaustive. In this regard, the leader must exercise the right balance between thoroughness and expediency. It is undesirable to cut off any member's comment by a sarcastic or preemptory remark. Curbing or ending a certain line of discussion should be done tactfully.

Unfortunately, some individuals have a tendency to use almost every kind of meeting for gripes and petty complaints, which has a deteriorating effect on a meeting's productivity. Such people are often excluded from important conferences or significant committee assignments because of this characteristic. When they are included, the discussion leader faces the challenge of letting them participate without allowing a disruptive or diluting influence.

The leader should avoid stifling a discussion when a disagreement is stated. In a situation in which the differences are so pronounced that the meeting does not seem to be progressing, it is helpful to work toward agreement on some generalities as stepping stones toward the specifics to be accomplished.

If the discussion is to be continued in a subsequent meeting, it is a good idea to end the meeting with a summary of what has been accomplished and where the next meeting will begin. The time and place of the next meeting should be set, and it should be clarified whether anyone is expected to do preparatory work prior to the meetings. If the purposes of the conference or committee have been accomplished, the leader should make this clear and should summarize the results and conclusions that have been reached.

Following are important characteristics of the conference or committee leader:

1. An ability to define the problem clearly and break it down into logical parts for analysis.
2. A quick mind to handle the interchange of ideas and keep them flowing along the right track.
3. Tact in eliciting or controlling discussion as needed.
4. An open mind and a sincere interest in the opinions and judgments of others.
5. A sense of timing to know when to conclude the discussion.

Additional Guidelines

1. It can be awkward to be a leader of a committee whose direct supervisor is a member. Other participants might tend to side with "the boss," thus reducing the effectiveness of the group.
2. If the leader has high authority over the other committee members, it has the potential of squelching the free exchange of information.
3. Each participant in a conference or committee should be qualified to make a contribution. It is a waste of time to include unqualified people unless their participation is viewed as a developmental experience for them.
4. From the standpoint of economy, the following should be avoided:
   a) Duplication of talents
   b) Involvement of those who cannot afford the time
   c) A larger number of participants than is necessary to accomplish the purposes
5. Simple rules of courtesy or protocol are important:
   a) Give sufficient advance notice.
   b) If a subordinate in another department is going to be involved, clear it with the head of that department.
   c) Give proper recognition and introduction to participants from outside your organization.
   d) When you must cancel a meeting, give prompt notice and an explanation to all participants.
6. Make a practice of starting on time so that everyone will learn to expect this and act accordingly.
7. Make a serious attempt to keep meetings within normal business hours.
8. At the outset of each meeting the leader should describe its intent, and the meeting should end with a summary statement that caps the discussion and states what is to happen next.

Divided Attention

Imagine yourself sitting in conference with an administrator for one hour and accomplishing only 30 minutes of work because the administrator spent the other 30 minutes conversing on the telephone or speaking with intruders. A coworker who is in the administrator's office on legitimate business deserves a reasonable level of, if not undivided, attention. This is especially true if the person is there by appointment. The administrator is guilty of inadequate attention when he or she reads paperwork, reviews the schedule book, or writes notes not related to the conversation. Allowing such distractions and interruptions shows lack of consideration for the person who has come to the office for a face-to-face meeting. It is inefficient as well as offensive to coworkers. An administrator will accomplish more overall by doing only one thing at a time and doing it well.

OFFICE MANAGEMENT

Day-to-day office activities are an important part of the overall functioning of the organization. Although a central office is usually needed, not all office work takes place in that location. Often, office work is decentralized in order to serve the organization efficiently. Some of the usual office procedures are discussed in this section.

Receiving Visitors

The manner in which people are received when they come to the office has a significant influence on their impressions of the organization. Since the receptionist has contact with many more people than does the administrator, the importance of this job function becomes apparent. Consistent application of these guidelines will help assure the effectiveness of the receptionist.

1. A person who comes to the office should receive a pleasant greeting and immediate attention.

2. The receptionist should be knowledgeable about how to deal with routine matters and where to direct matters that can be handled better elsewhere.

3. The receptionist should act on the principle that every visitor is important and every visitor's problems are important, at least to the visitor.

4. When possible, the receptionist should enhance the use of waiting time by providing information preparatory to the visitor's appointment.

5. The receptionist's function should not include idle conversation or entertaining of visitors. Reception is a business procedure that deserves the right balance of friendly formality and efficiency.

Organizing

Defining office employees' responsibilities and working relationships is a key to effective organization. Intraoffice business relationships and the particular duties of each person should be clearly understood; overlap or confusion reduces efficiency. Another concern is office space utilization. The placement of furniture and equipment in offices is important from several points of view: attractiveness, efficiency, traffic flow, and intraoffice relationships. The challenge is to use the available square footage for optimum effectiveness. A third aspect of organizing is the commitment of time. Appointments should be properly spaced to accommodate the necessary business without causing any waste of time. Appointment times should be clearly communicated so there is minimal chance of misunderstanding. Employees should maintain the established office hours; doing otherwise detracts from efficiency and contributes to a poor image.

Controlling Paperwork

As you go through your mail, deal with as many items as possible at the time. Pass appropriate items to the secretary for handling. Items that can be handled better by other employees should be sent on to them immediately. Certain items that require dictation can be answered right after they are read or can be grouped together if you have a scheduled time for dictation. Items that must wait for one reason or another should be filed by a method that will bring them to your attention again on the appropriate timetable. Following are some additional helpful procedures:

1. In the file drawer of your desk have a folder for each month of the year, and use them for filing materials that should come to your attention at a later date. For example, if you receive paperwork in June that should be handled in September, place it in the September file.

2. Also, have a tray on your desk labeled *immediate*, one labeled *tomorrow*, and one labeled *next week*, and place papers in the respective trays for timely handling.

3. Keep file folders labeled with the names of people with whom you often do business. When an item comes across your desk that should be handled through conversation with a particular person, drop it in that person's file; then make it part of the agenda for the next time the two of you meet.

4. Certain papers are marginal in terms of whether they should be placed in the office files. In order to safeguard against premature disposal and, at the same time, keep the files clear of unnecessary material, have a drawer that you view as a *temporary wastebasket*. Drop the items in the drawer and let them remain there for a safe period of time. After a certain period (e.g., once each month), move these papers to a location that you view as a *semipermanent disposal* and let them remain there for a longer time—several months. Any item that you have not needed during this time will probably never be needed. This system keeps unnecessary papers from cluttering up the files.

Processing forms is another aspect of paperwork. Every organization has certain standard forms to expedite management procedures. In most cases, the forms are prepared to furnish duplicate copies as needed. A purchase order is an example of a standard form. It is important to have an adequate supply of current forms, and office employees need to be knowledgeable about where to retrieve them and how to process the forms properly. Recording complete and accurate information on the forms is of obvious importance. It is desirable to keep a copy of each standard form that is processed.

Record Keeping and Filing

Questions as to which records to keep and the methods of keeping them require discriminating decisions. Even with the widespread use of computers, the need remains for filing hard copies of some records. Some administrators are guilty of "keeping everything and being able to find nothing," whereas others are too casual about keeping records. The following questions are pertinent to the record keeping of any organization, and the answers will serve as useful guidelines for establishing a simplified, streamlined, and crisp approach.

1. Should a secretary routinely prepare a duplicate of correspondence and file it, or should only an original be printed out? Should copies be filed on the computer or as hard copy?

2. By what system should office copies of standard forms be filed, and how long should they be retained?

3. Should all material kept on record be placed in file folders, or should some be placed in loose-leaf binders or kept as bound copy?

4. Which material should be filed in the secretarial office, and which should be in a different place in order to provide the right balance of accessibility and confidentiality?

5. How should confidential records be protected, and who should have access to them?

6. How often should the files be cleared of unnecessary and obsolete material, and who should be responsible for this?

7. Which kind of filing system should be used?

Certain decisions, responses, policies, and procedures should be in writing to reduce the chances of miscommunication or misunderstanding about what was said or decided. The written record can be referred to later, if the need arises. Examples that justify written records are financial agreements, contractual agreements, and policy statements. Usually, the minutes of a meeting furnish an adequate record of its proceedings. Sometimes decisions and agreements that justify a written record are reached outside of formal meetings. In these cases, it is a good idea to follow up with a letter or memo stating your interpretation of the important elements of the discussion. This helps to ensure that you and the others agree, and it provides a written copy for reference.

OFFICE AUTOMATION

Doing today's work with yesterday's tools is a pitfall that certainly applies to office management. Fortunately, many timesaving devices have been developed to enhance office work, such as computers, fax machines, scanners, and word processors.

Preparing and duplicating written materials are among the most common office tasks. These materials must be done on time, in correct form, and with acceptable quality. Frequent decisions must be made about which materials should be photostatically reproduced, duplicated, or printed.

A great deal of time can be saved if you learn to *dictate* correspondence and notes. Dictaphones are giving way to computer speech-recognition technology. Microsoft Office XP software includes a dictation mode compatible with word processing programs. Handheld recording devices are available that allow dictation to be played directly into computers via a scandisk external drive port for automated transcription by speech-recognition software; the transcribed text may be edited by the user or by office staff. The competent manager will plan for a state-of-the-art office and implement it on a timely schedule. It will be more efficient, more economical, and more effective overall.

Two developments have changed the way offices operate: electronic information processing and computer networking. In today's offices, most of the information is transmitted electronically. Networking computers for integrated office communications (intranet or LAN) allows instant transfer of information and work from office to office and from computer to computer, permitting several computers to share work on complicated problems for quick processing. As a result, management will have greater access to a "total information system."

Putting together an up-to-the-moment status report, complete with graphics, is greatly simplified by word processing and spreadsheet software. An administrator is able to call up such a report on the computer or make it available on a teleconference basis to several people inside or outside the organization.

All these technological changes impact the work habits and skills of administrators. They need to learn keyboard skills at least well enough to operate a computer. Improvements in voice-recognition systems someday may completely bypass keyboard use by administrators, but it may be several years before a system exists which is sensitive enough to handle large vocabularies. Administrators of the future will be able to dictate their correspondence directly into printed or electronically transmitted form.

The Paperless Office: Myth and Reality

Clearly, electronic data management is eliminating a large portion of the paperwork involved in administrative procedures. Personal computers with high-speed processors and at least 40-60 GB of hard drive memory have become the standard. CD-RW and DVD-RW with UDF reader/writer software allow backup storage of much more data (up to 4.7 GB) than floppy disks.

Computer software can provide verification of signatures and time of entry. These will be as foolproof as advanced technology can make them and will become admissible as legal documentation. Software programs exist with built-in verification that data have been received as sent without error. The dividing line between hard copy (print) and soft copy (electronic) may become blurred. Deciding when and how to make a record of a communication or data within an organization or between organizations will be a matter of administrative protocol.

Some paper will still be in use, although much less than is currently the case. The biggest change is that once paper has served its purpose, it will be dumped, shredded, or converted to electronic storage. Offices will store information, not paper. Much space now devoted to file storage will be free for other purposes, but files will still be necessary. Paper has several advantages over the electronic medium: It is tangible (can be held in hand), spatially flexible (can be flipped through), tailorable (can be annotated and scribbled on), passed back and forth, and spread out on a desk and visually scanned. Anyone who has proofread a long document for coherence appreciates the advantage of paper over a computer screen. Paper's major disadvantages: it can't be quickly searched, stored, or remotely accessed.

The reader should realize that any textbook on state-of-the-art office technology is near obsolete by the time it is published. Textbooks are not the best sources with which to stay abreast of technological developments. Trade journals and Internet Web sites such as ZDNet (see Online Sources at the end of this chapter) are much better sources. Administrators need to keep up with developing technology and utilize available resources. Most school districts and colleges employ a technical support staff who organize tutorials and workshops, as well as install new technology.

Physical Arrangements

Automation is redesigning the environment of the office. In some cases desks are giving way to "workstations" that resemble airplane cockpits. Voice communication, computers, and other freestanding equipment are built in or set into the workstation. Desks and chairs are positioned to fit the user at the touch of electronic controls, much like the seats in a luxury automobile.

Noise from telephones (with distinctive rings), copiers, and fax machines require office designers to come up with new ideas for acoustical shielding. The "open office" design is being replaced with small, private workstations with surrounding sound barriers.

Laptop and handheld computers are becoming part of the administrator's standard traveling equipment. These serve as workstations for the road with easy access back to the office via the nearest telephone or built-in modem in the hotel room or in one's home. Increasingly, satellite connections are becoming an option.

Applying Controls

An important aspect of office management is keeping a tight rein on matters that must be controlled. For example, issuing keys requires exact procedures, careful control, and accurate record keeping; otherwise the security plan will deteriorate (see figure 5-3). Office supplies should be controlled at the right level to avoid waste and still allow employees to feel that supplies are adequately available. Forms such as requisitions and budget adjustment forms should be available only to those who have official use for them. Student records should remain confidential as stipulated by law The right degree of control on the use of office machines—computers, word processors, typewriters, photocopiers, fax machines, and calculators—is necessary in order to avoid misuse and breakage. It is useful to have an access code and a fee schedule per sheet when the machines are used by personnel from other departments or for nonofficial purposes. This can help pay for supplies, maintenance, and replacement. Controls also are needed on the availability and flow of official information.

The Athletics Director's Office

Advances in computers and related technology have dramatically changed the operation of departments of athletics. In the director's office, calendars posted on the walls, notebooks holding various types of information, and numerous filing cabinets have been replaced with networked computers, personal digital assistants, fax machines, cell phones, and pagers. Several good computer programs are available that will keep track of schedules, officials, staff, and data on athletes. These programs are invaluable to the modern athletics director. Most of these programs employ e-mail to send contracts and reminders to schools, officials, and employees. The personal digital assistant (PDA) is making an impact on the entire athletics staff. These devices can send and receive e-mail, and programs are available for storing and retrieving statistics. The modern high school athletics program will need a major budgetary line for technology. Each head coach needs a PDA or portable computer, along with access to video equipment. The administrative staff will require access to networked computers, fax machines, copiers, portable computers, collaters, folding machines, cell phones, and pagers. Intelligent application of technology makes managing athletics departments infinitely easier than in the past (Denney, 2002).

REQUEST FOR ONE KEY

Issue one key to _____

Home Address _____

Phone _____ Date _____

I will not duplicate this key.

I will not loan this key.

I will return this key when my need
or employment terminates.

Faculty _____

Staff _____

Student _____

Other _____

(explain)

_____

Position

_____   _____

Office Address   Extension

_____

Signature of Person to whom Key is Issued

Key to fit _____

Building _____

_____   _____
Supervisor   Signature

_____   _____
Administrator   Signature

Do Not Write in
This Space

Hook No. _____

Date _____

Issued by _____

Posted by _____

Rec. by _____

Figure 5–3.   A key request form

POLICIES AND PROCEDURES

Policies

A *policy* is a governing statement that defines the appropriate action to follow under given circumstances. Policies are an important element in providing broad measures of control. They should be written to reflect the principles and values consistent with the mission of the organization. They are further attuned to the functions of the organization and to particular circumstances.

Many policies pertaining to a local situation are developed within a broad framework of higher authority. At the highest level, the Constitution of the United States sets forth certain conditions that affect policies in organizations throughout the nation. Within this context, educators must comply with: (1) equal rights legislation as it applies to differences in race, sex, and age; (2) the concept of separation of church and state; and (3) the various conditions inherent in due process. State codes contain various requirements that influence the policies and procedures of political subdivisions, including school districts and state colleges and universities. In addition, the State Board of Education establishes policies and procedures that apply to the local educational units, including such items as the number of days school must be in session, teacher certification requirements, subject requirements, and regulations pertaining to the school tax base. Within the framework established by higher authority, school district officials establish local policies.

The primary purpose of policies is to serve as guides in administrative actions. Some policies are strict in nature and not subject to exception, whereas others are considered guidelines and are subject to wide interpretation. A school district might have the policy of *keeping parents informed about student absenteeism*, but the exactness with which each school implements the policy might vary. Likewise, it may be the policy of a university that *faculty members who are newly employed will be issued contracts by August 1st*; however, unforeseen circumstances might necessitate exceptions to this policy. Nonetheless, if the basic policy is sound, it should be followed whenever possible. Generally, organizations should strive for consistency in the application of policies. Lack of consistency suggests arbitrariness, results in misinterpretation and confusion, and can have a demoralizing effect on staff and students. Policies and procedures should be viewed as dynamic. They should be kept current and undergo changes as the circumstances justify.

Standard policies and procedures are normally presented in writing, and the written form contributes toward their standardization. Nonwritten policies tend to be less standardized and may appear arbitrary or capricious. Consequently, setting policies in writing is recommended. Every organization of moderate size and complexity ought to have a set of written policies and procedures to serve as operational guidelines. Such statements can contribute to streamlining the organization and reducing confusion and misunderstanding. These statements should deal with matters of significance. If too many detailed issues are included, the organization becomes overstructured and loses the desirable amount of flexibility. Ideally, policy and procedure statements should be concise, easy to understand, and deal only with the matters of the operation that need to be standardized.

Other terms are employed in conjunction with policies, including: *regulations, rules, standard operating procedure (SOP)*, and *standard practices*. Some organizations draw a distinction between policies and regulations. Policies are broad statements that can be interpreted and applied widely over a number of similar situations. *Regulations* are more specific in addressing defined practices in particular situations. Regulations should be derived from policies and not be contradictory in intent.

Writing Policies and Constructing a Manual

As a general rule, the axiom that "many hands (or heads) make work easy" is true in developing policies and procedures. Involving several people in policy development has two potential advantages: it results in useful input, and people feel better about adhering to policies and procedures when they participate in developing them. It's a good idea to involve a variety of people with a wide range of perspectives.

Various approaches have been used successfully in creating policy statements. One method is to identify the topics for which written statements are needed, and then assign a committee to formulate the statements in draft form so that they can be finalized through approval of the administrator and/or employees. Alternatively, an administrator can draft policies and procedures and then go through the appropriate steps of having them reviewed by others in the organization. School policies are often adapted from those of other school districts or colleges where they have functioned successfully. This is a convenient strategy, as a number of schools now publish policies online (see the following section). Selective borrowing reduces the necessity to "reinvent the wheel" when writing policies

Outline and structure are indispensable when developing a policies and procedures manual. Two alternative approaches for organization are *alphabetical* and *functional*. Each has its advantages. Ease of updating is another consideration. Many manuals apply a decimal system of pagination to allow new pages to be inserted within a section without altering the overall pagination. For instance, pages in section one would be numbered 1.1, 1.2, 1.3, and so on. In addition, individual sections of the manual can be distinguished by a color format and tabbed so that they stand out clearly. A complete index of key words is recommended. It's important to build in several checkpoints during the compilation to identify any problems. Editing is crucial because policies can be cited during litigation. Most manuals are published in-house using desktop publishing. This can save money, but it may be worth soliciting bids from print shops for comparison.

Policies collected in an official manual can serve to answer crucial questions like what records or documentation employees are required to keep, and what the required procedures are under due process. Although the information contained in policies and procedures manuals varies among organizations, sections of manuals often are organized around the following general topics (Conn, 1991):

- General information about the organization, including organization charts
- Benefits available to employees
- Services provided to employees
- Salary and stipend issues
- Training and development resources
- Personnel issues, including explanation of due process

The policies and procedures manual should be distributed to all employees. New personnel should receive copies of the manual during their orientation and be advised to read it. Administrators should familiarize themselves with the manual to the extent that they can reference policies relevant to issues that require an interpretation. Linking an electronic copy of the manual to the school's Web page facilitates easy desktop reference and the updating of material.

Examples of Policy Manuals

Online policies and procedures manuals for physical education, intramurals, and athletics can be accessed through Internet search engines. Following are four sources (with Web addresses provided).

- The Division I manual published by NCAA offers an excellent example of a codified policy manual—including policies, regulations, procedures, and case studies. Access it online at <http://ncaa.org/library/membership/division_i_manual/>.

- Biloxi (MS) Public Schools provides an online version of their athletics policy manual at <http://www.biloxischools.net/athletics/athletic.htm>.
- The intramural policy manual of Michigan Tech University can be accessed at <http://www.mtu.edu/sports/handbook.htm>.
- Physical education policies for Ida Price Middle School in San Jose (CA) are available at <http://www.cambrian.k12.ca.us/price/shainc/policies&expectations.html>.

Sample Policies and Procedures

The following samples of policy statements that pertain to physical education and athletics are taken from a university policy manual (Jensen, 2000).

Policy Statement 1: Scheduling Facilities. The facilities of the department will be scheduled in accordance with the following priorities: 1) regular daytime classes and athletics practices and contests; 2) other activities sponsored by the department such as intramurals, extramurals, rehearsals, evening classes, and free play; 3) scheduled activities sponsored by other divisions of the school; 4) activities sponsored by student, faculty, and staff groups; 5) activities sponsored by groups not affiliated with the school.

Policy Statement 2: Grading. Because grades are important to students in several ways, the instructor is obligated to apply a sound system of grading. A responsible instructor will follow the steps listed below:

1. At the beginning of each course, clearly inform the students of the criteria for grading in that course.
2. Interpret for the students what each grade means in terms of quality and quantity of work or performance.
3. Be sure adequate information is obtained about each student and that accurate records are kept to support the grade.
4. Be especially careful in reporting grades in order to minimize the number of grade changes.

Policy Statement 3: Class Scheduling. Classes that need not be taught every semester should be placed on a rotation schedule, which results in an optimum combination of economy for the university and convenience for the students.

Policy Statement 4: Instructor Absences. When a teacher will be absent from class, both the reason for absence and the substitute arrangements must be approved by the department chairperson.

Specific Areas for Written Policies and Procedures

Each organization varies in terms of the areas for which standard policies and procedures are needed. The following list incorporates areas for which policy statements are typically prepared.

Class Administration
Attendance records and class make-up procedures
Student evaluation and grading
Class attire for teachers and students
Equipment and supplies for class use

Facilities
Priorities
Procedures for scheduling
Availability and restrictions for nonschool purposes
Maintenance of facilities
Facility security—key acquisition and accountability

Equipment and Supplies
Procedures relative to purchase
Maintenance and care
Checking out and in
Responsibility and accountability

Health and Safety
Safety regulations, particularly for hazardous areas and activities
Reporting accidents, injuries, and sickness
Availability and use of school nurse and physician
Care and treatment of injured persons
Student health services and insurance

Personnel
Handling of contracts
Tenure and dismissal procedures
Sick leave, official leave, and vacation time
Salary schedules and extra pay for extra service
Retirement
Definition of workload
Teacher evaluation procedures

Professional Development
Time and travel expenses for conferences, conventions, etc.
Advanced study and higher degrees
Professional development leaves
Additional remuneration

Regulations and Rules

Broad policy statements inform specific rules, regulations, and procedures, which in turn clarify and implement the stated policies. As mentioned earlier, the difference between policies and regulations/rules is their scope and degree of specificity. These distinctions are illustrated in the following example.

Policy. All students participating in physical education classes will dress in clothing that promotes uniformity, allows freedom of movement, ensures safety and hygiene, and is modestly revealing.

Regulation. Students are required to dress in white T-shirts (men)/blouses (women) and dark thigh-length shorts for activity classes.

Regulation. Students are required to wear white athletic socks and laced gym shoes with nonmarking soles.

Rule. Do not walk on the gym floor in street shoes.

To the extent that decisions and problems are repetitious, regulations, rules and procedures tell staff members how to respond without having to communicate directly with their superiors. Thus, rules and standard procedures save time. Staff can act routinely without having to inquire about every task or activity (Daft, 1998).

Standard Procedures

Procedures direct how something should be done (according to policy). In this regard, they are similar to (and complement) regulations. Proper administrative procedures are of crucial importance for effective management. Procedures that have been carefully thought out, properly refined, and tried and tested provide a distinct advantage toward overall efficiency. When they are clearly defined and applied consistently, procedures become *standardized*. Once standardized, a procedure is automatic and generally accepted by those concerned. (Not all procedures can be permanently standardized; some must remain flexible in order to accommodate changing circumstances.) Administrators often refer to what they label as "standard operating procedures" (SOP). It is highly desirable that the procedures are understood and supported by those affected by them.

Here are some procedural statements of the kind found frequently in policy and procedure manuals pertaining to physical education (Jensen, 2000):

1. Teachers of physical education classes for which students change into activity uniforms are expected to commence class five minutes after the regular beginning time, and end ten minutes prior to dismissal time to allow for dressing and showering.

2. In the case of an accident resulting in injury to a student, the instructor or supervisor in charge is required to administer emergency care in accordance with the procedures described in the American Red Cross First Aid Manual.

In addition to policies and procedures that are formalized in writing, an organization typically acquires a number of "practices" that have not been established as rules or regulations and are not formally enforced, but are nevertheless observed by reason of custom and practicality. Often the line between policy and practice is vague, unless the organization provides written confirmation of each.

References and Recommended Readings

Conn, James H. 1991. "An Open Book Policy." *Athletic Business* (December): 57–60.

Daft, Richard. 1998. *Organization Theory and Design*. Cincinnati: South-Western Publishing

Denney, Maurie. 2002. Director of Athletics, Jefferson High School, Lafayette, IN. Personal correspondence, January 21.

Goodman, Michael. 1998. *Corporate Communications for Executives*. Albany, NY: SUNY Press.

Jensen, Clayne. 2000. *Policy and Procedure Manual* (rev. ed.). College of Physical Education. Provo, UT: Brigham Young University.

Kitao, Kenji, & Kathleen Kitao. 2001. *Communication*. Albany, NY: SUNY Press.

Odgers, Pattie, & B. Lewis Keeling. 1999. *Administrative Office Management*. 12th ed. Cincinnati: South-Western Educational Publishing.

Quible, Zane. 2000. *Administrative Office Management: An Introduction*. 7th ed. Englewood Cliffs, NJ: Prentice-Hall.

Tyson, C. H. 1980. "Communicating with Self and Others." *Vital Speeches*, 47(16).

Online Sources

Robert's Rules of Order
http://www.constitution.org/rror/rror--00.htm
This is the public domain version, which has been updated in hard copy.

ZDNet
http://www.zdnet.com/
This site is an online electronic information service provided by Ziff-Davis Interactive, featuring product reviews on the latest technology, computing advice, forums, and a large library of downloadable shareware and freeware.

6

Management of Human Resources

People are the most valuable asset in an organization. This is particularly true in an educational setting, where the primary objectives have to do with the improvement of people. Schools are labor-intensive organizations. Between 75 and 85 percent of a school district's budget relates to personnel. Underlying human-resource management is the concept that if you can find and keep the right people in the right positions, and provide them direction and support, they will make the organization successful. Conversely, it is impossible to succeed in the absence of well-managed employees. Most educators are human-resource managers at some level. Department chairs and athletics directors manage teachers and coaches. Coaches manage assistant coaches and trainers. Teachers manage student teachers and paraprofessionals.

Human-resource management includes the following areas: job analysis and descriptions, selection and hiring, contracts, benefits, orientation, supervision, staff productivity, tenure, grievances, termination, personnel files, professional ethics, professional improvement, teacher unionization, and equal employment opportunity. Individuals in supervisory positions have to deal with most of these areas.

THE RECRUITMENT PROCESS

The need to fill a position in an organization usually occurs as a result either of growth or reorganization, or of retirement, termination or transfer of a current employee. The first task in filling a job vacancy is to establish a recruitment process. The purpose is to attract a pool of qualified candidates and to deter unqualified applicants. Managers should understand that the costs of a recruitment program could be substantial in terms of both time and money. A good recruitment system is one that is *effective* in attracting the best candidates, *efficient* in its procedures, and *fair* in terms of evaluation and selection of applicants. The first two steps in the process are conducting a job analysis and writing a job description (Webb & Norton, 1999).

Job Analysis and Description

Once the need for a position has been established, the next step is to conduct a *job analysis*. This involves identifying the requirements of the position: determining minimum education, certification, or licensure requirements; tasks and responsibilities; and the overall value of the position to the organization. Critical job analysis often results in modifications and realignment of job responsibilities.

Job analysis provides the basis for writing a *job description*. The description should be written carefully, as it not only describes the position, but also reflects on the hiring organization. A description may already exist for the current position but may need to be revised, based on the analysis. (Each current employee should eventually receive a job description that applies to his or her position.) Written descriptions of the various positions in an organization serve a useful purpose by clarifying the responsibilities of employees. Job descriptions should be written clearly and with enough specificity to cover the subject adequately, without spelling out every detail of the job. Some positions are easy to describe, whereas others are difficult due to the diversity or unusual nature of the responsibilities.

Jobs in schools and colleges that need to be described include principal, supervisor, specialist, dean, director, department chair, teacher, coach, facility manager, researcher, athletics trainer, secretary, receptionist, equipment manager, and issue-room clerk. The list is not exhaustive, nor do all of these positions exist in any one organization. Moreover, positions with the same title might include slightly different job descriptions among organizations. (Descriptions of the following administrative positions appear in chapter 2: superintendent, school principal, college dean, assistant dean, department head, and athletics director.) As a representative example, a description of the position of a swimming pool manager follows.

Title: Swimming Pool Manager
Requirements/certifications: Lifeguard, CPR, First Aid, Pool operator
Reports to: Area supervisor
Job Duties:
- Enforce safety rules and specific pool regulations.
- Secure swimming pool area.
- Institute access control.
- Assign and perform maintenance duties.
- Schedule lifeguards and train them in pool duties.
- Maintain chlorine/bromine and pH levels according to health department regulations.
- Backwash filter and maintain equipment in working order.
- Maintain professional relationship with building supervisor.
- Fill out all required forms.
- Assume responsibility for opening and closing pool on schedule.
- Attend bi-weekly managers' meetings.
- Attend weekly lifeguard training.

Selection and Hiring

An organization must have competent employees in order to function. Therefore, among the most important decisions made by an administrator are those involving the selection of personnel. If these decisions are consistently good, the organization will be staffed with well-prepared and effective people. The exact procedures for selecting and hiring personnel vary with different organizations, but certain procedures are basic:

1. Files are established and maintained on prospective employees so that when a position becomes vacant, these individuals will receive consideration.

2. A position vacancy form must be processed. This includes a description of the position, the required qualifications (and eventually a list of all to be considered and the candidate finally selected).

3. An announcement of the vacancy is posted in various venues and locations. The announcement includes the necessary information to attract the attention of potential candidates and an explanation of the procedures for applying.

4. Applications are received and handled in accordance with prescribed procedure.

5. After reviewing the applications, a few applicants are offered interviews, and from this group a final selection is made.

6. The terms of employment are agreed upon and the person becomes officially employed.

Employment of New Faculty Members

The following procedural steps are typical for the employment of a faculty member at a university. A similar process is implemented at the school district level.

- Submit a "Request for Appointment," including recommended salary and rank.
- Obtain authorization to make an official offer.
- Send offer letter to the selected candidate.
- Send a copy of the candidate's reply to the vice president.
- Initiate a letter of appointment (contract) to be sent to the new employee.
- After the signed contract is returned, have official notice of employment prepared and sent to the new employee along with information about orientation procedures.

In some cases, the decision of whom to hire is made by the chief administrator or several administrators. More commonly, a search committee and/or selection committee is involved. The committee recommends a candidate or submits a ranked list of candidates to the administration.

Sources of Applicants

Depending on the job position, it may be necessary to reach dozens or hundreds of potential applicants. Some thought should be given to the sources and methods used to generate a pool of qualified applicants. Some methods are more effective than others. The target pool for applicants can be divided into two general sources: external and internal. *External recruiting* is carried out by advertising in newspapers and professional journals, through employment agencies, by utilizing Internet job-posting sites, through referrals by employees and colleagues at other institutions, or by sending recruiters to college campuses and job fairs to conduct recruiting interviews.

External candidates are always somewhat of an "unknown quantity"; therefore, the organization may instead opt for *internal recruiting*, which has both advantages and disadvantages (Webb & Norton, 1999).

Advantages
- Saves time and money
- Provides employees with an incentive to seek the position
- Motivates employees to obtain additional training
- Facilitates evaluating the strengths and weaknesses of current employees
- Reduces time required for the new hires to become oriented

Disadvantages
- Limits choice of candidates
- Limits injection of new ideas and perspective, contributing to stagnation and complacency
- Can cause envy and jealousy among those not offered the position
- The organization is left with another position to fill

Most organizations seek a balance between external and internal recruiting. This provides a good mix of "new blood" and "old hands" in the workplace. When recruiting applicants, those involved in the process should be cognizant of policies against *nepotism* (i.e., giving preferential treatment to relatives). Most nepotism policies bar the one(s) doing the hiring or supervising from being a relative of the new employee.

Assessing Applicants

Much useful information can be obtained from a well-prepared job application form. Application forms allow the school to dictate the type of information sought from all applicants, yielding comparable data. These forms differ among organizations, but all should include essential information about educational background and degrees earned, previous work experience, special or unusual achievements, and contact information. Some school districts require a writing sample on application forms. School districts often include indemnifying statements on their forms that hold the district harmless for actions related to employment verification and reference checks (Castetter & Young, 2000).

The application becomes even more useful if it is accompanied by a personal résumé or vita. More attention is being paid to verification of information on applicants' résumés than in the past, following several high-profile cases of applicants fabricating credentials for college coaching positions. Verification is the responsibility of the school's human resources office or business office, whichever oversees hiring.

School districts use letters of recommendation or standardized reference forms to obtain additional information on applicants. Standardized forms can reduce some of the problems associated with letters of recommendation, as the forms can be tailored to address the specific criteria for the job position. Letters of recommendation may be provided by the applicant or solicited by the district. The information in these letters can be either confidential or nonconfidential. Assuming that an applicant's job performance in the future would be consistent with that of the past, recommendations from former employers are the most reliable sources of information. One should be cautious about accepting letters of recommendation at their face value. Such letters are usually written by individuals of the applicant's choice, and people are often more complimentary than is justified (Castetter & Young, 2000).

The best indication of potential success is immediate previous achievement in a similar position. Other indicators are less reliable but still useful. Figures 6-1 and 6-2 show examples of forms that can help to give structure to the selection process.

PROSPECTIVE EMPLOYEE RECOMMENDATION FORM

The below named applicant has given your name as a reference for a position in our school system. Your evaluation will be kept confidential. Please rate each category: (0) unobserved, (1) below average, (3) average, or (5) above average.

Date _____

To _____

RE: _____
(applicant)

Address _____

For _____
(position)

	0	1	2	3	4	5	Remarks
1. Appropriate model for students							
2. Overall appearance							
3. Scholarship in major teaching areas							
4. Scholarship in minor teaching areas							
5. Rapport with students							
6. Ability to effectively organize							
7. Ability to effectively communicate							
8. Ability to motivate students							
9. Ability to control students							
10. Competency in teaching (coaching)							
11. Acceptance of responsibility							
12. Overall professionalism							
13. Overall ability to succeed							

How long have you known the applicant? _____

In what relation have you known the applicant? _____

Would you hire the applicant for this position? _____

Additional comments _____

_____

Signature _____ Title _____

Figure 6–1. Example of a prospective employee recommendation form

INTERVIEW FORM

Name of Organization _____

After the applicant has left the room, complete this rating form. The rating categories are: (1) below average, (3) average, and (5) above average.

Interviewer _____   Date _____

Applicant _____   For _____
             (name)   (position)

	0	1	2	3	4	5	Remarks
1. Personal appearance							
2. Ability to communicate							
3. Personality appeal							
4. Interest in the position							
5. Positive professional outlook							
6. Knowledge of subject area							
7. Knowledge of student characteristics							
8. Attitude toward students							
9. Philosophy of teaching/coaching							
10. Degree of appropriateness for the job							

Strengths: _____

_____

_____

Weaknesses: _____

_____

_____

Comments: _____

_____

_____

Figure 6–2.   Example of an interview form for prospective teachers

Interviewing

One of the most useful processes in assessing job applicants is the selection interview. The interview should be structured with a "patterned approach" where the dialogue is guided by the interviewer, but the interviewee is encouraged to speak freely and in depth about relevant topics. The interviewer maintains control of the conversation to make certain that all relevant areas are covered systematically and within a reasonable time frame. Some schools use one-on-one interviews; however, more schools are employing panel interviews where the members of the group share in asking questions of the candidate. Whichever approach is chosen, it should be used consistently for all applicants.

Common errors by interviewers include lack of preparation, poor interviewing conditions, covering the content haphazardly, and asking improper questions. Regarding the latter, the safest way to steer interviewers away from asking problem questions is to provide standardized questions to be asked of all applicants. Improper questions asked of the applicant during a job interview can imply illegal discrimination. Figure 6-3 lists questions that can and cannot be asked by interviewers for the reasons given. Some questions are not discriminatory per se but might trigger scrutiny by the EEOC.

EMPLOYMENT CONTRACTS

A contract is an official written agreement between two or more parties with certain conditions described. It is common for teachers, coaches, and administrators to be employed under contract, whereas some other school personnel are employed under noncontractual arrangements, such as by time card at an hourly rate or by the month. A properly prepared and signed contract is a legal document, meaning that both parties are bound and are expected to adhere to the conditions of the contract unless the parties mutually agree to a contract release or modification. Certain points of information are essential in a teacher contract, such as job title and/or rank, duration of employment, salary, fringe benefits (unless explained elsewhere), and the basic responsibilities.

Both sides should live up to the conditions of the contract, regardless of whether the conditions are explicitly defined, implied, or reasonably assumed. Not all the conditions of employment can be described in a contract, but it should include the more important ones. In addition, it should be understood that the contract represents a "good faith agreement" between the two parties. Breech of contract without justifiable reason is both illegal and unethical and ought to be strictly avoided. On the other hand, both parties should be open to reasonable negotiation when there is sufficient cause to modify an agreement.

Figure 6-4 shows an example of a typical contract/letter of appointment. Most contracts include an attachment or enclosure that provides additional information about the specific conditions of employment.

BENEFITS

In educational organizations, employee benefits normally cost anywhere from 25% to 35% of an individual's salary. An employee should assume that the benefits are actually worth this much. Normally, the two major types of benefits program are retirement and insurance. Most professionals in education are enrolled in the state

ILLEGAL QUESTIONS AND THEIR LEGAL COUNTERPARTS

Subject	Illegal Question	Legal Question
National Origin/ Citizenship	Are you a U.S. citizen?	Are you authorized to work in the United States?
	Where were you/your parents born? What is your "native tongue?"	What languages do you read, speak, or write fluently? (This question is OK, as long as this ability is relevant to the performance of the job.)
Age	How old are you? When did you graduate from college? When is your birthday?	Are you over the age of 18?
Marital/ Family Status	What's your marital status? With whom do you live?	Would you be willing to relocate if necessary?
	Do you plan to have a family? When?	Travel is an important part of the job. Would you be willing to travel as required by the job?
	How many children do you have?	This job requires overtime occasionally. Would you be able and willing to work overtime?
Affiliations	To what clubs or social organizations do you belong?	Do you belong to any professional or trade groups or other organizations that you consider relevant to your ability to perform this job?
Personal	How tall are you? How much do you weigh?	Are you able to lift a 50-pound weight and carry it 100 yards? [if that is part of the job] (Questions about height and weight are not acceptable unless minimum standards are essential to the safe performance of the job.)
Arrest Record	Have you ever been arrested?	Have you ever been convicted of _____ ? (The crime should be reasonably related to the performance of the job in question.)
Military	If you've been in the military, were you honorably discharged?	In what branch of the Armed Forces did you serve? What type of training or education did you receive in the military?

Source: U.S. Equal Employment Opportunity Commission <www.eeoc.gov>.

Figure 6–3. Questions that can and cannot be asked in a selection interview

Name and address   Date _____

of faculty member _____

_____

On behalf of the Board of Trustees and the President, I am pleased to offer you a

faculty position for the year beginning _____ and ending

_____. Your salary will be $_____, to be paid to you in

twelve equal installments, commencing _____, and your rank or

title will be _____.

This letter is being sent to you in duplicate. If the terms or your appointment are
acceptable, please sign and return one copy within twenty days of the date of this letter;
otherwise this offer will lapse. We are required to set this time limit in order that we may
immediately recruit someone in your place if we do not receive your acceptance within
such period of time.

We are enclosing a memorandum relating to employment of faculty members of this
institution. Compliance with the terms of this memorandum is a condition of your
employment.

If you have any questions about this appointment, I suggest that you confer with your
dean, director, or other officer supervising your work. Questions concerning payroll
deductions, group insurance, and other fringe benefits should be directed to the Benefits
Office.

The Board of Trustees and the Administration are most appreciative of the loyal service of the faculty of the University. We trust that the continued spirit of cooperation and
devotion to duty will improve even further the quality of service to our students and our
performance in our other duties.

Sincerely,

Academic Vice President

I hereby accept the appointment stated above this _____ day of _____, 20____.

Figure 6–4. Typical letter of appointment for a college teacher

education retirement program, and these benefits are comparable from one state to another even though they differ in some respects. Well-established private education organizations usually provide retirement benefits that are comparable to those of the state schools.

When a teacher applies for a position, he or she should obtain complete information about the retirement and insurance benefits. This can be acquired from: (1) the written contract; (2) the school's policy and procedure manual; or (3) directly from the human-resource office.

The benefits package supplements, in a real way, the financial provisions of the employment contract. From a management perspective, a good benefits program helps in the recruitment and retention of competent personnel and enhances staff morale. The following comprehensive list of benefits is representative of those provided by more progressive employers.

Retirement—participation in the state retirement system and Social Security
Insurance—life, dental, medical, and disability
Workmen's compensation
Periodic medical examinations
Tenure

Leaves of absence—sick leave, maternity, emergency, professional development (conferences and workshops, travel study, sabbatical)

Credit union benefits
Legal services (under certain conditions)
Parking privileges
Free or reduced admission to special-interest and entertainment events

A disturbing trend, especially among institutions of higher learning, is to hire part-time or adjunct faculty members who are paid per number of courses taught without providing them any benefits. This practice should not be used to permanently replace full-time teaching positions that offer benefits.

TENURE

The rationale of tenure for teachers is predicated on protecting their freedom to teach and to advance knowledge, and the legitimate need to protect them from dismissal for unjustified reasons. Such reasons for dismissal might include personal differences with a school administrator, political activities, making room for an administrator's friend or favorite on the payroll, or for remarks made in the classroom which, although within the parameters of academic freedom, might not be acceptable to the administrative hierarchy. Tenure contributes to stability of the faculty and assures job security for individual employees.

New teachers are normally on a nontenured or probationary status for a period of 2 to 5 years in public schools (5 to 7 years in higher education) before becoming eligible for tenure. Tenure decisions should be made in consultation with a candidate's peers. Coaches may receive tenure as teachers, but coaching assignments normally are not tenured.

The concept of tenure is clearly justified, and the results are generally positive. However, the implied protection of lifetime employment can lead some teachers to become unmotivated, outdated, or incompetent. Rather than a guaranteed job, tenure implies job security in cases where no grounds for dismissal exist. Tenure systems routinely include provisions for removing tenured employees *for cause*, which covers such charges as incompetence, insubordination, and moral turpitude. The burden of proof normally falls on those making these charges. This led many state legislatures in the 1990s to implement versions of post-tenure review, which attempts to shift this burden to the employee. In any case, the initial remedy for below-standard performance by a teacher should be a mutually-agreed-upon development plan that incorporates corrective measures. School administrators should take awarding or revoking tenure as a serious responsibility.

Virtually all reputable colleges and universities award tenure. The practice is less universal among community colleges. Although most school systems in the United States operate under a tenure law, some do not. Where tenure laws do not apply, a "continuing contract" usually does. This means that a teacher's contract will be automatically renewed the following year unless the teacher is notified to the contrary before a specified date, in which case all aspects of required due process should be followed. Nineteen states employed the term "tenure" in legislation through 1998. according to the Education Commission of the States:

Alabama	Arkansas	California
Connecticut	Hawaii	Kansas
Kentucky	Louisiana	Maryland
Massachusetts	Michigan	Missouri
Montana	Nebraska	New Jersey
New York	Pennsylvania	Rhode Island
Tennessee		

Some states, including Wisconsin, offer tenure in selected school districts. Most other states without specific provisions for tenure have adopted terms for procedural protection, such as continuing contract status. In two states, Iowa and Nevada, collective bargaining determines contract terms.

A few states provide only renewable contracts for a specified number of years:

Arizona (three-year contracts)	Delaware (three-year contracts)
Mississippi (three-year contracts)	New Mexico (three-year contracts)
Oregon (two-year contracts)	South Dakota (three-year contracts)
Utah (five-year contracts)	Vermont (annual contracts)

ORIENTATION

Orientation involves familiarizing new employees with the organization and with their job responsibilities. Both the process and the content of orientation are important. Too often orientation consists only of filling out the necessary forms. New employees should be given the opportunity to ask questions and interact with other employees. Administrators and supervisors should take an active part in orientation of employees. Some organizations assign mentors to work with new employees.

The orientation of a new employee occurs through a combination of several phases beginning with the interviewing process. In addition, new employees usually go through formal orientation sessions. It is especially important that the sessions be well planned and effectively conducted. The manner in which the orientation is accomplished can have a significant influence on employees' early impressions of the organization and its administrators and can expedite the adjustment to their job.

Orientation sessions should cover the following areas (Noe, Hollenbeck, Gerhart, & Wright, 2000):

- Overview of the organization
- Policies, procedures, rules, and regulations
- Compensation and benefits
- Safety
- Tour of the facilities
- Job duties and responsibilities
- Performance expectations
- Introduction to other employees
- Orientation to the community and available housing

SUPERVISION

Supervision is a necessary function for diverse reasons within different educational contexts. Supervisors provide a link between upper-level administration and the instructional staff. Working at the school level, supervisors have two main purposes: staff development and evaluation. The focus of supervision ought to be on improving the teacher's overall performance. The analysis of obtained data and the relationship between the teacher and supervisor form the basis for strategies designed to improve the students' learning by improving teachers' effectiveness. In this sense, the supervisor can be seen as "a teacher of teachers" (Van der Linde, 1998). Some general principles apply to the supervisory process:

1. A supervisor should be viewed as both an educator and a motivator. The educational role involves teaching the employees how to perform their responsibilities more efficiently. The motivating function involves regular encouragement and meaningful challenge.

2. A congenial and businesslike relationship with teachers is recommended. This contributes to pleasant working conditions and yet prohibits staff from taking advantage of a relationship that might be too casual or chummy.

3. Each individual is different, and supervision therefore must be flexible. While some individuals may require little direct supervision due to their insight, understanding, and work ethic, others may have little feel for the circumstance or the assignments and therefore require constant guidance.

4. When a person has a split assignment (e.g., teaching and coaching), that involves work responsibilities mixed with limited supervision, care must be taken to insure that the individual will perform with efficiency in both assignments.

A major responsibility of instructional supervisors is to observe classes and evaluate instructors and programs. This task must be carried out systematically and fairly, as a supervisor's observations may constitute part of teachers' performance evaluations (see chapter 11). The clinical supervision of teachers is a process encompassing three phases: the planning conference with the teacher, the classroom observation, and the feedback conference (Van der Linde, 1998).

The following guidelines are suggested for conducting a *planning conference*:

- A suitable time for the upcoming classroom visit should be agreed on by both parties.

- Good rapport should be established with the teacher; the person being evaluated should be made to feel at ease and encouraged to do his or her best.

- The aim and nature of the follow-up (feedback) discussion should be clearly outlined at this time.

- Any queries and questions on the part of the teacher should be answered to prevent uncertainty. The supervisor should adopt an attitude of helpfulness.

- In order to plan the classroom visit properly, the evaluator should be informed about the teaching-learning situation, including the characteristics of learners.

During a formal classroom visit (the observation period), the evaluator should follow established guidelines. Economical use of time is important for both the teacher and the supervisor. The supervisor should arrive before the lesson commences and remain until it has been concluded.

Staffo (1993) offers guidelines for supervisors (or principals) who observe physical education classes and evaluate programs as part of their responsibilities. He suggests avoiding unreliable methods such as eyeballing, anecdotal evidence, intuitive judgments, and even rating scales—all of which are influenced by subjective opinion. Following is a list of objective standards:

- The teacher should appear organized and have a clear, concise lesson plan.

- Appropriate warm-up activities should be included in the lesson plan.

- There should be proper use of class time.

- Skills/drills/activities should be progressive in their sequence.

- Evidence of teaching/learning should take place.

- The teacher should offer corrective feedback to students and utilize positive reinforcement.

A classroom visit is basically without value if there is no feedback. The *feedback conference* should be a natural outflow of the classroom visit, as it may be the last step in the evaluation. The follow-up discussion should take place as soon as possible after the completion of the visit. This conference is indispensable for a mutual understanding and for solutions to problems. Teachers are now in a situation where they are able to explain their behavior in the classroom. Both the positive and the negative factors should be discussed. As for the negative, the supervisor should focus on mistakes or deficiencies in the act of teaching rather than on the person of the teacher. Consensus should be reached by both parties as to the need for change and how to bring it about. The emphasis should be on instruction and on guidance toward improved teaching. Remedies that are realistic should be discussed, and steps to promote continuing professional growth should be outlined.

The supervisor observes instruction on a continuing basis. During the course of the school year, a variety of teaching methods and a range of appropriate activities should be evident. Moreover, it should be apparent to the supervisor that the physical education program is staying abreast of current developments in the profession and is responsive to changes in society. Gender equity and health-related fitness are two current areas of emphasis that should be evident in the program.

PERSONNEL FILES

A file is kept on each faculty member and usually on other employees. It is important that personnel files contain as much useful information as possible, but not become filled with trivia or irrelevant material. The files should be available only to those who rightfully have access to them, and they should be treated with a high level of confidentiality. Most administrators would probably agree that individuals should be able to examine their own files upon request; however, this practice is usually governed by an organizational policy that spells out the handling and accessibility of personnel files. In large organizations, the trend is toward keeping personnel records on computer rather than hard copy.

JOB PERFORMANCE

The total environment of the school should be conducive to high-level productivity. There ought to be adequate facilities, equipment, and supplies, along with challenging assignments and encouragement. Friction, apprehension, and frustrations should be minimized.

The administrator is usually in a strong position to influence employee *morale*, which is an important factor in performance. If good morale is to be attained, the administrator must have concern for the staff members and their welfare and must maintain an atmosphere of trust, honesty, and openness. Clearly stated objectives, consistency of procedures, and overall administrative efficiency have positive influences on morale. Since the crucial factor in education is student progress toward the educational objectives and teachers are the primary facilitators of student progress, teacher effectiveness must be viewed as the primary objective of administration.

Practically all administrative efforts either directly or indirectly influence productivity of school employees. The recruitment and retention of talented teachers and coaches is an initial step. Licensing and certification requirements along with

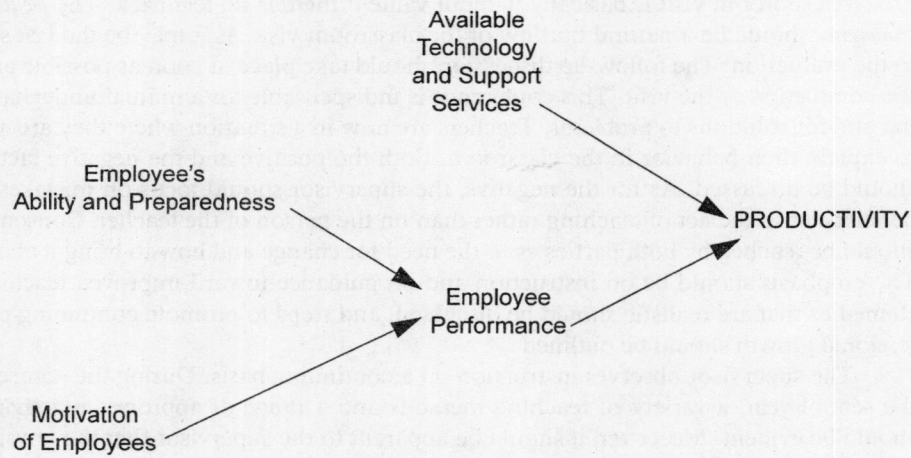

Figure 6–5.   Factors contributing to employee productivity

upgrades play an important role in maintaining performance levels. Performance evaluation is a crucial factor (see chapter 11). Incompetent teachers and coaches must be removed expeditiously and fairly. Likewise, good teaching and coaching should be recognized and rewarded (Pitch, 1996).

Stress Management

Workplace stress leads to poor performance, absenteeism, and high turnover. Consequently, school systems are following the lead of private corporations in their efforts to alleviate employee stress. Benefits of reducing stress among employees include lower healthcare costs and insurance premiums.

The first step is to identify workplace *stressors*. Stressors may include workload, coworkers, performance goals, or the environment (noise, confusion, and so on). Coaches are particularly susceptible to unrealistic workloads and performance expectations. Creating a low-stress work environment in schools is a major goal. In addition, stress in employees' personal lives can carry over into the job. Employees should be trained to recognize the signs of stress before "burnout" occurs. Burnout may suggest reassignment in particular instances. *Stress testing* (e.g., blood pressure monitoring) is the second step in managing stress. This preventive measure can be carried out in conjunction with employee preferred provider organizations (PPOs). *Treating job stress* is the third step. Employers increasingly are providing therapeutic resources to supplement medical treatment. These include stress counselors, nutritionists, exercise physiologists, and on-site gymnasiums, ergonomic specialists to redesign work areas, and employee recreation programs.

PROFESSIONAL DEVELOPMENT

Professional development (also known as staff or human-resource development) historically has focused on remediating deficiencies. The recent trend has been toward proactive development. The purpose of professional development is to provide the learning necessary to (1) enable the employee to perform at the level of competence required for present and future assignments, and (2) provide opportunities for personal and professional self-fulfillment. Professional development should be viewed as a valuable supplement to an individual's basic preparation and professional experience. It should be approached with enthusiasm and high expectations and should be viewed as an opportunity to improve both the individual and the organization (Webb & Norton, 1999).

Professional development is based on the precept that "the best way to get rid of a poor teacher is to make him or her into a good one." Although this sounds idealistic, it should be the first consideration and receive an honest effort. The steps in professional development are first to assist teachers and coaches in diagnosing and determining development needs, and then to design a development plan to meet those needs. In order for professional development experiences to be beneficial, they must be well planned (this usually involves a written proposal); approved and funded with enough lead-time for adequate planning and preparation; and include an oral or written report of sufficient detail. In the case of professional development leaves, interim reports are also often required.

Methods and strategies for professional development include: (1) enrolling in graduate study; (2) attendance and participation at conventions, clinics, and workshops; (3) meaningful research and writing; (4) working with a mentor; and (5) visiting other schools and programs. The National Association for Sport and Physical Education (NASPE) provides professionals who will deliver workshops at local schools in curriculum planning, assessment, and teaching practices.

It is essential for administrators to hold a long-range view of staff development. Although a return on investment may not be immediately forthcoming, over time a well-planned program will produce substantial benefits, including better-qualified staff, improved motivation and morale, improved job performance, and increased organizational productivity.

In allocating funds for professional development, administrators must make sure that the optimum improvement is gained for the amount of funds expended. In this regard, it is not good enough simply to respond to employees' ideas. It is also important to aggressively generate and implement development opportunities. In connection with each employee, especially the faculty, the concept of the individual development cycle presented in figure 6-6 should be applied.

Retrenching Personnel

Sometimes an educational organization finds that the curricular emphases have shifted away from the competencies of certain faculty. Solutions to this problem are either to replace the faculty members who have become outdated or to have them *retrench* to fit the present and future curricular needs. Retrenching is usually done through various professional development opportunities over a period of time. Occasionally, the need for new competencies is so pressing that, in order to meet the need, a person must go through extensive retraining. Administrators should carefully monitor the need for professional retrenchment and try to provide attractive and timely opportunities to help employees keep current with the needs of the program. A teacher who falls behind in educational methods and subject matter certainly cannot function effectively on the job, and this is a situation where prevention is better than cure.

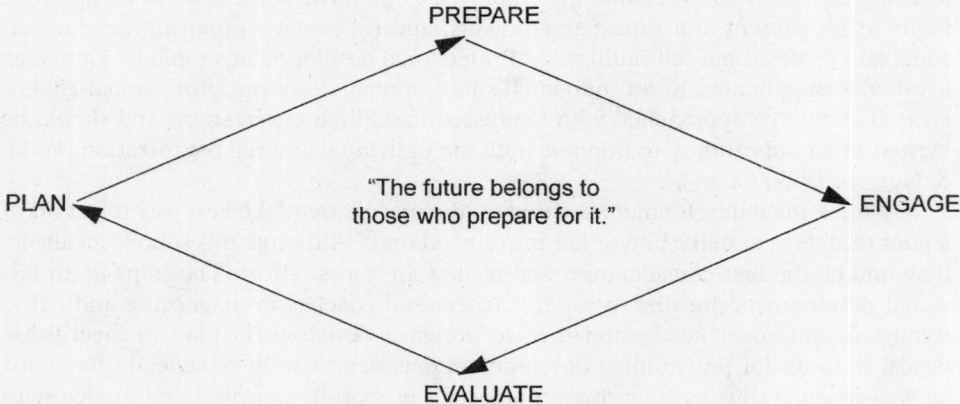

Figure 6–6.   Diagram of an individual development program cycle

MANAGING PROBLEM EMPLOYEES

Dealing with difficult employees is a time-consuming, worrisome problem for managers. Generally, the person's immediate supervisor is the one most intimately involved with the situation. The problem may result from the unilateral behavior of an employee, from interactions with other employees, or interaction with the employer. Understanding the common types of behavioral problems is helpful. Ivancevich (1992) has identified the following major categories of problem behaviors:

1. Those whose work is unsatisfactory due to lack of ability, training, or motivation.

2. Those whose personal problems off the job affect job productivity (e.g., alcoholism, drug abuse, and poor family relationships).

3. Those who commit illegal acts while on the job (e.g., stealing, abuse of fellow workers).

4. Violation of company rules (e.g., chronic tardiness; carrying weapons).

Corrective Measures

The employer should establish rules and communicate them to employees. Employee behavior is then assessed and modified where necessary. Employees must be aware of the rules and consider them fair. It is useful to seek employee input when periodically revising workplace rules. When rule violations occur, the administrator must conduct a fair investigation. The employee who is in violation of the rules must be confronted about his or her unacceptable behavior. If the problem is one that would benefit from therapy, such as alcoholism, this remedy may be required.

The first step in the disciplinary process is to counsel the employee, which may include a warning. If the problem continues, the next logical step is to repeat the warning and inform the employee that the incident will be entered in his or her personnel file. Organizations often apply a hierarchy of increasingly serious forms of reprimand. Eventually, the options narrow down to the employee being transferred (if appropriate) or terminated. Discharge of an employee is accompanied by costs. In addition to creating a position that has to be filled, this action may result in a grievance being filed or in litigation.

Termination

A sound bit of advice is: "Hire carefully, dismiss sparingly." Termination of employment should be the last step in a process of progressive discipline. Effective discipline programs have two central components: documentation and progressive punitive measures (see Noe et al., 2000). Termination can occur by a natural process such as expiration or retirement, or by initiative of the individual or the organization. Termination procedures should be spelled out in the policy manual, and they must be enforced in order to ensure that terminations occur in an orderly manner. The requirements for due process in disciplinary matters leading to termination include: (1) clear and unambiguous notice; (2) timely, thorough, and fair investigation; (3) proof of failure to perform; (4) record of equal treatment; and (5) establishment that termination is a fair penalty (Pulhamus, 1991). Some organizations utilize peer reviews, in which a panel of colleagues acts as mediators or arbitrators to determine whether management's decision to terminate the employee should be sustained. Employees are usually allowed at least one internal appeal to the decision. If a termination is actuated, it is important that administrators process a standard termination form that ensures proper checkout and clearances in terms of keys, equipment and library materials, and stipulates the final financial settlement.

Negotiated departures work well when an employee is willing to leave. This makes the separation easier on everyone. The context is that a mismatch exists between the employee and the job. A considered strategy is to assist employees so they can actively engage in seeking other employment. Severance packages or "buyouts" often are a part of this process. For example, an employee may be offered a year's salary to leave willingly. This approach suggests to the terminated employee that the organization values them as a person, even though their performance on the job is below standard (Leavit & Johnson, 1998).

Grievances

For the overall good of the organization, it's important to keep gripes and complaints in perspective and in the proper channels while still giving them appropriate attention. The organization should implement a prescribed procedure for handling grievances. The procedure should be simple but effective, and should involve the following steps: (1) Discuss the grievance with one's immediate supervisor and attempt to solve it at that level. (2) If this effort is unsuccessful, take the grievance to the next higher administrative level. (3) If the grievance remains unresolved, it probably will be placed in the hands of a grievance committee composed of three or more staff members. This committee will study the situation and attempt to bring the matter to a satisfactory solution.

Grievances ought to be kept to a minimum because they use up time and energy that could better be spent on more productive matters. Effective control of grievances closely relates to the administrator's sensitivity and responsiveness to potential tension-causing situations, along with personality, good judgment, and implementation of sound procedures. However, it is not possible to avoid grievances entirely; therefore, an organization must have a fair procedure for dealing with them. In handling grievances, the administrator should always focus on finding the best solution, not on proving a point. An effective grievance system discourages employees seeking relief through litigation. This saves the school and the employee a great deal of time and expense.

Some schools and colleges maintain an Ombuds Office. (The term derives from the Swedish language and refers to a people's representative.) The campus Ombudsperson provides problem-solving assistance for work-related issues, investigates reported internal complaints from staff (and/or students), reports findings, and helps to achieve equitable settlements of grievances.

PROFESSIONAL ETHICS

The assumption that educators and the schools are guardians of the nation's cultural and moral values suggests that schools should maintain and strengthen the ethics of those they employ. Teachers and coaches often are placed on a pedestal by the local community, and they act as role models for young people. Moreover, being a member of a profession itself suggests high standards of conduct and practice.

Professions adopt codes of ethics as recommended guidelines for the conduct of those who practice. The education profession has a code of ethics set forth by various associations, including the National Education Association (NEA) and the American Association of University Professors (AAUP). Most coaches' associations, including the American Football Coaches Association, also have developed codes of ethics. The American Alliance for Health, Physical Education, Recreation and Dance (AAHPERD) has prepared a code of ethics (see Appendix B). All administrators, coaches, and physical education teachers should familiarize themselves with the contents of this code.

TEACHER UNIONIZATION

Many school administrators traditionally opposed unionization of teachers because it reflected a "labor versus management" image. Unions, they argued, created a barrier between teachers and administrators that strained a congenial working relationship. Others believed that unions tended to damage the public image of the education profession. These perceptions have faded as teachers' unions have demonstrated their positive impact and gained acceptance among enlightened educators and the public.

Most teachers join unions at the school-district level. Through collective bargaining or other strategies, local unions have improved working conditions and won financial gains for the profession. Two national groups are at the forefront of teachers' efforts to organize professionally: the National Education Association and the American Federation of Teachers. Surveys indicate that over 80 percent of teachers belong to one of these two groups, both of which maintain chapters in individual states.

Unions confirm teachers' legitimate role in shaping the work conditions and learning environment in schools. Administrators need to develop appropriate and effective bargaining skills when involved in negotiations with teachers' unions. Bargaining models that promote "win-win" strategies are gaining acceptance (Webb & Norton, 1999).

EQUAL EMPLOYMENT OPPORTUNITY

Human-resource management must function within the legal environment. Legal requirements affecting the workplace derive from the Fourteenth Amendment of the U.S. Constitution, which states that all individuals are entitled to equal protection under the law. Equal employment opportunity (EEO) refers to the government's attempts to assure that all individuals are treated equally in employment compensation, terms, conditions, and privileges. To accomplish this goal, the federal government has resorted to constitutional amendments, legislation, and executive orders. School administrators must pay particular attention to laws and regulations affecting the management of people in the workplace (Noe et al., 2000). Figure 6-7 lists the major EEO laws and regulations.

The Equal Employment Opportunity Commission (EEOC), a division of the U.S. Department of Justice, is responsible for enforcing most of the EEO laws. The commission has three major responsibilities: investigating and resolving discrimination complaints, gathering information, and issuing guidelines. Employees who feel they are victims of job discrimination may file complaints with the EEOC. Various time lines and requirements apply when filing discrimination complaints. Failure to file a complaint within 180 days of the alleged discriminatory act generally results in the case being dismissed.

Law or Regulation	Requirements
Equal Pay Act of 1963	Discrimination on the basis of sex is prohibited on pay scales and benefits. Equal skill, effort, responsibility, and similar working conditions are the criteria of determination.
Title VII of the Civil Rights Act of 1964	No person shall, on the grounds of race, color, religion, sex or national origin, be excluded from participation in, be denied the benefits of, or be otherwise subjected to discrimination under any program or activity receiving federal financial assistance.
Age Discrimination in Employment Act of 1967 (amended in 1978)	Discrimination on the basis of age is prohibited against those in the age bracket of 40 to 70 in hiring, promotion, discharging, and benefits.
Executive Order 11246 Amended by EO 11375 of 1968	Regarding Title VII discrimination, requires the elimination of all existing discriminatory conditions. Affirmative Action requirements include setting goals in achieving employment mixture.
Title IX of the Education Amendments of 1972 (Public Law 92–318)	Designed to eliminate discrimination on the basis of sex in any education program or activity receiving federal financial assistance.
Rehabilitation Act of 1973—Sections 503 and 504	Section 503 requires that organizations shall take affirmative action to employ and advance in employment qualified persons with disabilities. Section 504 prohibits discrimination against qualified disabled persons in employment and in the operation of programs and activities receiving federal financial assistance.
Veterans Readjustment Act of 1974	Employment openings are to be listed at Job Service, veterans should be given priority, and veterans are eligible for readjustment points.
Americans with Disabilities Act of 1990	Prohibits job discrimination against individuals with disabilities. Applies to all organizations with more than fifteen employees.
Civil Rights Act of 1991	Amends the laws enforced by EEOC (i.e., Title VII, ADEA of 1967, ADA of 1990) to provide for damages in cases of intentional employment discrimination and to clarify provisions regarding disparate impact actions.
Family and Medical Leave Act of 1993	Organizations with fifty or more employees must permit employees to take up to twelve weeks of unpaid leave per year for family and medical reasons.

Figure 6–7.   Equal employment opportunity laws and regulations

Employment discrimination per se is not illegal. Employers make numerous discriminations concerning their employees—on suitability for promotion, in awarding merit pay, and so on. A legitimate basis for discrimination might be an employee's job performance. What an employer cannot do is discriminate on the bases of the factors prohibited in the laws and regulations listed in figure 6-7. In addition to the protected classes of sex, age, religion, ethnicity, national origin, and disability, several states have implemented laws forbidding discrimination based on sexual orientation. In 1998, President Clinton issued Executive Order 13087, which prohibited job discrimination based on the sexual orientation of federal employees. This order would apply to Department of Defense schools, for example.

Theories of Discrimination

How does one determine if an employee in a protected class has experienced illegal discrimination? Complaints are based on three common theories of discrimination: disparate treatment, disparate impact, and reasonable accommodation (Noe et al., 2000).

Disparate treatment occurs when employees with similar qualifications and in similar situations are treated differently based on protected status. For example, if a school district fails to hire women with young children (assuming mothers will be more frequently absent) but hires men with young children, the applicants are being treated differently based on sex.

In cases where disparate treatment is charged, the plaintiff (the party filing the charge) has the burden to make a *prima facie* case against the employer by showing that: (1) they belonged to a protected class, (2) they applied for an employment benefit and were qualified, (3) they were rejected despite being qualified, and (4) after the plaintiff was rejected, the employer offered the employment benefit to someone with similar qualifications.

Once the plaintiff has established a *prima facie* case for discrimination, the *burden of proof* shifts to the employer to demonstrate that there was a legitimate reason for rejecting the plaintiff. The employer may argue that for this particular job, the protected factor (e.g., sex) was a *bona fide occupational qualification* (BFOQ). (For example, the position of equipment manager in a particular situation required access to the men's locker room.) BFOQs must be directly related to an employee's ability to do the job to be acceptable by the courts.

Disparate impact occurs when a seemingly neutral employment practice disproportionately excludes a protected group from employment benefits. The exclusion may not be intentional, but nevertheless a protected group is affected disproportionately. For example, a school district might require applicants for security jobs to be a certain height. However, height qualifications would likely have a disparate impact on women and on male applicants from some ethnic groups. The burden would be on the district to show that height is somehow related to successful performance of the job—an unlikely assertion in this case. Where protected groups historically have been excluded, employment decisions based on job seniority can have a disparate impact. Normally, the courts expect the mix of employees hired and promoted to adequately reflect the pool of available applicants.

Reasonable accommodation places a special obligation on the employer to accommodate an employee's protected status, often relating to religion or disability. In the case of a wheelchair-bound employee, the employer may be required to provide wheelchair ramps, access to upper floors of a building, and restroom facilities that meet ADA requirements. Employees may voice religious objections to being required to work on the Sabbath or holy days.

Title VII of the Civil Rights Act of 1964 protects employees from retaliation by their employer for opposing a discriminatory practice or participating in a complaint.

Affirmative Action

Affirmative action is a policy that makes special efforts to hire and promote minorities, women, and other disadvantaged groups as a way to compensate for past discrimination. The goal of affirmative action is to institute employment practices that will result in a workforce that reflects the racial and sexual composition of the population. Disadvantaged applicants must be recruited and personnel practices revised to meet the special needs of these applicants. In 1965, Executive Order 11246 required all public and private employers to establish affirmative action plans (Chandler & Plano, 1988). Since January of 1973, all educational institutions (both public and private) with one or more federal contract(s) of $50,000 or more and 50 or more employees have been required to maintain a written affirmative action plan.

The EEOC may impose consent decrees on employers with records of past discrimination, requiring them to hold a certain number of job positions for disadvantaged groups. So-called "set asides" and quotas have become increasingly controversial, however. The imposition of these legal remedies has in some cases led to charges of "reverse discrimination." Affirmative action has become conflicted terrain in the courts and the political arena. Human-resource managers should work closely with the school district's legal counsel in drafting affirmative action plans.

Sexual Harassment

Sexual harassment refers to unwelcome sexual advances. It can occur in two ways. *Quid pro quo* harassment refers to a benefit (e.g., promotion) or punishment (e.g., dismissal) being made contingent upon the employee submitting to sexual advances.

A more subtle form of harassment is a *hostile working environment*. This occurs when someone's behavior in the workplace makes it difficult for an employee of a particular sex to do his or her job (e.g., sexually explicit language or offensive pictures on the wall).

The court may hold an employer liable for the actions of an offending employee. In making this determination, the court attempts to discover whether the employer had (or should have) known about the harassment, and if the employer tried to stop it. Most harassment complaints involve male harassment of females; however, men have filed complaints against women, and same-sex harassment has been charged by employees. Schools should take four steps to create a workplace free from sexual harassment (Noe et al., 2000):

1. Develop a policy statement making clear that harassment will not be tolerated.
2. Train employees to identify inappropriate workplace behaviors.
3. Develop a procedure for reporting sexual harassment.
4. Implement prompt disciplinary action to protect victims of sexual harassment.

Women in Management

Women have faced barriers to advancement to higher management positions (as have certain ethnic groups). Studies have found that no significant differences exist between women and men that would limit the capacity of women to perform effectively in administrative roles. Furthermore, the legal framework has supported equal opportunities for women and men in the form of the Civil Rights Act of 1964 and the Title IX legislation of 1972. Yet, women continue to be underrepresented in administrative positions in schools, particularly in physical education and athletics. More women today are interested in managerial positions than in the past. This desire, and more open attitudes, will undoubtedly contribute to greater numbers of women administrators in physical education and athletics—but appreciable progress toward this goal has yet to be made.

References and Recommended Readings

Castetter, William, & I. Phillip Young. 2000. *The Human Resource Function in Educational Administration.* Columbus, OH: Merrill.

Chandler, Ralph, & Jack Plano. 1988. *The Public Administration Dictionary.* 2nd ed. Santa Barbara, CA: ABC-CLIO.

Ivancevich, John. 1992. *Human Resource Management: Foundations of Personnel.* Boston: Irwin/McGraw Hill.

Leavit, William, & Gail Johnson. 1998. "Employee Discipline and the Post-bureaucratic Public Organization: A Challenge in the Change Process. *Review of Public Personnel Administration,* 18(2): 73–81.

Noe, Raymond, John Hollenbeck, Barry Gerhart, & Patrick Wright. 2000. *Human Resource Management: Gaining a Competitive Advantage.* Boston: Irwin/McGraw-Hill.

Pitch, M. 1996. "In Defending Teachers, Clinton Calls for Help in Improving Quality." *Education Week* (Sept. 18), p. 22.

Pulhamus, A. R. 1991. "Conflict Handling—A Common Sense Approach to Appraising Supervisory Performance." *Public Personnel Management,* 20(4): 485–92.

Staffo, Donald. 1993. "Guidelines to Assist a Principal or Supervisor in Evaluating a Physical Education Lesson/Program." *Education,* 114(1): 74–76.

Van der Linde, C. H. 1998. "Clinical Supervision in Teacher Evaluation: A Pivotal Factor in The Quality Management of Education." *Education,* 119(2): 328–33.

Webb, L. Dean, & M. Scott Norton. 1999. *Human Resources Administration: Personnel Issues and Needs in Education.* Columbus, OH: Merrill.

Online Sources

Association for Supervision and Curriculum Development
http://www.ascd.org/
This site provides a forum in education issues and professionalism and shares research, news, and information concerning supervision and curriculum development.

American Federation of Teachers
http://www.aft.org
This site covers current issues and news affecting teachers, lists conferences and meetings, and offers a portal to AFT publications, surveys, and reports.

National Education Association
http:www.nea.org
Click on the site map hyperlink for membership information, current events, hot topics, news releases, and publications.

U. S. Equal Employment Opportunity Commission
http://www.eeoc.gov
This site offers links to press releases, statistics, laws, regulations and policy guidance, training information, and a list of publications.

7

Financial Management

To paraphrase a former U.S. Senator, "A million here, a million there, and pretty soon you're talking about real money." We are in an era where $50-million school district operating budgets and $15-million college athletic budgets have become common. We are building $25-million recreational gymnasiums and $50-million high schools. Today's school administrators are indeed managing "real money."

Each administrative unit and the programs within it must be managed in a financially sound manner. Directors, teachers, and coaches who make budget decisions must maintain credibility with those who control the school's finances. This means that budget requests should be based on accurate information and sound rationale. Expenditures should be made wisely and in accordance with the budgetary plan. Accurate financial records should be kept, and the program conducted in a manner that is cost-efficient. If these procedures are applied consistently, they will win financial support. Conversely, there are very few things that can cause an administrator more grief than the mismanagement of financial resources.

PRINCIPLES OF FINANCIAL MANAGEMENT

The following basic principles serve as guides to sound financial management of instructional programs, intramurals, and athletics.

1. To the extent that values are accurately applied in budgeting, the budget truly represents the relative significance of the activities within the organization.

2. Administrators are often judged by their competence in financial management. Meticulous preparation and thorough justification of budget proposals and careful control of the expenditures reflect heavily on an administrator's image.

3. Regardless of who else is involved, the administrator of the given budgetary unit is immediately and ultimately responsible for proper financial management.

4. As a general rule, those who furnish money for any enterprise have the right to know how it is spent. This is particularly true of the tax-paying public.

5. Funds for athletics should be allocated and accounted for in generally the same way as funds for other aspects of the school program. The revenue should go into designated funds, and the program should operate on an allocated budget, which is subject to regular audits.

6. Maintain an adequate operating budget, but invest any surplus funds on deposit either within the organization or in an approved investment program. Certificates of deposit (CDs) with a staggered payoff schedule work well for this purpose.

7. In financial management, saving money is important, but spending money is also important. The goal is to obtain and spend the right amount in the right manner for the good of the program and to get the most for the money spent.

8. One aspect of financial efficiency is to stop doing some things in order to have additional resources to do other things that are more important.

PLANNING THE BUDGET

Budget planning and preparation are more complex processes than most people realize. Budget planning impacts all facets of program planning, because the budget must be designed to finance the programs. In turn, the programs must be administered in accordance with the budget. Following are some guidelines for budget planning:

1. The budget should accurately represent the financial needs of the program in relationship to the stated program objectives.

2. Honesty in budget preparation is the recommended approach. Padding of the budget will become apparent to financial officers. This practice damages the credibility of the administrator and may result in budget cuts.

3. Accurate determination of enrollment, student credit hours, or other related factors is fundamental to effective budget planning.

4. The planning process should take into account anticipated income from all sources (both appropriated funds and revenue) and balance this against the anticipated expenditures.

5. The possibility of financial emergencies should be anticipated and included in the budget. Funds may be placed in contingency to cover unforeseen needs.

6. The budget should have the right balance of rigidity and flexibility—enough rigidity to give adequate structure and enough flexibility to accommodate unforeseen changes.

7. The budget should be prepared on a timetable that leaves ample opportunity for thorough analysis, evaluation, and review (see the section on the budget calendar later in this chapter).

8. The budgeting process ought to involve all those in the organization who have administrative responsibilities.

9. The budget categories and the amount in each category should be correctly aligned to fit the financial needs of the different programs.

Budget Preparation

Although the exact steps in budget preparation vary with different budgetary approaches, certain steps should be included in all approaches: (1) clearly determining the objectives of the program the budget is to finance, (2) determining as accurately as possible the real financial needs of the program, (3) preparing the budget in adequate detail so that it can be clearly understood, and (4) presenting the budget through the prescribed administrative channels.

Each organization has its own budgetary forms designed in accord with local or state regulations. However, all budget forms share similarities across organizations. Figures 7-1 and 7-2 are examples of budget preparation forms, showing the different categories. Although budget forms may be similar, the procedures followed in arriving at the information requirements vary significantly with the budgetary approach. The following section includes descriptions of four approaches to budgeting: incremental, PPBES, zero base, and cost analysis.

Incremental Budgeting

Incremental budgeting uses a historical base for estimating needs and revenues during the next budget cycle. The assumption is that each budget entity will receive the same "increment" as a percentage increase or decrease from the previous year. This form of budgeting is rather superficial and mechanical, and yet, due to its simplicity, it has its proponents. Despite the weakness of this method, it is probably the most widely used approach. It is relatively simple, requires minimal work, and helps to preserve the status quo. With this approach, the person preparing the budget might be provided with a guideline figure—say, a 5% increase over last year's budget— which then becomes the starting point. The additional money is applied and other adjustments are made in accordance with informed judgment of how the money should be used. Under this system, the most mechanical approach is to apply the percentage increase to every category. However, based on changes in circumstances and projection of needs, different budget categories are increased by varying amounts. Each line item is examined in relation to overall expected revenues. Some line items may be decreased so that others can be increased disproportionately.

This approach has one clear drawback. The present status of the budget tends to be basically perpetuated without thorough analysis of productivity as related to expenditure (i.e., return on the dollar). It is assumed that last year's budget is accurate and adequate, and increases are tacked on each year based on available revenue. Thus, unnecessary programs and inefficient procedures may be retained year after year. It is conceivable that a budget item that should have been eliminated or drastically reduced years ago is still in the budget and has received standard annual increases, while some other item that deserves considerably more emphasis has also received standard increases. This does not necessarily happen, but it is possible with this approach to budgeting.

Inherent to the incremental approach is a need to control wasteful spending, because the approach encourages a tendency to make needless purchases toward the end of the fiscal year. This is because departmental budget managers often feel that if the entire budget allotment is not spent, next year's budget will be reduced to match this year's spending level. This is a foolish and wasteful approach to budgeting, given that money is often spent on items that otherwise would not be purchased. In this type of budget climate, the budget manager is given little incentive to save money.

DEPARTMENTAL BUDGET REQUEST SUMMARY SHEET

Department _____

College/Division _____ Signature of Dept. Chair

Account Code _____

Signature of Dean

Column	A	B	C	D	E
Object of Expenditure	Category Code #	Present Budget Amount FTE	Request for Next Year Amount FTE	Program Improvement Request Amount FTE	Total Budget Request (C + D) Amount FTE*
1. Administrative salaries		$	$	$	$
2. Full-time Faculty salaries					
3. Full-time Faculty supplemental earnings					
4. Part-time Faculty salaries					
5. Full-time Staff salaries					
6. Student Assistant salaries					
7. Student wages					
8. Non-student part-time or temp. wages					
9. Payroll taxes, insurance, and retirement					
10. Equipment repairs and maintenance					
11. Rented-leased equipment					
12. Computer Services					
13. Capital Equipment					
14. Travel					
15. Supplies and other expenses					
16. Miscellaneous					
17. TOTALS					

* Full-time equivalent personnel

Figure 7–1. Budget proposal form for an instructional department, showing the different categories

ATHLETIC BUDGET REQUEST SUMMARY SHEET

Account Code _____   (Sport) _____

Object of Expenditure	Code #	Last Year	Present	Request Next Year
1. Administrative salaries				
2. Faculty salaries (coached)				
3. Full-time Staff salaries				
4. Student Asst. Salaries				
5. Student Wages				
6. Payroll Taxes, Ins., & Ret.				
7. Telephone				
8. Vehicle Operations				
9. Repairs and Maintenance				
10. Capital Equipment				
11. Travel				
12. Supplies				
13. Postage and shipping				
14. Printing				
15. Programs				
16. Laundry				
17. Advertising				
18. Films				
19. Gear				
20. Award Jackets				
21. Recruiting (off campus)				
22. Coaches Clinic				
23. Scouting				
24. Grant-in-aid				
25. Grant-in-aid—Spring				
26. Books				
27. Training Table				
28. Tutoring				
29. Officials				
30. Guarantees				
31. Game Management				
32. Facilities				
33. Dues and Membership				
34. Insurance				
35. TOTALS				

Approvals:

_____   _____
Administrator   Director

                                                      _____
                                                      Coach

Figure 7–2. Budget proposal form for an athletics program, showing the different categories

Ways in which this annual ritual of unneeded spending can be controlled include the requirement that purchase orders be approved by the immediate superior and business officer during the last month of the fiscal year. Another method is to establish a reward system referred to as a "percent return rule." It simply means that a certain percentage of the total amount of the original budget (say 2%) can be carried over to next year's budget and placed in the unallocated category. This would mean that for a departmental budget of $300,000 as much as $6000 could be carried over. This encourages tight control of funds and provides for some flexibility in next year's budget. Unfortunately, few educational organizations permit this procedure.

Program Budgeting

School budgeting moved into a modern phase with program budgeting, an approach that features several advantages over the earlier form of fiscal decision making. The primary advantages of program budgeting are: (1) it places emphasis on specific programs supported by a set of funds, and (2) it makes a greater amount of data available to monitor the relationship between programs and funds. The primary change over traditional approaches is that budgeted funds are internally organized according to their objective or purpose. This functional approach offers a more sophisticated method for internal decision and control processes. Conceptually, program budgeting moves the budget process into a conscious recognition of the relationship between money and programs.

The trend in program budgeting has been toward standardizing the account code structure in order to facilitate compliance with state and federal reporting requirements. Adopting standards set by the U.S. Office of Education or the National Center of Educational Statistics, most states have implemented account codes. Local districts develop their own fund structures and control processes for internal decision making (Thompson, Wood, & Honeyman, 1994).

This budgeting approach utilizes object and function codes (commonly four digits) for revenues and expenditures. *Object* codes identify the sources of revenue (e.g., state appropriations) and classify expenditures by what specific goods or services (e.g., training-room supplies) are being purchased. *Function* codes identify the type of activity or services (e.g., instruction) on which funds are spent (Drake & Roe, 1994). Following is an example of function codes for expenditures:

1000   *Instruction*
    1100   Regular program
        1110   Elementary
        1120   Middle school
        1130   High school
    1200   Special programs
    1300   Adult programs

2000   *Support Services*
    2100   Pupils
    2200   Instructional staff
    2300   General administration
    2400   School administration

3000   *Community services*

4000   *Nonprogrammed charges*

5000   *Debt service*

Local districts often utilize a *program* code extension to identify subject area budgets. For example, the physical education budget might be assigned the code 028. Thus, expenditures for middle-school physical education instruction (adhering to the above format) would be coded 1120-028.

The development of the concept of program budgeting was a historic breakthrough that laid the groundwork for more sophisticated approaches to budgeting—notably the planning, programming and evaluation budgeting system discussed in the next section.

Systems Approach (PPBES)

In order to coordinate the various steps involved in program budgeting, a systems approach emerged, labeled PPBES (originally PPBS), which stands for *planning, programming, budgeting,* and *evaluation system.* The systems approach to budgeting has been utilized by a spectrum of public agencies. The most widely known PPBES system, developed by the U.S. Department of Defense during the 1960s, is still used by that agency in its revised form. By the late 1960s, the merits of this approach were recognized by other public agencies, particularly in the field of education. As demands by the public for accountability increased, many school systems across the country began using PPBES.

The *planning* component of PPBES involves preparing goals (objectives) and defining expectations in terms of student accomplishments. The *programming* phase involves the selection of program activities and events that will be included for the purpose of accomplishing the goals. The *budgeting* component entails the preparation of a financial proposal that will enable the sponsorship of the planned program. Of course, this must include all financial needs such as facilities, equipment, and personnel. The *evaluation* phase of the system involves a careful analysis of how adequately the budget fits the program, how well the program leads toward the goals, and how valuable the accomplishments are as related to costs. Stated outcomes are linked to assessment criteria. When systems management is applied to the educational process, the student is the product.

One of the main advantages of the systems approach is that it forces one to give adequate attention to all the important tasks of budget preparation and administration. With a large, comprehensive program each step becomes more involved, but regardless of the size of the program, the system still uses the same four steps. In this approach to budget management, the components cannot be isolated from the total system. The four phases constitute a complete cycle, with each step having a logical sequence and a close relationship to the others. The final step of the cycle (evaluation) logically lays the groundwork for the first step of the new cycle. A review of goals and objectives should take place every three to five years to reflect changes in society and technology in the operation of the schools. The goal process serves a double purpose. It not only replaces individual goals with system goals, but by involving individuals in the process it tends to precondition these individuals toward acceptance of the established goals (O'Donnell, 2002).

Major drawbacks of PPBES include organizational resistance to adapt the conceptual thinking and technology necessary to implement this radical change in budgeting philosophy. Often, adoption has been followed by discord as programs and staff compete for limited resources. Moreover, PPBES requires significant investment of time and energy to gather and interpret information required by the system. Consequently, PPBES works best in larger school districts with adequate resources. The initial popularity of PPBES has waned because of its complexity and its incompatibility with the historically nonaccountable culture of schools. Inherent in the PPBES approach was the idea of zero base budgeting (Thompson, Wood, & Honeyman, 1994).

Zero Base Budgeting

Zero base budgeting (ZBB) was developed in the private sector in an effort to find a better way to relate input to company objectives. Later, it was introduced into the public sector by President Jimmy Carter. Zero base budgeting was implemented by many local governments in response to taxpayer unrest. ZBB continues to enjoy popularity among constituents supporting tax reform and strict accountability. Schools began adopting ZBB in response to fiscal problems that began to mount in the inflationary decade of the 1970s. Although its popularity has waned, some school districts continue to use this approach.

Zero base is a system of budgeting that focuses on the complete justification of all expenditures each time the budgeting exercise takes place. It does this by assuming that every element of the organization starts from zero and must justify its existence by its relative contribution to the organization's objectives. Its attractiveness as an approach to budgeting can be attributed to three basic characteristics: (1) the participative nature of the method by administrators at different levels, (2) the merging effect it causes between budgeting and planning, and (3) the crisp influence it has on decision making.

Perhaps the foremost feature of ZBB is that it provides an avenue for budgeting and planning to occur jointly. Too often these functions are performed independently of one another. When program planning and financial planning proceed together from a zero base, it becomes easier to coordinate the decisions about programs and finance.

In effect, ZBB results in a thorough review of functions from the ground (zero) base and proceeds in an upward direction. The method was developed to offset some of the undesirable features of conventional budgeting techniques. The most important difference is that ZBB avoids the practice of considering last year's budget as the base; therefore, it moots the question of how much of an increase is required next year to continue with a particular function. Rather, the question is whether funding should be allocated at all. This method denies the assumption that each budgeted item should be automatically continued and funded.

Some schools have adopted *modified ZBB budgeting*. One modification utilizes a procedure to build a new budget on a reduction over the previous year wherein restoration of the prior funding level requires extensive justification. The assumption is that all budgets can cut waste. In another modification, budget officers prepare three scenarios for their unit: one based on a reduced budget, one based on a sustained level of funding, and one incorporating a stipulated increase in funding (Thompson, Wood, & Honeyman, 1994).

The specific steps to be carried out when implementing ZBB vary from one application to another. However, in a general sense, the following are common aspects of ZBB approaches (American Management Association, 1990):

Defining the goals. Stating the goals (objectives) of the program is an important step. The goals influence the decisions involved in preparing the budget.

Planning the program. The program must be planned in light of the goals. In the program planning phase, the following questions are important: What is the reason for implementing any particular aspect of the proposed program? What would happen if it were not done? Is there a better and more economical approach? Where does it fit in terms of priority? After the current aspects of the program have been scrutinized in this manner, new activities and approaches can be considered by the same procedure.

Identifying alternative decisions. The previous step should result in a careful analysis of the relative value of the various aspects of the program. The purpose of this step is to identify and evaluate alternative approaches to accomplishing those program objectives that are considered important (for example, having the school cooperate with the city on the construction of new tennis courts and having them lighted so that they serve both the school and the community, or closing certain facilities during low use periods so that money can be diverted to other facets of the program).

Cost analysis of the alternatives. This step involves cost comparisons of the alternatives. It would include questions such as the following: What are the relative costs of a team traveling with two vans as opposed to one bus? What would be the relative cost of playing more games at home and fewer games away, taking into consideration all of the financial factors, gate receipts, travel costs, guarantees, and so on?

Making the decisions. The information from the previous four steps brings the situation to the decision-making point. It is now time to answer two important questions: (1) Which functions should receive funding and which ones should not? (2) For those activities that will be funded, which of the alternative approaches is the best? The decisions must be realistic and in accordance with the information obtained from the previous steps.

If the ZBB approach is used properly, it will result in funding that fits the best plan for the accomplishment of the stated objectives. With the ZBB method, each aspect of the program has to be justified on a cost-benefit basis, and this has to be done for each budget cycle. Precedent, tradition, personal preference, and bias have no place in the ZBB method. It is a straightforward calculation of how much money can be justified in support of each function, when they are weighed carefully against each other in light of the stated objectives.

The ZBB method has specific advantages. For instance, it allows new programs to compete for funds against established programs. It increases decision flexibility with discretionary funds. It is especially useful in high-growth situations where program expansion is prevalent. Although cost reduction is not a required objective of ZBB, this budget approach is oriented toward cost reduction because the process naturally identifies and eliminates areas of the program that are overfunded or relatively ineffective in goal accomplishment.

ZBB is not a panacea, but it can be especially valuable to an organization that needs a "face lift" in terms of budgetary procedures, or one that needs to add rigor to its decision-making process. On the other hand, ZBB should be avoided by any organization whose budgeting approach is adequate and yet simpler and less time-consuming than the ZBB method. In other words, apply the adage that "if it isn't broke, don't try to fix it." Also, ZBB ought to be avoided if the organization is involved in major changes in personnel, procedures, or organizational structure.

If zero base budgeting is used, an interesting question is how often the process should be repeated. The advantages and disadvantages of using ZBB as an annual exercise are summarized as follows (American Management Association, 1990):

Arguments for annual ZBB:

1. A different system would have to be used in off-years if ZBB were not used annually.

2. ZBB provides alternatives for budget modification during changing circumstances.

3. Administrators will become more proficient at using the system if it is done annually.

4. If the system is effective, then it would seem that annual implementation would be an advantage.

Arguments against annual ZBB:

1. The greatest benefit of the ZBB system accrues the first year and diminishes thereafter.

2. There would be a tendency to repeat the same decisions every year.

3. The process is time consuming.

4. Good managers review programs critically on a continuous basis without ZBB.

5. ZBB is not needed every year, since programs and activities change little from one year to the next.

The drawbacks of zero base budgeting are similar to those of PPBES. ZBB incorporates a vast amount of analysis that is not normally a part of the budget exercise. Although it may create an atmosphere of rethinking that can have a renewing effect on an organization, it can also lead to competitiveness and overfunding. Some schools have concluded that the increased effort required to make ZBB work is not worth the savings that resulted.

Cost Analysis

In education, cost analysis is a procedure whereby all expenses of the program are related to the total number of students served. In this regard, enrollment projection is the single most important task in budgeting because it shapes all other aspects of the budget. Enrollment is the key factor driving revenues and expenditures, with obvious implications for programs. The interdependency of the various steps in the budget process is most evident in enrollment estimation. Various methods exist that can be used to project enrollment, such as cohort and trend analysis. This discussion will be limited to methods for calculating current enrollment.

In carrying out cost analysis, projected costs are computed on a per-student basis. The total number of students is figured in one of two ways: the average daily *student membership* or the average daily *student attendance*. The membership represents all of the students enrolled in the school; whereas on any given day the attendance is usually less than the membership. Most experts agree that the membership is the better measure of the two because teachers' salaries must be paid, whether 95% or 100% of the pupils are in attendance. Also, school facilities, furnishings, books, and other costs remain approximately constant and do not fluctuate with average attendance. School districts tend to utilize average daily attendance (ADA) where truancy has been a problem. This provides schools with a funding incentive to improve attendance.

If cost analysis is employed, a different formula (amount per student) applies at the different school levels, because the nature of physical education and athletics varies at the different levels. The formula might allow for a certain dollar amount per student to cover the total expenses of physical education and athletics (and related programs); or it might provide specific amounts per student for physical education instruction, for athletics, and for intramural sports. In some cases, the school as a whole receives a set amount per student, and the principal assigns the amount for each phase of the total school program.

Although some room for variation exists within budgeting methods, the concept is consistent: the amount of allocation is based on the student population (either enrollment or attendance). This approach is predicated on the idea that a close relationship exists between educational costs and number of students. Under this method of budgeting, the best way to obtain more money in support of a program is to get an upward revision of the formula.

Sometimes (e.g., at the college level), cost analysis is based on the production of student credit hours rather than on the number of students. In such cases, noncredit activities such as athletics and intramurals have to be budgeted on a different basis.

PRESENTING THE BUDGET

Several steps may be involved in presenting a budget, depending on the particular situation. In a high school, the proposal would logically go from the program director to the principal, to the superintendent, and to the school board. In a university, the budget would pass from the department chair to the dean, to the vice president, to the president, and to the board of trustees. At each step along the way, the budget proposal must be reviewed and approved. At any stage before final adoption, modifications might be required. These guidelines will help in the presentation process:

1. Provide sufficient copies of the proposal on time.

2. Be sure that the budget proposal is in the recommended format and that it is sound, accurate, and complete.

3. Have available all the information to justify the budget, such as (a) the objectives of the program, (b) evidence of previous success, (c) rate of student participation, and (d) statistical data about the program that would be useful to the reviewers.

4. Include a cover sheet that summarizes income and expenditures, so that the total budget picture can be seen at a glance.

5. Consider involving other persons who can speak in favor of certain aspects of the proposal. The circumstances of the review indicate whether this would be appropriate.

6. Be thoroughly familiar with the content of the budget proposal and the rationale behind it.

7. Be precise and forthright in your presentation and responses. Avoid being vague or evasive.

SOURCES OF FUNDS

Public schools operate with a combination of external (publicly allocated) and internal (self-generated) funds. The most reliable source of funds is the *budgetary appropriation* from general funds. These funds provide for personnel costs, supplies, and fixed expenses. This money is doled out to schools on a state-mandated basis. Sometimes it barely covers expenses, but it does provide stability and consistency to the budget. For this reason, it is clearly the preferred funding source.

Special fees constitute another source of income. Public schools are requiring students to pay more and more fees as a result of tight budget situations. Students often pay a registration fee, a transportation fee, or an athletics fee. Sometimes a physical-education uniform fee or a facility-use fee is also paid by each student. Colleges often charge a fee for use of recreational gymnasiums.

Government or private *grants* can fund program areas and facility development. Look for usable funds that fit your goals. School administrators should identify staff who have talent for writing grants and should compensate them for their successful efforts. The goal is to bring in as much self-generated income as possible (Morgan, 1998).

Several other sources of income are pursued with varying amounts of success by different educational institutions. These include: *donated funds* from foundations, companies, and individuals; *annual giving* by individuals through a fundraising program or athletics booster club; *athletics guarantees* paid in connection with contests played away from home; revenue from *concessions, auto parking,* and *program sales* at athletics contests; and *special fundraising* events, such as walkathons, marathons, and novelty athletics contests.

For some athletics programs *gate receipts* provide a source of revenue. In the public schools, gate receipts are relatively small, but in certain university programs the gate receipts are substantial. It is strongly recommended that gate receipts go into the general fund, and that budgeted funds be used to finance the athletics program. This puts athletics on the same financial basis as other school programs, and it is a protective measure against abuse that promotes stability in budgeting athletics.

Facility construction normally is financed through *general obligation bonds* (tied to local taxation) or through *revenue bonds* (secured through revenue generated by the facility). *State appropriations* and *federal grants* are also utilized to finance school facilities. Administrators are routinely involved in funding for capital improvements, but rarely are teachers and coaches concerned with this type of funding.

It should be noted that in some states the law prohibits the expenditure of tax dollars on athletics. In this situation, the athletics department should have the benefit of student athletics fees or help from boosters (or other sources) to stabilize funding to ensure a healthy program.

EXPENDITURE OF FUNDS

In public education, taxpayers have a right to expect a full measure of return on the money spent. In order to control expenditures properly and ensure accountability, it is necessary to be explicit about who is authorized to expend funds from a particular budget. The use of an authorized signature card, such as the one shown in figure 7-3, gives a clear definition of who can expend funds.

In the case of educational budgets, the great majority of expenditures are for the payment of personnel and the purchase of equipment and supplies. Occasionally, large amounts are paid to contractors in connection with construction projects. Relatively small amounts are expended for travel, leasing of equipment, student awards, and other ongoing expenses.

Payment of Personnel

Personnel are hired on *regular contract, special contract,* or *time-card* arrangement. Full-time faculty and administrative staff are normally hired on an annual contract.

AUTHORIZED SIGNATURE CARD

Account Code _____

_____ for
Authorized Signature

Account Name _____

_____ whose signature appears
(Name—Print or Type)

above is authorized to expend from the above account code.

Restriction (if any): _____

_____

Approved by:_____
(Dean or Director)

Date: _____

Figure 7–3. An authorization form for expenditures against a particular financial account

Important elements in the contract are the title, the faculty status, the salary, the conditions of payment (such as monthly over the contract period), and at least a general description of responsibilities. Some contracts are lengthy and definitive, whereas others cover only the essentials. Contracts may be accompanied by supplementary information impinging on conditions of employment.

Special contracts are used for the employment of game officials and the accomplishment of special projects (see figure 16-2 in chapter 16).

Clerical and support services personnel may be hired on either a contract or a time card arrangement. This involves processing a form that establishes a record of employment and spells out the essential conditions. For those hired on time cards, the agreement includes hours per week and the hourly pay rate. Once employment is established, time cards (indicating the number of hours for which each person is to be paid) are submitted on the prescribed schedule.

Personnel must be hired and paid in accordance with the approved budget. Monthly budget reports are invaluable to the administrator as a method of checking that the amounts being paid each employee correspond with the agreements and the amounts that were allocated.

Purchases

The normal procedure for making purchases is to complete the purchase form(s). Sometimes a different form is used if the purchase is to be made from another division on campus or within the school district, as opposed to a purchase from a supplier or vendor. Figure 7-4 shows a purchase order, and figure 7-5 is an example of a purchase requisition. Following is a discussion of some guidelines relating to purchasing.

When completing a purchase form, provide all pertinent information, including a complete and correct description of the item(s), the number desired, color (if this applies), sizes, model number, and the recommended source of purchase, if there is reason to make such recommendation.

Place orders far enough in advance to make sure that the items will be available when needed and to allow the purchasing agent ample opportunity to properly investigate the various sources of purchase. Avoid rush orders, which are inefficient and tend to be more expensive.

Order the correct amount. Too much of a certain item in stock occupies valuable storage space, costs interest on the money invested in the items, and presents the risk of potential loss through fire, theft, or deterioration. Conversely, with orders that are too small, you sometimes lose the advantage of bulk rates, pay a high price on the next order due to inflation, and decrease efficiency because of frequent purchases and deliveries.

As a manager of public funds, you are obligated to search out and accept the best bargain. Of course, the best bargain does not always mean the cheapest price. Other important considerations are involved, such as quality of the product, quality of service, and convenience. However, when all else is essentially equal, the vendor who can provide the best price is the one that should be selected. Avoid accepting substitutes by vendors unless you feel satisfied with the product and the price. Also, when items arrive, be sure they go through the proper receiving steps and are properly inventoried.

There are two kinds of organizational plans for purchasing: decentralized and centralized. With a decentralized plan, each school or department makes its own purchases directly from the vendors. The disadvantages of this system include having too many school personnel dealing directly with vendors, which creates the inability to buy in quantity, and the lack of control due to numerous people involved in the various aspects of the purchasing process. The greatest advantage is the convenience of quick buying to fill unexpected needs.

With the more commonly used centralized approach, all purchase are made through a central purchasing office. It is economical because it allows quantity buying and involves purchasing specialists who often have more leverage with vendors and experience negotiating with them. For these reasons, the centralized method is usually the better of the two.

The normal procedure in purchasing is to obtain bids from different vendors whenever the purchase is large enough to justify doing so. The purpose in *bid buying* is to obtain the best price possible for the specified product or for some acceptable substitute. In preparing a package for bid, one must provide exact specifications of the materials to be purchased. Incomplete or vague descriptions can contribute to inconsistencies among bids from the different vendors. The description must also indicate whether substitutions will be considered. An organization can consistently save money through the bid-buying process because it creates competition among vendors and affords the purchaser the opportunity of accepting the "best deal."

PURCHASE ORDER

Date _____   Ship When _____

Ordered
from   ⌐   ⌐   ⌐
Vendor   Ship to

                  ∟   ∟   ∟

Bill in Duplicate to: Requested by: _____
                                                      Authorized Representative
                                                            or Principal

                                         Approved by: _____
                                                           Program Director

                                         Approved by: _____
                                                          Purchasing Director

Account Number	Vendor No.	Purchase Order No.	

Invoice Number	Total Amount	Discount	Partial	Full	Net Amount	Approved for Payment

Quantity Ordered	Quantity Received	Description	Unit Cost	Est. Cost
		TOTAL AMOUNT		

Figure 7–4.   Example of a purchase order (courtesy of Alpine School District)

School ____________________

Street ____________________ A/C No. ____________________

Date ____________________ City ____________________ ____________________

Quantity	Units	P.O. #	Index #	Description	Supp. A/C #	Code	Unit Cost	Total Cost

Approved By: ____________________ Received By: ____________________

Figure 7–5. Example of an intraschool district requisition form (courtesy of Alpine School District)

Often, large organizations have *direct buying* sources from which large purchases are made. Direct buying means making purchases from the wholesaler of the product and eliminating the retailer. Normally a substantial price advantage is gained with this method, while there is the disadvantage of the loss of service and convenience provided by a local vendor. Naturally, retail suppliers oppose this procedure because it eliminates their opportunity to do business. In each situation, the advantages and disadvantages of direct buying must be carefully weighed. If possible, information should be kept current as to which items can be bought from a local retailer at the same price as from the wholesaler. This can help to maintain a good relationship on the local level while saving money on larger items. An organization should not be coerced into doing business with local retailers at a higher cost unless sufficient reasons exist for doing so.

The following system for ordering equipment used by the Beaverton School district in Oregon can used as a model.

1. The school district establishes a comprehensive physical education and athletics request catalog by developing a master file with item specifications and estimated costs. This catalog lists the most frequently requested supplies and equipment for the physical education and athletics program.

2. The catalog is distributed to physical education teachers and athletics coaches through the school principals.

3. Individual coaches and teachers make requests from the catalog by specifying the identification number and quantity of each item. The requests are reviewed and totalled by department for each school.

4. Each school's request is then submitted to the central office where the requests are reviewed for accuracy and entered into the computer with all data necessary to produce a purchase request.

5. The bid document is reproduced in sufficient quantities to supply each vendor with two copies. Vendors indicate bid prices on items as specified. Substitutes must meet original specifications. Bid documents are returned to the central office by bid deadline.

6. Completed bid documents supplied by vendors are scanned into the computer to obtain updated vendor file names, addresses, vendor numbers, item numbers, and prices of all submitted items. The completed bid documents are processed to provide a vendor bid report, which lists each item, in ascending bid price, by vendor.

7. Selection is made of successful bids by product and unit price from the bid report.

8. Itemized purchase orders are produced and processed, and when the orders are filled, inventory records are established.

BUDGET REPORTS AND ADJUSTMENTS

In order to manage a budget effectively, it is necessary to receive periodic budget reports from the financial services division. Normally, these reports are provided on a monthly basis. The reports should include all the financial transactions that have occurred during the reporting period. These reports can serve two purposes: (1) as a check to make sure that only the correct entries have been recorded against the budget, and (2) to allow the administrator to evaluate whether the expenditures are on the proper time schedule or whether the rate of spending is too rapid. The budget must fit the approved plan, both in amount and kind of expenditures and in the time frame over which the expenditures are made.

Each budget report should be carefully reviewed soon after it is received. Also, the report ought to be kept for a reasonable time for possible future reference. Sometimes it is necessary to transfer funds from one category to another, although this is not recommended as a common practice. For example, suppose a certain amount of money to pay the salary of a full-time secretary is in the staff salary category. Part way through the year, the secretary leaves, and a decision is made to finish out the year with part-time student secretaries. In this case, the remainder of the money would have to be transferred from the staff salary category to the student wage category. Ordinarily, educational organizations have a special form on which to process budget adjustments (see figure 7-6).

The Budget Calendar

School districts operate on a budget calendar to assure that budget steps are completed by specified dates. A calendar reduces the tendency to procrastinate and assures that adequate time has been allotted to complete the process in an orderly and thoughtful manner. The following calendar is used by a district where the budgeting process begins in November of the school year (Gross, 1996).

November
Prepare budget calendar.
Establish budget principles.
Collect and review budget-related data.

December
Prepare and compile budget.

January
Prepare initial budget proposal.
Submit proposal to school board for review and comments.

February
Revise budget proposal, based on initial review.
Resubmit budget to school board for initial approval.

March
Revise budget as needed.
Receive approved final budget.

April
Implement bidding process.
Set employee contracts.
Communicate budget to staff, parents, and public.

In most states, the fiscal year normally commences on July 1; however, several states operate on biennial (two-year) budgets. The fiscal year for the federal government begins on October 1.

```
From _____   To: Budget Office
_____   _____

             BUDGET ADJUSTMENT REQUEST

                   For School Year 20___

I hereby request a transfer of funds for the following budget accounts:
   DEPARTMENT          ACCOUNT NUMBER      INCREASE        DECREASE
_____     _____     $_____     $_____
_____     _____     _____      _____
_____     _____     _____      _____
_____     _____     _____      _____
_____     _____     _____      _____
_____     _____     _____      _____
_____     _____     _____      _____
_____     _____     _____      _____
_____     _____     _____      _____
_____     _____     _____      _____
_____     _____     _____      _____

                                  TOTAL  $_____     $_____

Approvals (authorized signatures):     Date:
_____   _____
_____   _____
```

Figure 7–6.   A form for transferring funds from one budget category to another

COMPUTERIZED BUDGETING

Software is available that allows users to monitor revenue sources and track various funds when preparing school and department budgets. The better software incorporates spreadsheets, bookkeeping packages, and records of receipts and payments, and it can create an auditing trail. In addition, budgeting software can interface with inventory software. Most school districts and colleges have adopted some form of computerized budgeting. This technology minimizes the time spent entering budget data and assists in preparation of reports and end-of-the-year statements. Among the specific advantages of computerized budgeting are:

1. Accuracy is increased and errors are more easily detected.
2. Budget report forms are readily generated, modified, and transmitted.
3. The process is both fast and economical.
4. Numerous copies of reports can be produced, which is convenient and time saving.
5. The financial management process can be streamlined by eliminating certain steps.
6. By-products of the process assist in other aspects of administration, such as shortcuts in purchasing procedures, improved inventory tracking, and availability of computer online budget information.

Computers have facilitated the trend toward school site budgeting, incorporating decentralization and increased accessibility of information. Advanced Budget Accounting (ABT) is a promising development in school finance that incorporates the latest Internet linking capabilities. In some school districts, each school's budget is linked electronically to the district office's budget information. School principals use an *electronic workbook* to draw up their budgets and then submit them to the district office via the Internet. Schools can decide at which point during the budget process they will make the information accessible online (Olson, 1997).

ACCOUNTING

Financial accounting is one of the most important aspects of management, as it provides the base from which almost all educational decisions involving money are made. Accounting is concerned with the school's property. This responsibility is met by: (1) documenting, analyzing, recording, and summarizing budgeted and actual transactions; and (2) providing adequate safeguards to preserve property through the use of audit procedures (see the section on auditing later in the chapter). All financial transactions are initiated by documents to indicate by whose authority and for which budget unit they were transacted. Accounting is more than just bookkeeping, which concerns the recording of financial data on forms or in electronic format. Accounting implies a broader purview including evaluation and interpretation of data. Costs, when compared to services performed or benefits gained, may indicate waste and excessive expenditures. It is the responsibility of the accountant to assist school administrators in getting 100 percent value out of every dollar spent. Effective accounting procedures promote efficiency.

Generally, the basis for accounting is either the accrual or the cash method; however many schools use modified versions of these two approaches to meet their unique needs. *Accrual accounting* is a system that recognizes revenues and expenditures when the school's money is committed. The commitment is noted whether or not the receipt or payout of funds actually has occurred. For example, a commitment is formed when teacher contracts are issued. The purpose of accrual accounting is to reflect the actual financial position of the school at any given point in time. Generally, the benefits of this approach are deemed to outweigh its greater complexity.

In contrast, *cash basis accounting* records only those transactions where an exchange of funds actually takes place. The checkbook of a private individual is an example of cash basis accounting. The checkbook balance is adjusted when a bank deposit is made or a check is written. No adjustment is made for the house payment due in three weeks. Cash basis accounting is much simpler but doesn't reflect unpaid commitments. Although most small nonprofit organization rely on cash basis accounting, most school systems are not small organizations and, in addition, face special demands in handling public funds.

Theoretically, accrual accounting is a better option; however, in practice most schools have implemented a hybrid form of accounting that contains elements of both accrual and cash-basis systems. This is because schools need to account for future revenues to implement their budgets, yet many state statutes require them to operate on a cash basis. Standardized accounting procedures have acknowledged *modified accrual accounting* as the appropriate method to be used by government agencies such as schools (Thompson, Wood, & Honeyman, 1994).

Special Funds

The fund system of accounting gives schools and colleges more latitude in carrying out their responsibilities and meeting their objectives. Funds are independent fiscal and accounting entities that maintain a separate record of assets and liabilities and are self-balancing. Schools employ both *public funds* and *trust funds*. One example of public funds is the *general fund* through which schools obtain the major portion of their operating budget. Trust funds normally are restricted to certain uses (for instance, many schools create a special fund for transportation). Schools also utilize contingency or petty cash funds for emergency expenditures. Examples of *special revenue funds* are gate receipts from athletics events and donations from booster clubs. Special funds may be defined by the source of revenue and/or the purpose for which they are used. As a practical example, parents and students may ask the school to initiate a fund to send the cheerleaders to summer camp, with money coming from fundraising events and private donations. Administrators must assure that special funds are deposited in a timely manner in a separate account and are used only for their stated purpose.

Audits

Financial audits are part of the normal procedures of almost all budgeted agencies. Audits involve the examination of financial *procedures* and financial *expenditures*. Many people think of audits in a negative sense, but they shouldn't be seen this way under normal circumstances. An audit is a useful and necessary procedure for both the larger organization and the budgetary units on which it is performed. On the unit level, an audit can be an educational experience for both the administrator and the staff. Recommended procedures and guidelines relative to expenditures that might otherwise remain vague can be clarified. An audit usually has a sharpening effect on the financial process, which contributes to overall efficiency.

Audits serve three purposes: (1) to detect errors in the accounting process, (2) to recommend changes in accounting procedures toward improved fiscal operations, and (3) to demonstrate to public funding agencies that the organization is operating according to statutory requirements. Several different types of auditing procedures have been devised, with each serving a particular purpose. *Internal auditing* is a set of procedures that occur within an organization to provide a system of self-checks. Internal audits can be periodic or continuous and may include *pre-auditing* protocols designed to assure proper accounting measures in advance of transactions. *External auditing* is a formal examination of an organization's financial records by an independent (outside) auditor, such as a certified public accounting (CPA) firm, to verify accuracy and compliance with regulations. This service often is referred to as a *general comprehensive audit*. In the case of public schools and colleges, *state audits* are carried out to assure compliance with statutory regulations involving the distribution of state or federal funds.

Moreover, there are *routine* audits and *special* audits. Routine audits occur periodically for the reasons just explained. They are part of the overall financial procedures of most organizations. A special audit may be initiated by a higher authority inside or outside the organization or by the unit administrator. One reason for a special audit is to investigate a suspicion or accusation about misuse of funds or improper procedures. Also, special audits are sometimes done in preparing for a change of administrative personnel or reorganization of the unit. An audit should be viewed as a process that will enhance financial management and reassure that the financial records and procedures are correct.

ATHLETICS BUDGETING

The budgeting process in athletics is sufficiently distinct from that of other school programs to merit further comment. As noted above, athletics departments have additional sources of revenue in addition to the general appropriation. The director of athletics manages income from such sources as gate receipts, concessions revenue, broadcast media fees, participation fees, and booster club donations. On the college level, some NCAA Division I schools have operating budgets in the tens of millions of dollars. In public schools the numbers are smaller, but the principles of good management are much the same.

The great majority of expenditures in the athletics budget go to coaches' salaries and transportation costs. Other major expenditures are equipment, supplies, and uniform replacements. Additional funds are spent on awards, hosting contests and tournaments, paying officials, and office supplies. The school district budget should cover most of the expense for maintenance and repair of facilities, unless this line item has been transferred to the athletics budget. Some school districts maintain separate transportation accounts to cover travel. This relieves the director from having to compute costs and apply formulas for team trips. Directors in public schools usually are freed from the responsibility of determining coaches' salaries, as this task normally is carried out at the district level. Coaches' salaries may be negotiable or determined by a salary schedule. Salary requests usually are compared with neighboring districts' pay scales to determine equitable and competitive figures.

Athletics directors on limited budgets must establish some type of rotating system for purchasing large-ticket items, like team uniforms. A chart should be constructed with all sports listed, indicating what year uniforms were last purchased and when they are scheduled to be replaced. Different types of uniforms have varying life spans, and this should be factored in. This system prevents the budget from being overburdened by having to replace uniforms for several sports at once.

The budget process for athletics is similar to that of instructional programs. The first step in the process is to set the goals for the program. Goal statements should be tied to activities and expenditures. Many school districts require an annual report from the athletics department, in which goals should be included. The director should meet with coaches individually, and as a group, in developing goals and priorities. Coaches have three main tasks: (1) take a complete inventory of equipment and supplies, (2) prepare a request for essential items, and (3) furnish a written explanation for any extraordinary items. After coaches list their specific needs, they can be given the opportunity to explain and defend their budget requests. These meetings provide coaches with some idea of how much money will be available.

The next step is to secure prices from manufacturers and suppliers. Once this has been completed, the director presents the athletics budget to the building principal. The director should be prepared to explain and defend the budget at this point. It's a good idea to have copies of coaches' requests and quotes from vendors on hand. Rarely do these meetings involve serious disagreements if the director has kept requests within the amounts normally allocated. The budget then goes to the school board for final approval.

Once the budget has cleared all reviews, it is returned to the athletics director as the *approved budget* for the coming year. The director now can began the purchasing and bidding process. Relatively small purchases (usually under $500) normally don't require bids. For major purchases, bids usually are solicited from at least three vendors. Athletics directors are obligated to apply the same strict guidelines and procedures to their budgets as those that govern instructional budgets in the schools.

AVOIDING FINANCIAL PITFALLS

Administrators stumble into pitfalls usually because of carelessness, and rarely because of dishonesty. The following guidelines for avoiding pitfalls should be helpful:

1. Keep adequate records of financial agreements and transactions. The following paperwork must be kept on file: (a) personnel contracts or other employment agreements; (b) purchase requests; (c) a record of the essential information submitted on time cards; (d) contracts for athletics contests; and (e) memos, letters, and other written agreements that explain financial terms. It can be embarrassing and awkward to be unable to produce the necessary paperwork in support of financial commitments or transactions.

2. Money can be collected only by designated personnel. This includes the sale of tickets and collection of student fees. Those authorized to collect funds usually have clear instructions on how to handle and process the funds; and if much money is involved, the individual is bonded. (A surety bond insures the school against loss caused by the person handling money.)

3. All collected funds should be deposited in a timely manner in the appropriate account, so that an official record is made. The funds, in turn, are utilized by the approved procedure and are subject to financial audit.

4. Avoid having an unauthorized flow of petty cash. If you have an authorized petty cash account, be sure it is tightly managed in accordance with correct procedures.

5. Avoid having cash in obvious places and never leave cash unsecured (where it offers the potential for theft). In cases where a large amount of money is collected, deposits should be made frequently.

6. Follow the prescribed financial procedures of the organization in all cases, and have financial records subject to regular audit. Any deviation from this practice has real potential for getting an administrator into difficulty.

References and Recommended Readings

American Management Association. 1990. *Financial Management*. New York: AMA.

Drake, Thelbert L., & William H. Roe. 1994. *School Business Management: Supporting Instructional Effectiveness*. Boston: Allyn and Bacon.

Gross, Michael. 1996. "The School Budget: Blueprint for Success." *Momentum*, 22:52–54.

O'Donnell, Susan. 2002. "Systems Education." *The E-Files*, No. 10 (11/01–2/02). Online EdLoop post: <http://www.arthurhu.com/2002/03/system.htm>.

Olson, Lynn. 1997. "Electronic Workbook Bares Budgets in Seattle." *Education Week*,16(26): 3.

Sperber, Murray. 1991. *College Sports, Inc.* New York: Henry Holt.

Thompson, D., C. Wood, & D. Honeyman. 1994. *Fiscal Leadership for Schools: Concepts and Practices*. New York: Longman.

Online Sources

American Education Finance Association
http://www.ed.sc.edu/aefa/>
AEFA serves as a forum and information network for the exchange of ideas concerning education finance issues among academic researchers, program administrators, and policymakers. The site provides a hyperlink to the *Journal of Education Finance*.

ERIC Clearinghouse on Educational Management. "Trend and Issues: School Finance."
http://eric.uoregon.edu/trends_issues/finance/index.html
This site includes links to material on funding disparity, finance equity, cost effectiveness, cost-cutting trends, and fundraising strategies.

8 Public Relations

Public relations begins with two fundamental premises: (1) schools cannot function without public support, and (2) the public has the right to know what is going on in schools. Public schools rely on the public's money. Members of the public want to know what is being done with their children and their money. If they feel dissatisfied with the efforts of the schools, they have a right to bring about desired changes. This is one of the basic reasons for the necessity of open and forthright communication between schools and the public. Without it, a school cannot serve its purpose in a democratic society.

Public relations has two dimensions. The political dimension (discussed above) lays the groundwork for adequate public funding for the schools. The second dimension addresses the school's relationship with students, parents, and others in the community separate from funding. A higher level of public support is required if schools are to be outstanding. This will become evident in the discussion that follows.

Educators often give too little attention to public relations. Putting an effort into public relations allows educators to improve their effectiveness and to win broader and more favorable public support for their programs. Public relations can no longer be a sporadic function, relied on only when a tax referendum is imminent. It must be a continuous effort. An effective public-relations program must have a strong element of consistency. It should involve an ongoing effort that keeps ahead of potential problems, as opposed to a crisis or "brushfire" approach— "an ounce of prevention is better than a pound of cure." By the same token, public relations is more than a media campaign; the process needs to incorporate two-way communication with the public. The most important element is credibility. Straightforward honesty is crucial; the consequences of any amount of dishonesty are usually disastrous in terms of public relations.

School personnel involved with physical education and athletics have significant public-relations opportunities and responsibilities. Athletics are of particular interest to the public and attract relatively large public support and media coverage. Physical education is somewhat less visible to the public, but most parents are concerned about the health and fitness of their children and want to know what the schools are doing in this area. Despite their differences, these two areas share the following distinction: the public invests considerable resources in physical education and athletics by way of funding, facilities and the like. If adequate support is to be maintained, the programs must be good enough to justify such a large expenditure, and the public has to be convinced of the value of physical education and athletics. Ultimately, they must be willing to pay the bill.

DEFINITION AND SCOPE OF PUBLIC RELATIONS

People frequently interpret public relations as coverage in the print and electronic media. The media is certainly one important aspect of public relations, but an overall public-relations effort is much more inclusive. For example, in a school setting, public relations at a minimum includes the following: internal relations among school personnel, including both students and employees; relations with parents; and relations with the public in general.

The avenues through which an effective public-relations program is launched include: (1) the mass media (radio, television, and newspapers); (2) direct conversation (face-to-face or by telephone); (3) written communications (letters, notices, reports, bulletins, and report cards); (4) the perception and image of the program that is communicated by students and faculty; and (5) the effective conduct of school programs and demonstrations, athletics contests, and other public events. In short, public relations involves everything that influences the image of the school in the minds of the total public.

Organized public relations may range from the "hard-sell" campaigns of business or politics to the "soft" approach of nonprofit organizations. An effective public-relations program can be surprisingly simple. It need not contain "show business," gimmicks, or dramatic events. It can be a well-planned, normal, ongoing exchange of useful information among school employees, students, and members of the community. It should be a naturally integrated part of the school's overall efforts. This does not mean that there is never a need for special emphasis. Sometimes a reason exists to inform the public about issues or matters of special concern, to convince the public to support a new idea or program, or to recognize a particular need of the schools.

Public relations can be thought of as both an attitude and a process. As an attitude, it involves realistic optimism. As a process, it involves open and effective communication, with emphasis on the positive.

PURPOSES AND IMPORTANCE OF PUBLIC RELATIONS

The purpose of public relations is to bring about a harmony of understanding between any group and the public it serves and upon whose good will it depends. School public relations has two fundamental components: administering excellent programs, and effectively communicating that fact to the public. Certainly, the first component is more important, but both aspects deserve adequate attention. Among the specific purposes of school public relations are the following:

1. To inform the public as to the work and accomplishments of the school.
2. To establish confidence in the schools.

3. To rally adequate support for educational programs.

4. To develop awareness of the importance of education.

5. To improve the partnership that should exist among the schools, the students, the parents and other members of the public.

6. To correct any misunderstandings regarding the aims and activities of the schools.

U.S. opinion polls show that the majority of parents are basically satisfied with the schools their children attend; however, public perceptions of the quality of schools has been in a steady decline since the 1990s. Americans traditionally assign higher grades to the schools in their own communities than to the nation's schools as a whole. Given the opportunity to choose between reforming the existing system or finding an alternative system, a large majority say they favor improving the existing system. The public feels that the biggest problem facing local public schools is lack of financial support, yet they often are unwilling to provide the level of support that schools need (Rose & Gallup, 2001). Historically, education has been a local concern in which communities, not the federal government, have assumed the primary responsibility for the schools. This has obvious implications for public relations.

PLANNING PUBLIC RELATIONS

A good public-relations program is not haphazard; it must be well planned and wisely implemented. Taking time to plan public-relations activities greatly increases their chance of success. Lew Armistead (2002), former public-relations director for the National Association of Secondary School Principals, has suggested the following ten-step planning process for a public-relations campaign:

1. Identify the specific public-relations challenge or opportunity (e.g,. to demonstrate that student activities are important to the students' development).

2. Determine the key audiences for this communication project. Realize that there is no "general public" but rather numerous specific publics, each with unique characteristics.

3. Find out what your targeted audiences already know or suspect about the subject of the campaign. Remember that perceptions are more important than reality, and don't communicate to audiences what they already support.

4. Determine how members of your audiences receive information, and which sources have the most credibility (e.g., do they read mailings or throw them away without reading them?).

5. Set measurable objectives you hope to achieve with each audience (e.g., financial support for intramurals will increase by 50%).

6. Develop message points for what you want your audiences to understand and believe (e.g., student activities plus academics equal a complete education).

7. Develop a list of specific communication activities that will deliver your message (e.g., videos, student testimonials, question-and-answer sheets).

8. Determine what resources are necessary to complete the communication activities. Do you have the available time and funds?

9. Create a time line to make sure everything is done when it should be (e.g., a campaign in support of a school bond issue that must be completed before election day).

10. Evaluate the successes of your communication campaign upon its completion. This is crucial so that you don't make the same mistakes year after year.

Methods

A well-balanced public-relations program includes a variety of approaches, several of which are described in this section. Of course, the total public-relations effort extends far beyond these methods. It involves every kind of information and impression given to the public by school personnel, whether intentional or unintentional.

A good place to begin is the school's *physical facilities*. The buildings should have a welcoming appearance to visitors. They should be clean, attractive, well lighted, uncluttered, and in good repair. The grounds also should be attractively landscaped, fenced securely, well lighted at night, with the lawn mowed and the shrubbery trimmed. Signs should be posted strategically to make it easy for visitors to get around both indoors and outdoors.

Parent conferences allow parents to feel more connected to the school. Most often conferences are held at the beginning of the school year or at the end of a grading period (when performance reports are distributed). Conferences have been commonplace at the elementary-school level but underutilized by middle and high schools. Some schools give teachers a staff development day (with the morning off) and extend conference hours through the afternoon into the evening to accommodate working parents. The school library, auditorium, and gymnasium make good meeting rooms to handle large crowds. Scheduling five-minute conferences allows time for parents to meet all the teachers. Both teachers and coaches should take the time to hold conferences with parents (Voors, 1997).

Another valuable approach is to take students out into the community on *field trips;* special-interest tours; and educational visits to recreational venues, museums, historical sites, industrial and manufacturing plants, and the like. This is one way to make education more visible in the community and make the community more familiar to the students.

It is important to publish periodic *reports* relative to the school's well-being (financial and otherwise) about its personnel, programs, or special achievements. Reports about school deficiencies and special problems that need public assistance also can be helpful. Many school districts now place their annual reports online.

One of the most useful means of public relations is through *handouts and notices* to students and parents, including report cards as well as various informational materials and notices of programs and school events.

Information can be transmitted by use of school-generated *bulletins* or *newsletters* that are mailed to alumni and staff or carried home by students. Such publications should contain information that keeps people informed of programs and current events. They also can be used as a management tool to remind readers of policies and procedures (Baker, 2001). It is important that they are written in an appealing manner and with a positive image, so that the targeted audience will be motivated to read them. Figure 8-1 shows the front page of a newsletter distributed by a college Department of Health, Kinesiology, and Leisure Studies.

Report cards sent home to parents that include more information than the simple basics can serve as a good method of communicating important facts about student

HKLS UPdate

Purdue University

Newsletter of the Department of Health, Kinesiology, and Leisure Studies, Purdue University - Vol. 2, No. 1 - Summer 1998

Editor: Kim Lehnen

Klehnen@purdue.edu

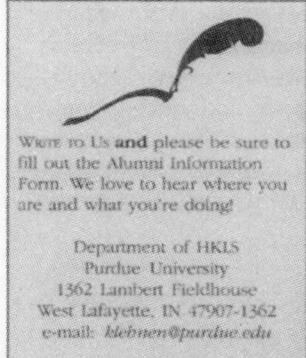

Write to Us **and** please be sure to fill out the Alumni Information Form. We love to hear where you are and what you're doing!

Department of HKLS
Purdue University
1362 Lambert Fieldhouse
West Lafayette, IN 47907-1362
e-mail: *klehnen@purdue.edu*

Also check out our web page at:
www.sla.purdue.edu/academic/hkls

FROM THE DEPARTMENT HEAD

On behalf of the Department of Health, Kinesiology, and Leisure Studies, I am pleased to provide highlights from the 1997-98 academic year. The year can be summarized as one of continued reflection and growth for the department. We have achieved much collectively, and individual successes have been numerous. With assistance from the university, our school, faculty and staff, students and alumni have worked hard toward the continued improvement of the department.

By creating various improvement faculty teams, the department has continued its review and modification of undergraduate and graduate curricula and related recruitment and retention issues and policies. Equally, service programs have been reviewed with recommendations for the future of our Physical Education Service Program, the Adult Fitness Program, and other outreach activities. Approval and support for the creation of the A.H. Ismail Fitness and Nutrition Research and Education Center is a major highlight for the year. Fund raising has been successful and will continue for the Center in the coming year.

Research, publication, and grant and contract activity appears to be on the rise in the department as faculty continue to be productive on all fronts. Our faculty have served in many prestigious professional service roles, and Professor Joan Duda spent eight weeks abroad as the International Visiting Scholar, which was bestowed by the Australia Sport Psychology Society.

The department renewed its relationship with alumni and former faculty by mailing its first newsletter in many years. Beyond sharing departmental events, the newsletter assisted in soliciting over $6,000 in gifts. We thank everyone for their support.

As noted in this newsletter, the department celebrated the success of a number of students on April 19. As a part of the ceremony, three awards were named in honor of Professors Emeriti Anthony Annarino, Hal Veenker, and Carol Widule.

The department has made strides in the upgrade of Lambert through painting, the Wastl lab renovation and dedication, and again, the creation of the Ismail Center. Various curricular, personnel and facility issues face HKLS in the future, but our new administrative structure, team efforts, and sense of community will meet these challenges successfully. There are many challenges and opportunities ahead, and we pledge to continue our goal to be one of the best departments in the country. Thanks again to everyone for your continued support and best wishes!

Thomas J. Templin

Reprinted with permission of Department of HKLS, Purdue University.

Figure 8–1: Newsletter (front page) of a Department of Health, Kinesiology,

achievement. Physical education grades tend to be distinct from classroom grades and may call for a different format or additional comment. Sometimes, other information like fitness test scores can be enclosed with the report cards. This insures the information will receive more attention than if distributed separately.

Student performances on *youth fitness tests*, with regional and national norms, can be highly meaningful to parents and the public. The awarding of fitness certificates and other forms of special recognition can reinforce the message of student achievement.

Exhibits, presentations, demonstrations, and *open houses* have been used to provide parents and others in the community a firsthand view of successful programs. Parents like to see their children perform or participate in these programs. Such events as gym circuses, dance performances, and sports skills demonstrations have been used successfully in physical education. Whenever programs of this kind are conducted, they should portray educational aspects of the activity as well as entertain the public. An effort should be made to engage the community in special events. For example, someone from the mayor's office can be asked to present an award (Baker, 2001).

Community service by teachers and coaches is an effective method of public relations. Examples include conducting fitness workshops, opening the gymnasium on weekends, or coaching in youth sports programs. Well-kept *bulletin boards* can serve an excellent purpose. Useful information can be communicated via bulletin board displays to other staff, to students, and to parents when the visit the school. *Trophy cases* are useful in connection with athletics and other extracurricular competition. Much interest and pride can be instilled through the proper display of trophies.

Exhibits in shopping malls are another good way to publicize school programs. Shopping malls have become the modern "town square." Booths can be staffed by students or teachers to disseminate information. Schools can put on demonstration activities in malls with staging areas (Baker, 2001). In addition to newsworthy items that are routinely provided to the news media, special *feature stories* can be developed about new programs, outstanding teachers, or unique student achievement.

Online *Web pages* provide a worldwide venue for publicizing one's school and programs. Departments and programs from the college level to elementary schools have Web sites on the Internet to describe programs and announce upcoming events. They can include images and hyperlinks to related information, as well as written information. Professional consultants can help with the design of a Web page, and a Webmaster should be assigned to keep it up to date. Figure 8-2 shows the Web page of a public school physical education program

Valuable information for use by the media can be obtained from state and national organizations. For example, every state has a Department of Education with one or more *specialists* in the area of health, physical education, school recreation, and athletics. This state agency is capable of providing useful information in the form of news releases, newsletters, and information bulletins.

National organizations can also be a helpful source. For example, the President's Council on Physical Fitness and Sports provides publications, facts and statistics that are of interest to the public. Educational *films* may be acquired on free loan or at reasonable rental rates from such agencies as AAHPERD and the President's Council. Films can help members of the public better understand the objectives, values, and concepts of physical education. The Centers for Disease Control and the Surgeon General's Office put out reports on health and fitness that can promote the need for physical education classes.

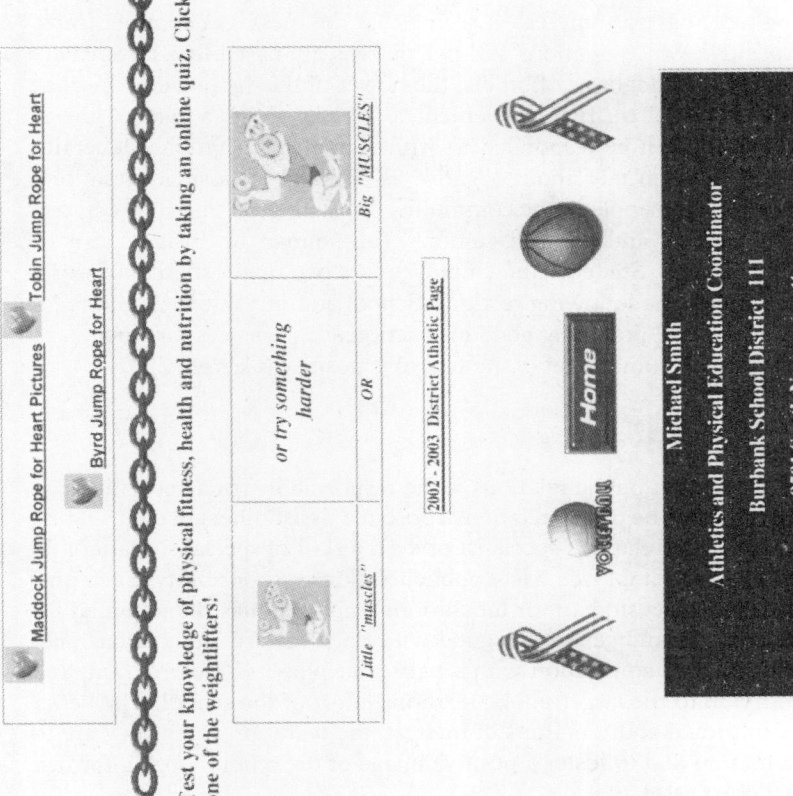

Reprinted with permission of Burbank (IL) School District 111.

Figure 8–2. Web page for a public school physical education program

Word-of-mouth communication is a crucial but often unrecognized medium for public relations. Schools have an interest in seeing that the messages communicated are accurate and positive. Since opinions and perceptions are based to a large extent on comments and recommendations of others, the power of the "grapevine" must be acknowledged. Many school districts implement a "rumor line," a phone line in which gossip and rumors can be responded to with factual information. Generally, individuals are more prone to talk about negative as opposed to positive situations. Schools need to know what people in the community are saying, so that they can correct inaccurate impressions and squelch rumors. Talk spread by insiders can be extremely helpful or harmful. Students and employees are two major sources of word-of-mouth impressions (see the following section). School administrators should focus on employee satisfaction and providing good instructional experiences for students to assure that the information coming out of the school is positive (Carroll, 2002).

Who Should Be Involved?

Some people think that public relations is the responsibility of a specialist who can give overall direction to the public-relations effort and assist others in contributing their fair share; but a public-relations specialist or even a staff of specialists cannot do the job alone. Every school employee has a public-relations responsibility. Each one, whether a coach, secretary, custodian, or lunchroom employee, has the potential for presenting a positive image and generating good will. Each faculty member has public-relations opportunities in connection with a particular phase of the program, and each needs to contribute to the larger public-relations effort of the school. The better the employees are informed about matters of interest, the better prepared they are to give accurate information and to foster a positive image of the school. Well-informed employees can be a great asset.

The idea that "quality attracts quality" applies to education. The first consideration of quality should be personnel. If a school is staffed with high-quality people, they will build quality into the other aspects of the program. An administrator should consider the public-relations qualifications of each prospective employee, and this should play its rightful role in the selection of those to be hired.

Each individual in physical education and athletics has a responsibility to improve the public image of the profession. This can be partly accomplished by all of us giving attention to the following personal standards:

1. Are you well groomed and appropriately dressed, or do you belong to the sloppy sweatshirt group? Sweatshirts have their place, but not in public venues.

2. Check your personal habits. Do you smoke in front of your students? Do you use inappropriate language? Have you let yourself become flabby and overweight through lack of exercise and improper diet?

3. What about the quality of your teaching? Do you have your courses planned in advance for the semester, with objectives you want your students to reach, or do you improvise from day to day and on the spur of the moment?

4. Do you treat your students with consideration and understanding, but at the same time maintain discipline?

5. Are you well-read in your area, or do you find the scholarly life a bore? Are you seen at public cultural events in addition to athletics contests and school events?

Students can be effective ambassadors of the school. How they represent the school and what they say about it can certainly impact the school's image. Attention is focused on certain students because of outstanding achievements in academics or athletics. These students have a greater-than-normal influence on the school's image.

Often, what happened at school on a given day becomes the topic of discussion in the home. The impressions communicated by the students have a significant influence on the minds of parents. Parents accept the views of their children as reliable indicators of school programs, often at face value and without seriously questioning them. For these and other reasons, the first front on which the public-relations battle should be fought is the classroom. If the students believe in the program, the teachers or coaches, and the learning process, they can be a major factor in maintaining a positive image and garnering public support. Students can be your greatest allies in public relations.

The idea that "people listen better when others toot your horn" is an important consideration. Even with the same set of facts, what you say about your own program has less impact than what someone else says. Therefore, it is often useful to involve nonschool personnel in public-relations efforts. For example, it could be beneficial to have a news article by a prominent physician about the health instruction program, the need for encouragement of good health practices in the home, or emphasis on physical fitness. If a prominent citizen states that a facility or program is needed, the majority of others in the community will likely accept what is said.

A school–community public-relations council can be a highly useful instrument if it functions properly. Such councils can be a valuable source of ideas and can help implement them. In addition, councils can serve as a barometer to measure effectiveness and provide useful feedback for evaluation of the public-relations effort. Clearly, the school public-relations program depends on the efforts of many people, including employees, students, and nonschool personnel.

Identifying Publics

A *public* is a group of people drawn together by common interests; located in a specific geographic area; or characterized by some other common feature such as age, religion, sex, occupation, race, nationality, politics, income, social status, affiliation, or educational background. The more that you know about a particular public, the easier it is to direct an effective public-relations approach (Boone, et. al., 2002). In order to gather information about a public, various techniques may be used including surveys, questionnaires, opinion polls, interviews, and face-to-face conversations. Prevailing opinions of a particular public are often the result of experiences in early life, but they are also influenced by current experiences and by the images projected in the mass media.

Among the several categories of publics that must be considered in a school's public-relations effort, Newsom (1996) lists the following:

- *Media publics*—Print, broadcast, local, regional, national
- *Member publics*—Professional organizations, accreditation organizations, athletic conferences
- *Employee publics*—Labor unions, teacher organizations
- *Community publics*—Civic clubs, chambers of commerce, realtor associations
- *Investor publics*—Private foundations, benefactors, charities
- *Government publics*—Federal, state, and local; legislative bodies, regulatory groups

- *Consumer publics*—Students, parents, sports fans
- *Special publics*—Watchdog organizations, activist groups

Until recently, each neighborhood and/or community had its own school. The neighborhood school concept enhanced parental awareness and support. However, recent trends, including school consolidation, busing of students to meet integration guidelines, and demographic changes, have altered the reality of the neighborhood school in many communities. These developments make it more difficult for school administrators to identify the publics with whom they need to communicate, and it makes the publics less accessible. Consequently, it is important to maintain a list of relevant publics.

The effectiveness of the methods of communicating and the content of the communications vary with different publics. It is important for the schools to have a sound plan that gives adequate attention to all segments of the population the school serves. Sometimes the target group for certain information constitutes a portion of the larger public, such as sports fans, members of the PTA, particular ethnic groups, or parents with children of a certain age. It is important to define the audience toward which the public-relations effort is pointed. Targeting audiences improves effectiveness.

COMMUNICATION IN PUBLIC RELATIONS

Communication provides the thread that binds an organization and its clientele by ensuring common understanding. Effective communication is essential in good public relations, as persuasive communication can sway public opinions. Chapter 5 contains a major section covering communication. Certain aspects of that section apply to communications with the public; therefore, the reader is encouraged to review it in connection with the present topic.

The following "Six Cs" of communication have provided a model among public-relations specialists. They can be effectively applied to the field of education and to athletics. They include:

- *Content.* Communication should accurately state the intended message as clearly as possible. Accurate content is fundamental to effective communication.
- *Clarity.* Communication should be stated in a manner that will be easily understood. This will reduce the chances for misinterpretation, misunderstanding, and confusion.
- *Credibility.* Communication should be based on information from appropriate sources.
- *Context.* Communication should fit the situation or the environment. It must be appropriate in terms of kind, style, and content.
- *Continuity and Consistency.* Communication should flow with consistency and regularity in order to maintain open channels and mutual trust.
- *Channels.* Appropriate channels of communication should be used.

News Media

One effective channel of communication is through the news media. This includes both electronic and print media. News items, regardless of form (written, oral, images), are generally of four types: pre-event, post-event, feature, or editorial.

All these can be used effectively in the school's public-relations program, and each should receive consideration. News stories often are generated by direct contact with news personnel. Try to establish personal relations with as many newspeople as possible. News media occasionally are invited into the school for athletics contests, science fairs, and other events. Someone on the school staff should be assigned to see that media representatives are accommodated appropriately.

Photographs attract special attention to news articles and can be highly communicative in and of themselves. Ready access to useful photographs can result in the publication of pictures along with printed material. Keep updated pictures of yourself and staff in personnel files—both serious photos and cheerful photos—for appropriate occasions.

To establish good working relationships with the news media while protecting the integrity of your program, consider applying the following strategies and administrative guidelines (Potter, 1997):

- Establish a protocol for media information and share it with staff members.
- Designate a person to handle media inquiries who can be available within a reasonable time period.
- Keep your supervisor and the school's chain of command apprised of media issues and information provided to the media.
- Stay informed about board policies on media procedures, as well as state and federal regulations on sunshine laws (requiring that deliberations of boards and public agencies be conducted publicly and reports be made available to the public) and confidentiality requirements. Consult the school attorney about the extent to which you can legally share information.
- Provide in-service training for staff who have media responsibilities.
- Notify staff of inquiries about them, and notify parents of inquiries about students.
- If media want to interview students, obtain parents' permission and accompany the media crew to make sure the interview is conducted properly.
- Construct the message you want to disseminate before a media interview. Plan what you want to emphasize and work it into the conversation. Be cautious in making remarks during interviews, as excerpts may be used out of context.
- Anticipate press inquiries; plan together with staff as to possible questions and appropriate responses. Employ brainstorming and role playing, if helpful.
- Be informed of news deadlines and see that information reaches the media in time to ensure that it will be used.

Don't underestimate the power of the news media. They have an unparalleled role in shaping public opinion. Always be honest with them—tell the truth. Don't tell the media more than you want them to know; be careful about volunteering information. If you don't know the answer to a question, say so. You might respond, "Sorry, I can't answer that." Keep the language simple and avoid educational jargon. Finally, keep in mind that what *you* consider news may not be news to the media (Potter, 1997).

Responding to Negative Incidents

Negative incidents are inevitable. Sooner or later, every school administrator is going to be confronted with one: a star athlete is arrested, a student brings a weapon on campus, or a teacher employs an inappropriate method of discipline. Most large school districts have a public-relations professional who can take the heat while you get your thoughts together. This person also can give you suggestions on the best way to handle the incident. The worst thing you can do is say, "No comment." This response has two drawbacks: it implies that school is "stonewalling," and it suggests there is something to hide. Instead, admit what you can and then make a positive statement about your school's beliefs and policy, not about the incident. You may be justified in stating that the information has to be kept confidential under the law; or you may comment, "There has been an accusation, but I haven't been briefed on the situation yet." Realize that the public expects an explanation, but also that they understand that negative things can happen to anyone. Explain your position in clear and straightforward language; avoid jargon and "waffling." Finally, assume that there is no such thing as "off the record." Anything you say may end up in the public record (Million, 2000).

Handling Complaints

Every organization that deals with the public will receive some complaints. This is especially true with education that involves people's children and is financed by public funds. A complaint involves one of two circumstances: a weakness or fault that justifies the complaint, or a misunderstanding. A complaint left unresolved results in a continuing problem, one that is likely to grow and lead to other complaints.

From the public-relations point of view, complaints should be handled in an open-minded and gracious manner. The person dealing with the complainer should avoid being defensive and abrupt. Try to understand the complainer's point of view and help him or her to understand yours. If the complaint is justified, then admit it and make the necessary corrections. If it is not justified, help the complainer to recognize that. This type of objective approach will help soothe feelings and win support.

Administrators should structure meetings to provide opportunities for staff, students and the public to vent their feelings or concerns, especially when a controversial issue arises. Unfortunately, many people are better talkers than listeners. Listening is as much a part of good public relations as talking. Listening not only can be highly informative, it also represents a show of respect for another person's point of view.

Publicity

Publicity (attracting public attention or public acclaim) involves drawing attention—either intentionally or inadvertently—toward something such as a person, an event, or a program. It is usually done intentionally, for promotional reasons. Publicity is one important aspect of public relations, but it is not synonymous with public relations. Normally, school athletics seeks and receives much more publicity than other school programs. Consequently, publicity about athletics events is significant in the school's public-relations effort. However, there may be other activities and events that warrant publicity, including dance concerts, science fairs, physical education demonstrations, and scores on national fitness tests.

Publicity differs from advertising in that it obtains exposure for your program without paying for it. In this regard, it has more credibility than advertising because it has not been bought. Publicity is obtained primarily by creating reasons for journalists to believe that their readers/viewers want to know about your program—that it is newsworthy. The good publicist can come up with ideas ("hooks") to persuade journalists to cover a program or event. In the commercial sector, this is called a "pitch."

However, if you maintain good relationships with journalists, they may inadvertently serve as unpaid spokespersons for your programs. For these reasons, publicity can be a very powerful part of your marketing effort.

The major disadvantage of publicity is that because you are not paying for the coverage, you do not have control over it. There may be a lot of coverage, or there may be little or none. It may come too early or too late. It may be positive, or it may be negative. But when it is positive and widespread, it can be tremendously effective. Publicity should be an essential part of the marketing mix within the total PR effort (Hauser, 1999).

Marketing Physical Education and Athletics

The idea of strategic marketing, which developed in the corporate sector, has been exploited by the public sector including schools, and especially by college athletics programs. Marketing should be a component in the school's comprehensive public-relations effort for promoting athletics, intramurals and physical education. The basic rationale for strategic marketing is that the public may not understand or appreciate the goals, objectives and benefits of these school programs. Marketing has as its purpose the persuasion of the public to endorse, support, purchase or utilize a product or service—in this case, the curricular and extracurricular programs of physical education.

Marketing must be sensitive to the changing environment in the community in which schools operate, along with the contingencies that accompany change. Internal contingencies—such as goals and priorities of programs—are generally subject to control; while external contingencies—such as technology or the economy—are normally beyond control. Identifying these contingencies precedes a successful marketing plan.

A first step in marketing is to conduct market research of a targeted public (students, parents, taxpayers) through tools such as surveys, opinion polls or interviews. The collected data might identify attitudes, perceptions, needs, or interests of the public. This information is then used to shape a positive image for the program. At this point, the task shifts to promotion of the program through a publicity campaign. This may encompass replacing misconceptions and redesigning the program to garner public support. The final step is the control phase of the marketing strategy—the ability to determine whether your marketing plan achieved the desired results. Evaluation is the process of comparing the end results with the stated objectives of the marketing effort. This should be designed as cyclical feedback with improvement built in (McFarland, 2001).

SPORTS INFORMATION DIRECTORS

The position of sports information director (SID) has become standard at the college level and exists at a few large high schools. This title may be a stand-alone position or one combined with the position of assistant athletics director (e.g., assistant AD for public relations or media relations). Occasionally, the SID is housed in the college's public-relations department. The profession is growing rapidly. The College Sports Information Directors of America (CoSIDA) now lists over 1,800 members. Most sports information directors have a degree in journalism. In addition, graduate programs in sports management typically include a course in sports marketing. Some schools offer internships for students interested in a career in this area (McCleneghan, 1995).

Sports information directors serve as the press agents for the athletics department, and at the same time they take on myriad other duties. In addition to writing press releases, they may coordinate promotional events and sponsorships, conduct fan surveys, handle licensing and merchandising, plan and implement advertising, be in charge of press-box seating and credentials, be responsible for newsletters, and work on special projects. Their media responsibilities include contact with TV sports directors and reporters; newspaper sports editors, columnists and reporters; the school's newspaper sports editor; live TV game producers and directors; and booster club officials. Internally, they work closely with the athletics marketing director, the school photographer, the public affairs office, and the recruiting director, as well as with coaches and the director of athletics (Neupauer, 1997).

Currently, sports information directors are spending less time with media duties and taking on more marketing responsibilities, while assisting the AD with fundraising. The typical SID at a large university will have an office staff of two or three full-time assistants plus several student interns. At this level, the office may distribute over a hundred publications and thousands of press releases to media representatives every year. In carrying out their responsibilities, sports information directors spend a lot of time on the phone talking to people.

Public-Relations Channels for Athletics

In most high school athletics departments, the director or coaches handle public relations. Various promotion methods can be employed. Schools can maintain an athletics bulletin board in the main lobby with press items and schedules on it. Commercially produced season schedules should be placed throughout the school and in public venues around the community. Before each season, the department should supply all media with rosters, schedules, and other needed information. Some departments produce a weekly highlights sheet distributed to students. Also, a weekly press release keeps the media up to date. Coaches should let the director see any press releases before giving them to the media. To avoid embarrassment, always check spelling and grammar on materials to be made public. In addition, coaches can hold preseason meetings with athletes' parents to promote communication (Hoch, 1999).

Following are several special public-relations opportunities that exist in connection with school athletics:

1. The manner with which athletics events are presented—public address procedures, adequacy of seating, ushering, cheerleading, half-time performances, and the conduct of coaches and athletes.

2. Athletics game programs—content, design, distribution.

3. Newspaper, radio, and television coverage—amount, type, and content.

4. Booster clubs—potentially helpful, if they have valid and clear objectives and are well managed. (Conversely, a booster club that is poorly managed can become a problem.)

5. Ticket promotion and sales, and parking provisions.

6. Public appearances by coaches at banquets, on television shows, in radio interviews, and at other public settings.

7. Awards presentations and ceremonies.

GUIDING PRINCIPLES AND STRATEGIES

Based on the collective experiences of successful programs, several fundamental principles and strategies serve as guides for the public-relations process. They help to summarize the material presented in this chapter.

1. A school system cannot avoid public relations. Relations with the public will exist in some form—good, bad, or indifferent.

2. The most fundamental aspect of good public relations is a superior program, for which there is absolutely no substitute. Educators should not expect public support of programs that produce inferior results.

3. The public-relations program should be clearly defined, including purposes, procedures, responsibilities, and expected results. The better the program is understood, the better chance it has for success.

4. The public has a right to information pertaining to the school and all of its various programs. Educators should try to avoid operating in a world of their own, isolated from the public (to whom they owe their existence). However, school officials must stay in control of the program and reserve the right to make appropriate management decisions regarding public relations.

5. A vital function of public relations is to establish an open channel so that the public will be sufficiently informed to properly influence decisions.

6. While good public relations requires a well-conceived plan, flexibility and adaptability are necessary. It is important to adjust to changing circumstances and also to capitalize on unexpected opportunities.

7. The public is entitled to know the truth about school policies, finances, and programs. Hidden weaknesses fester and sooner or later erupt. If granted full knowledge of the facts, the public often can help in the solution of problems.

8. The way in which something is done is often as important as what is done. The particular approach can make a great deal of difference in how a decision or a circumstance is accepted.

9. Public education is a social institution financed and maintained by the total public, not just those with children of school age. Therefore, educational programs must be designed and managed with the benefit of all citizens in mind.

10. One excellent means of public relations is through the students. A teacher may have 200 students, or a coach may have 80 athletes—all of whom can be agents for public relations.

11. Public-relations whitewash wears off sooner or later, and then the facts become painfully obvious. Don't try to cover up embarrassing issues.

12. Every well-planned public-relations program is pointed toward a defined population. Identify the population and subpopulations that deserve special attention.

13. News stories and other reported information should be factual, relevant, free from bias, and appealing.

14. The public-relations program should be a continuous, integrated phase of the overall school effort. Poorly planned, hit-and-miss efforts are only partially successful at best.

15. One of the most effective public-relations approaches is person-to-person, teacher-to-student, student-to-parent, or teacher-to-citizen contact.

16. Some aspects of public relations need funding. If funding is inadequate, the success of the program is jeopardized. Adequate funding is just as necessary for public relations as for other school programs.

17. Effective public-relations specialists usually stay in the background instead of seeking the limelight; they keep abreast of the factors that affect the program, develop a wide circle of acquaintances, and make helpful contacts.

18. A highly reputable image can serve several useful purposes, one of which is public relations.

19. A sense of humor, especially when some of it is directed at yourself, can be helpful in public relations. Good, wholesome humor is a universal language.

20. Public relations, like other phases of the school program, requires both near- and long-term comprehensive planning.

21. School personnel who make sharp or abrasive statements or who criticize frequently or react defensively must gain control of these dysfunctional behaviors in order to serve as positive forces in public relations.

EVALUATION

Regular, critical evaluation of the overall public-relations effort can be a useful and purifying process. The organization's mission statement (see chapter 1) is the starting point for such evaluation. The public-relations program should be designed to contribute to the organization's mission.

Evaluations of public-relations programs should examine two outcomes: (1) what the organization did in terms of PR and how much it cost, and (2) what were the results, tangible and intangible. One of the principle methods for evaluating the public-relations effort is to survey the various publics to see if the targeted messages were received and were effective in changing the level of support, attitudes, or beliefs. You want to know who heard or saw what, and what they thought of it. This may require constructing questionnaires or conducting interviews with members of these groups. It may mean assessing changes in behaviors (e.g., attendance at PTA meetings or an increase in season ticket sales). The overall intent of evaluation should be to identify collective perceptions, as these are what constitute the organization's image. Following your assessment of the public-relations effort, you should have a sense of what works, what doesn't work, and what is worth the cost and the effort (Newsom, 1996).

References and Recommended Readings

Armistead, Lew. 2002. "10 Steps for Planning PR." *The Education Digest*, 67(6): 57–61.

Baker, Ken. 2001. "Promoting Your Physical Education Program." *Journal of Physical Education, Recreation, & Dance*, 72(2): 37–40.

Boone, Edgar J., R. Dale Safrit, & Jo Jones. 2002. *Developing Programs in Adult Education: A Conceptual Programming Model*. 2nd ed. Prospect Heights, IL: Waveland Press.

Carroll, David. 2002. "Make the Grapevine Work for Schools." *The Education Digest*, 67(7): 56–60.

Hauser, Cliff. 1999. "Marketing Overview: Publicity." *Marcomm Wise Knowledge Bank*. Online article: <http://entrepreneurs.about.com/gi/dynamic/offsite.htm>.

Hoch, David. 1999. "Handling the PR." *Scholastic Coach & Athletic Director*, 68(8): 4–5.

Kowalski, Theodore. 1996. *Public Relations in Educational Organizations*. Columbus, OH: Merrill.

McCleneghan, J. Sean. 1995. "The Sports Information Director—No Attention, No Respect, and a PR Practitioner in Trouble." *Public Relations Quarterly*, 40(2): 28ff.

McFarland, Allison J. 2001. "Developing a Strategic Marketing Plan for Physical Education." *The Physical Educator* (early winter): 191–97.

Million, June. 2000. "No Comment? No Way!" *The Education Digest*, 65(9):59–60.

Neupauer, Nick. 1997. "Sports Information: The Most Coveted, Ignored Profession." *The Public Relations Strategist* (fall issue). Online article: <http://www.prsa.org/_Resources/profession/6k039735.html>.

Newsom, Doug. 1996. "Evaluating Public-Relations Programs." Chapter 15. In T. Kowalski (Ed.), *Public Relations in Educational Organizations*. Columbus, OH: Merrill.

Potter, Les. 1997. "Getting the Media on Our Side." *The Education Digest*, 62(5): 22–24.

Rose, Lowell, & Alec Gallup. 2001. "33rd Annual Phi Delta Kappa/Gallup Poll of the Public's Attitudes Toward the Public Schools." *Phi Delta Kappan*. Online article: <http://www.pdkintl.org/kappan/k0109gal.htm#1a >.

Voors, Rob. 1997. "Connecting kids and parents to school." *The Education Digest*, 62(5): 20–21.

Online Sources

College Sports information Directors of America
http://cosida.ocsn.com/
CoSIDA provides a directory of members, offers workshops for SIDs, maintains some 20 active committees, and publishes a monthly digest.

National School Public-Relations Association
http://www.nspra.org/entry.htm
NSPRA provides communication products, maintains resource and research files, and has contacts and resources within the corporate communication industry. NSPRA maintains chapters throughout the country that provide local networking opportunities for members, and it also provides workshop assistance to school districts, state departments of education, and state and national associations.

The President's Council on Physical Fitness and Sports
http://www.fitness.gov/
The President's Council promotes, encourages, and motivates Americans to become physically active and participate in sports. The site provides numerous resources for teachers, coaches, and fitness professionals.

9 Legal Liability and Risk Management

Murphy's Law states, "Anything that *can* go wrong *will* go wrong." This may seem like an overly pessimistic point of view, not the mindset that should prevail in schools. As a preventive measure, however, acknowledgment of Murphy's Law can alert educators to potential liability problems. Administrators, teachers, and coaches should develop the ability to anticipate things that can go wrong and correct such situations before they cause harm.

Legal liability is based on the principle that every person has a right to justice and fair treatment under the law. Among the rights enjoyed by citizens is the right to personal safety. This right and other individual rights are receiving greater protection by our legal system, while traditional government immunity has eroded. Public school employees are no longer held immune from liability under the law simply because they are employed by a state agency. Educators are obligated by law to pay particular attention to the protection, health, and welfare of students. This chapter is designed to help potential administrators, teachers, and coaches better understand the legal concepts and principles that pertain to physical education and athletics.

Although physical education and athletics are among the more hazardous activities in the school, there is no reason to be apprehensive about providing a well-rounded program in these areas. Generally, the courts are aware of the inherent risks in these activities and acknowledge that the potential values exceed the risk expectancy, thus falling within the parameters of "reasonable risk" (see the section on risk management later in the chapter). It is assumed that programs are based on adequate knowledge of student needs and are conducted by competent instructors. One must recognize, however, that liability suits against educators are not uncommon. Therefore, school personnel must: (1) be well informed about the laws that pertain to educators, (2) use adequate caution and control in all activities within their responsibility, and (3) always try to act prudently. Educators can meet their legal obligations by carrying out job responsibilities in an informed and careful manner.

IMPORTANT LEGAL TERMS

Following are definitions of some common legal terms that may be encountered by educators who manage personnel, facilities, and programs.

Act of God	An extraordinary event due to forces of nature that could not have been foreseen or prevented.
Assumption of risk	Involvement in an activity or situation in which the participant freely and knowingly assumes the risk of injury, relieving the defendant of a duty otherwise required to be performed.
Attractive nuisance	A hazardous facility or situation that attracts participation, especially that of young children.
Battery	Intentional, unpermitted, and offensive touching of the person of an individual by another (not necessarily resulting in an injury).
Civil law	Area of law that addresses a noncriminal infringement on the rights of a person, agency, corporation, or other. Tort and contract disputes are examples of civil actions.
Common law	Law determined by the courts or custom, as opposed to statutory or legislative-made law.
Comparative negligence	A method (distinct from contributory negligence) for apportioning relative degree of responsibility for an injury, in which the award is reduced or denied based upon fault assigned to the plaintiff.
Contract	An agreement between parties that contains three elements: an offer, an acceptance, and consideration (something of value contributed by each party such as money for a service rendered).
Contributory negligence	Conduct on the part of the injured party contributing to the harm suffered, which falls below the standard required for his or her own protection.
Criminal law	That portion of the law which deals with violations of the criminal statutes, generally falling into two categories of felonies and misdemeanors.
Damages	Monetary compensation awarded in a civil suit as reparation for the loss or injury for which another is liable.
Defendant	The party against whom action is brought and from whom relief or recovery is sought in a lawsuit.
Due process of law	Formal (as in judicial) proceedings, which normally include the right of notice, opportunity to be heard and defended, and the right of appeal.
Equal protection of the law	The right of equal treatment by the law and law enforcement agencies for all persons under similar circumstances.
Foreseeability	The degree to which a person responsible for a negligent act could have (and reasonably should have) foreseen the danger.
Immunity	Exemption from a duty or liability granted by law to a person, to a group of persons, or to the state (as in sovereign immunity).

Injunction	A ruling issued by a court directing a person or agency to perform or refrain from performing a specific act.
In loco parentis	Acting in the place of the parent or guardian. Teachers generally are considered to stand *in loco parentis* during the time students are under their supervision. (Status is influenced by the age of the students and the nature of the school.)
Liability	Accountability and responsibility to another person, enforceable by civil remedies or criminal sanctions (e.g., liability for injuries caused by one's actions or a by product).
Malfeasance	A wrongful or unlawful act involving or affecting the performance of one's duties.
Misfeasance	The performance of a lawful duty in an improper manner.
Negligence	Failure to exercise due care or to follow procedures taken by a person of ordinary prudence under similar circumstances to protect another from harm (an act of commission or act of omission).
Nonfeasance	Failure to perform an act that one is under an obligation or duty to perform.
Plaintiff	One who initiates legal action or a claim against another party.
Proximate cause	A situation or factor that is the near and contributing cause of an injury.
Prudent person	One who acts in a careful, discreet, and judicious manner in view of the particular circumstances.
Respondeat superior	The responsibility of an employer for the negligent acts of his employees.
Statutory law	Law governed by statutes enacted by legislative bodies (as distinguished from common law).
Tort	A civil wrong or violation of duty (independent of breach of contract) that produces an injury or damage to another person or to property.

TORT LAW

Tort law addresses the rights, obligations, and remedies applied by the courts to provide relief for persons who have suffered harm from the wrongful acts of others. The law of torts derives from common-law principles supplemented by legislative enactments. (Traditionally and constitutionally, it has been left to the states.) Tort law is distinct from criminal law. Tort actions are brought by private citizens in civil proceedings, whereas criminal actions are initiated by governments.

The law of torts has four objectives. First, it seeks compensation for injuries caused by the culpable action or inaction of others. Second, it seeks to shift the cost of such injuries to the person(s) legally responsible for inflicting them. Third, it seeks to discourage careless and risky behavior in the future. Fourth, it seeks to vindicate legal rights and interests that have been compromised or diminished. These objectives are served when tort liability is imposed for intentional wrongdoing or for negligence. Most tortious acts by school employees are negligent torts (see the following section).

However, intentional torts do occur in school settings. These include assault, battery, trespass, false imprisonment, invasion of privacy, misrepresentation, and fraud. Mere reckless behavior, sometimes called willful and wanton behavior, does not rise to the level of an intentional tort. Moreover, tort actions aren't dependent on an agreement between parties as in contract disputes (West, "Torts," 1998).

Where legal liability exists under tort law, a person is liable only when that person intentionally or negligently causes or contributes to an injury. The injured person (plaintiff) must prove that the injuring person (defendant) committed a tort in order to collect damages.

NEGLIGENCE

Most injuries that result from tortious behavior are the product of negligence, not intentional wrongdoing. *Negligence* is the term used by tort law to characterize behavior that creates unreasonable risks of harm to persons and property. A person acts negligently when his or her behavior departs from the conduct ordinarily expected of a *reasonably prudent and careful person* under the given circumstances. A prudent person is expected to use common sense and experience to determine the proper degree of care and vigilance to avoid imperiling the safety of others. The fact that a teacher is certified by the state indicates a level of expected behavior. When one falls short of these expectations and this failure contributes to injury to another person, the teacher can be declared negligent. Good intentions alone are no safeguard, because negligence denotes an unintentional failure to do what is reasonably expected (West, "Torts," 1998).

Not every accident-producing injury gives rise to liability for negligence. Some accidents cannot be avoided, even with the exercise of reasonable care. Negligence is often difficult to prove, and many individuals who are inclined to sue become discouraged when the legal grounds are weak or insufficient. In order for a person to be declared negligent, the following four elements must exist (Van der Smissen, 1996):

1. The service provider (defendant) must have a *duty* toward the injured person(s) (plaintiff) to protect them from unreasonable risk. Teachers, coaches, and administrators clearly have such a duty toward the students entrusted to their care.

2. The service provider must have *breached* that duty by an act of commission (misfeasance) or act of omission (nonfeasance), meaning that one who does nothing when something should have been done is often as liable as one who responds inappropriately.

3. The breach of the standard of care (duty) must be directly related to the injury that occurred. In other words, the breach of duty must be the *proximate cause* of the harm. To be considered the proximate cause, an act need not be the immediate or even the primary cause, but only a substantial contributing factor in producing the harm.

4. The plaintiff must have been *harmed* by the wrongful act committed by the defendant, in the form of physical or emotional injury (or damage to property or reputation).

These four elements pertain to legal negligence, and all of them must exist in order to justify legal action. Sometimes nonlegal negligence is committed by a teacher or coach, but if no tort results (due to the lack of one or more of the four elements such as "harm"), the teacher's behavior might be the rightful subject of administrative action but is not a matter for the courts.

Situations of Potential Negligence

Several conditions frequently are found in teaching or coaching that, if unattended, could injure a student and cause the teacher to be found guilty of negligence. It is important to understand potential sources of negligence and deal with them effectively.

Unsafe Facilities

Poorly maintained recreational facilities are a leading cause of personal injury lawsuits. The legal obligation to provide a safe environment for students is the primary responsibility of school administrators but extends to teachers and coaches, who are defined as *premise operators* during their supervision of an area. The premise operator must protect users from known risks. The obligation for ordinary care includes the following duties (Maloy, 1996b):

1. To inspect regularly for hazards or dangers.

2. To maintain the premises and correct defects.

3. To warn participants and spectators about hazards that are not readily apparent.

4. To implement a plan for reasonable supervision and security.

5. To have an emergency medical plan.

Faulty Equipment

Acting both as instructor and supervisor, the teacher or coach must prevent students from using defective equipment. For example, requiring or even permitting students to climb a dangerously frayed rope would be grounds for negligence, if the rope should break and cause injury. The same would be true of permitting students to use faulty gymnastic apparatus or defective playground equipment. In addition to preventing the use of defective equipment, the duty exists to inspect all equipment periodically, with the intent of discovering any defects.

Attractive Nuisances

The presence of hazardous facilities or equipment that may be attractive to children and that the owner has failed to secure is defined in tort law as an *attractive nuisance*. Equipment like climbing ropes and gymnastic apparatus must be safely secured or made inoperative in order that students not be attracted to using it in the absence of proper supervision. A swimming pool left unsecured or lacking supervision is an attractive nuisance. Whether a situation would be legally declared an attractive nuisance is influenced by the laws of the particular state, the age and competence of the injured person, and the various circumstances surrounding the incident.

Inadequate Supervision

Lack of proper supervision is the cause of most school-related negligence lawsuits. Schools have an inherent duty to supervise participants who are taking part in sponsored activities. Administrators would be wise to establish written policies and procedures governing the supervisory responsibilities of employees, and to ensure that teachers and coaches follow these procedures. Two levels of supervision are distinguishable. *General supervision* implies that a responsible adult has the entire area under supervision but is not expected to have every individual under supervision. This level of supervision should always be in place with children present. *Specific supervision* implies that the teacher or coach provides constant and continuing supervision in the immediate area where an activity takes place. This level of supervision is prudent when high-risk activities are being conducted, such as dismounts from parallel bars or springboard diving. The level of supervision required is also based on such factors as age, mental ability, and skill level of participants. Requirements for the number of supervisors needed are based on the characteristics of the participants and the nature of the activity. Specific ratios of supervisors to users have been applied to playgrounds and natatoriums.

Educators have a duty to supervise all school grounds during normal operating hours. Supervision may also be required before and/or after school hours, depending on circumstances surrounding use of the property. Only in a *sudden emergency*, a life-or-death situation, can a school employee be relieved of normal supervising duties. Notwithstanding, an employee's absence from the scene of the injury does not automatically constitute negligence. When a plaintiff alleges that a supervisor's absence was the proximate cause of the injury, the law requires proof that the supervisor's presence would have prevented the injury (Gaskin, 1996).

Administration of Student Discipline

From an administrative standpoint, school discipline has two goals: to ensure the safety of students and staff, and to create an environment conducive to learning. All school personnel should be familiar with state laws and district policies involving discipline. Complaints and lawsuits by parents routinely arise from the administration of disciplinary measures by school personnel. Imprudent discipline can constitute an intentional tort. Charges of battery have resulted from the overzealous application of corporal punishment, and several states have prohibited this practice. The American Academy of Pediatrics has recommended that schools abolish corporal punishment in deference to alternative forms of student behavior management. Teachers and coaches should have formal training in discipline strategies in a school setting, including an understanding of the basic psychological principles of behavior modification. The school's focus should be on prevention of discipline problems.

The following risk management tips relative to disciplining students should help administrators to preempt litigation in this area (Sawyer, 2002):

- Establish written policies regarding appropriate discipline of students.
- Establish written procedures to follow when disciplining students, including documentation of action taken and names of witnesses.
- Notify parents in writing of all disciplinary policies, procedures, and actions.
- Establish a policy that no teacher is allowed to hit a student.
- Establish the fact, in writing, that teachers and coaches are employees of the school district, and that part of a teacher's and coach's duty is to maintain discipline.

Physical education teachers and coaches have been more inclined than other teachers to impose physical forms of punishment. This has included not only corporal punishment but also the use of calisthenics or running laps in an attempt to discourage undesirable behaviors. It is arguable that mandatory physical exercise in a punitive context may cause students to become less inclined to exercise voluntarily. This runs counter to the goal of physical education. California's state physical education association, CAHPERD, has adopted position statements opposing the use of any form of physical activity as punishment in a school setting.

Other Hazardous Situations

Poor selection of activities can lead to charges of negligence where there is a likelihood of collisions among participants, being hit with objects, or inappropriate rough-and-tumble activities. Low-organized games like bombardment and dodge ball need to be kept under control. Trampoline and minitramp activities have been eliminated by many schools (see below). Inadequate provision for individual differences in size during combative activities also can result in injuries. Lack of provision for protective equipment could be declared an act of negligence. Most "hand-me-down" equipment eventually reaches a marginal state when it should be discarded. A general rule to follow is: if you lack adequate equipment, don't conduct the activity. Permitting students to participate in tackle football, fencing, hockey, or lacrosse without safe and proper equipment is particularly dangerous.

Interpretations of Negligence

Jury trials prevail in tort cases. A jury decides whether a defendant in a given situation acted reasonably in performing a duty to protect from harm. The law requires jurors to use their common sense and life experience in determining the proper degree of care and vigilance expected of school employees to avoid imperiling the safety of others.

The duty of care is the major question decided by the courts. The duty of care required to avoid tort liability is that which a reasonable person of ordinary prudence would exercise for the safety of others under the circumstances. (In a few jurisdictions, the standard of duty owed is lowered to liability only for willful or wanton misconduct.) The test to be applied in most situations is what the particular individual should have reasonably foreseen and done.

This standard, often referred to as the *reasonable person* test, allows courts to adapt society's views of reasonableness as they change from time to time and place to place. It is important to recognize that the law does not require that all possible harm be removed, only that which can be done so reasonably. What an educator thinks is reasonable might be considered unreasonable, and thus negligent, by a jury of noneducators. Of course, jury decisions may be appealed, and lawsuits often are settled without adjudication. The best strategy, however, is to avoid actions that appear to be a breach of duty and result in unnecessary injuries to others.

Defenses to the Charge of Negligence

The mere fact that a school employee is sued doesn't necessarily mean that the defendant will be found guilty of negligence. Several defenses are available against a charge of negligence. The best defense is if one of the elements required to prove negligence (duty, breach, proximate cause, or injury) is not present. In addition, the following defenses should be considered in response to a charge of negligence (Cotten, 1996).

Lack of Legal Duty or Lack of Proximate Cause

Occasionally, a plaintiff will attempt to establish liability against an individual or organization that has no legal duty related to the damage. In the absence of legal duty, liability normally does not exist. For example, courts have ruled that colleges no longer are in an *in loco parentis* relationship with students for certain types of activities and forms of protection.

In order to satisfy the criterion of *proximate cause*, the plaintiff must show that the negligent conduct was the "substantial factual cause of injury" for which damages are sought. Even though an individual breached a duty and was involved in the chain of events associated with injury to another person, this individual might be free of liability due to lack of proximate cause. A clear and direct linkage must exist between the action of the defendant and the injury to the plaintiff. Evidence must show that the defendant's action or inaction substantially contributed to the cause of injury. For example, there may have been inadequate inspection of playground equipment, but the failure to inspect did not contribute to the cause of injury (Van der Smissen, 1996).

Contributory/Comparative Negligence

In some cases, the person who sustains an injury is a party to the problem. For instance, the plaintiff may have taken more risks than were necessary. Examples might include disobeying warnings or not following safety regulations. In such cases, the injured person contributed to his or her own injury and therefore may be liable under the doctrines of contributory or comparative negligence.

The concept of *contributory negligence* asserts that the conduct on the part of the injured party contributed as a legal cause to the harm suffered. This defense is currently utilized in some half dozen states. Under this doctrine, any contributory negligence on the part of the plaintiff, regardless of how slight, bars recovery in a negligence claim. This doctrine has come to be viewed as overly harsh, as it leaves the plaintiff with no legal recourse even if the defendant was negligent. Consequently, the vast majority of states have abandoned the doctrine of contributory negligence in deference to a form of comparative negligence.

Comparative negligence (comparative "fault" in some states) is a method for apportioning the relative degree of responsibility for an injury. In states using *pure* comparative negligence as a defense, the award to the plaintiff is reduced by the percentage of fault assigned to him or her. Other states employ *modified* comparative negligence, which operates on the theory that the plaintiff is not entitled to recovery unless the negligence of the defendant was substantial—often defined as equal to or greater than 49 or 50 percent. Where the plaintiff's fault exceeds the designated percent, the plaintiff would receive nothing. Some states also utilize the doctrine of *joint and several liability* to apportion liability when more than one party is at fault.

As to whether students can be found to have contributed to their injuries, the general rule is that children over the age of fourteen are capable of negligence. Those under seven cannot be held negligent. Children between the ages of seven and fourteen can be held negligent depending on the circumstances. A student over age seven is held to a standard of "self-care" expected of someone of similar age, intelligence, and experience (Cotten, 1996).

Assumption of Risk

In some states, assumption of risk has been subsumed under the doctrine of comparative negligence. Elsewhere, assumption of risk implies that when individuals voluntarily expose themselves to known and appreciated risk, they cannot recover for injuries resulting from that risk. Schools most often are subject to *primary assumption of risk*, which involves consent of the participant and relieves the defendant of a duty that the defendant might otherwise owe the plaintiff. Two types of primary assumption exist. *Express* assumption of risk involves a verbal or written agreement in which the participant agrees to accept the inherent risks of the activity. The signer assumes the risks caused by ordinary negligence. Schools may utilize an "agreement to participate" form that explains risks in writing. *Implied* assumption of risk is presumed when an individual voluntarily participates in an activity that involves inherent, well-known risks, in the absence of a formal agreement. Notably, primary assumption of risk does not apply as a defense in product liability claims.

Three requirements are necessary for primary assumption of risk to be used as a defense: (1) the risk must be inherent in the activity, (2) the participant must voluntarily consent to the risk exposure, and (3) the participant must appreciate the risks inherent in the activity. *Inherent risks* are those risks that cannot be eliminated without changing the nature of the activity—such as the risk of breaking a collarbone while blocking in football. Voluntary consent would not apply where physical education classes are required or certain activities are required of all students enrolled in physical education. A degree of voluntary consent is implied when a student elects one activity among a range of choices within a physical education requirement (Cotten, 1996).

Spectators, as well as participants, assume inherent risks at sporting events. Fans at baseball and hockey games are presumed to be aware of balls or hockey pucks entering the stands, and therefore have assumed the risk of being struck. They may exercise the option of sitting behind a screen or Plexiglas barrier. The doctrine of assumption of risk in conjunction with the doctrine of limited duty has held sway in court cases involving most spectator injuries. Generally, the courts have ruled that there is no duty to warn spectators of such inherent risks as being hit with balls or pucks. Nevertheless, some ballparks are posting warning signs in the dugout seating areas. When distractions are present in the spectator areas during a game—such as vendors or mascots—the proprietor may be held to a higher standard in assessing liability. In addition, when fans are "priced out" of protected seating areas, the courts may reinterpret assumption of risk (Fried, 2002).

Acts of God

Some events or conditions of nature cannot be foreseen or controlled by humans, and no person can be held responsible. These *acts of God* (also known as acts of nature, or *Vis major*) constitute a defense against negligence only if the injury is directly and exclusively caused by the natural event. Examples are a student being struck by lightning or injured in an unforeseen windstorm. Notwithstanding, certain events foreseeably lead to predictable consequences. For example, thunder may precede a lightning strike. Thus, the concept of foreseeability can be a factor in determining liability when a natural event causes injury. Advisedly, lifeguards remove patrons from outdoor swimming pools upon hearing thunder.

Immunity

Immunity confers an exemption from a duty or liability that is granted to a person or class of persons. While immunity may provide a defense against negligence under specified conditions, it does not extend to behavior that is judged reckless, wanton, willful, and intentional. Several types of immunity are recognized in law.

Sovereign immunity refers to the immunity of government and its agencies from being sued. Generally, it does not extend to the employees of state agencies, unless they have been indemnified by state statutes. Historically, sovereign immunity afforded governments broad protection from liability for negligence. However, over the last four decades government immunity has been greatly restricted by legislatures and the courts. Several states have passed laws by which schools and their employees can be sued under tort claims acts.

The distinction drawn between governmental functions and proprietary functions of state agencies as a basis for immunity has given way to a theory based on discretionary acts versus ministerial acts. Under this test, the nature of the employee's act determines governmental liability. While discretionary (decision-making) acts are generally protected, activities that are operational in nature, such as teaching, supervision, and facility management, are considered ministerial and not protected by immunity. Public school teachers, coaches, and administrators are generally liable for their actions in carrying out their duties.

Statutory immunity from tort liability is conferred by legislation and differs from state to state. Most states have instituted recreational-use immunity covering the use of undeveloped rural land open for recreational purposes. In some states, this immunity is limited to private property; in others, it may extend to school-owned parks or playgrounds. *Good Samaritan statutes* provide another form of immunity. Enacted in some form by all the states, these statutes provide immunity from liability to those individuals who voluntarily come to the aid of an injured person. Teachers and coaches in a supervised activity setting have no immunity under these statutes, since they owe a duty of emergency care to their students (Maloy, 1996a).

If a school district is not protected by immunity, the doctrine of *respondeat superior* ("Let the superior give answer") holds the district liable for negligent actions by its schools, employees, and volunteers. The individual who was negligent (ordinary negligence) also remains liable. A school is expected to utilize reasonable care to discover the moral unfitness of an employee, and it may be held liable if it hires, retains, or recommends an unfit employee. This doctrine has been broadened recently to include intentional torts such as sexual harassment and sexual abuse by school employees (Cotten, 1996b).

RISK MANAGEMENT

Today's schools are faced with risks that were rare or unknown a decade or two ago. The most salient development has been the increase in incidents of violence on school campuses along with the accompanying escalation of security measures. Schools also are being held to higher standards of care for protecting students and staff from accidental injuries. In response, school administrators must operate in a proactive mode. Rather than waiting for an incident to occur and then reacting to it, schools must anticipate and move to preempt potential problems. This approach employs the concept of risk management.

Risks are viewed broadly to include physical injury, potential litigation, and financial loss. *Risk management* is defined as the coordinated efforts to protect a school's human, physical, and financial assets. This is accomplished by identifying risks to which a school may be exposed, analyzing the probable frequency and severity of such risks, and implementing control measures to reduce or eliminate them. The goal of this effort is to control loss of assets. *Loss control* falls into three categories: loss avoidance—avoiding or abandoning risky activities; loss prevention—reducing the frequency of losses through maintenance, supervision, and instruction; and loss reduction—reducing the severity of unavoidable losses by minimizing damage through salvage actions and by providing counseling after a school tragedy.

Risk management implies systematic planning and implementation of policies and strategies, communicating them to all concerned, and enforcing them consistently. These policies must be supported by ongoing monitoring and adjustments. Inspection of buildings, equipment, and playfields will reveal conditions needing repair or alteration and will identify potential sources of liability. Administrators should develop standard checklists to aid inspection of specific areas such as physical education and athletics facilities. School districts also may request inspections by their insurance carriers or by state and local agencies. In addition, rules governing the use of school facilities by community groups should be part of community-use agreements, and user groups should be required to provide insurance unless the district is willing to assume liability (Gaustad, 1994).

Physical education, intramurals, and athletics are among the more risk-prone areas in schools because of the nature of the facilities, equipment, and activities. Wet areas such as locker rooms and swimming pools pose special risks. Gymnastic apparatus and playground equipment carry intrinsic risks, as well. Activities in which students run at full speed, rapidly change directions, jump, climb, throw and catch various objects, or manipulate bats and sticks in close contact while on a wide variety of surfaces expose them to inherent risks. Table 9-1 lists the number of injuries from recreational activities and equipment that occurred in the United States in 1999. The table records the top ten recreational causes of reported injuries by number but doesn't indicate numbers participating in each activity or severity of injuries.

School administrators must determine what are acceptable risks, eliminate unacceptable risks, and manage those which remain. One example of schools assessing a risk as unacceptable has occurred with the use of trampolines and minitramps. Following the 1977 statement (reissued in 1981) by the American Academy of Pediatrics (AAP) that "trampolines be barred from use as part of the PE programs in grammar schools, high schools, and colleges, and also be abolished as a competitive sport," many school districts across the nation began eliminating trampoline and minitramp activities from the physical education curriculum and the extracurriculum. This decision was based on the incidence of catastrophic injuries documented by the AAP coupled with the trend among insurance companies to exempt activities on trampolines and minitramps from liability coverage (Lahey, 2000).

Springboard diving provides a case for assessing acceptable levels of risk. For the past several years some colleges and schools have been removing diving boards from

Table 9–1
Estimated Number of Injuries from Selected Recreational Activities/Products (Numbers rounded to nearest 1,000)

Activity/Product	Estimated injuries (1999)
1. Bicycles	627,000
2. Basketball	600,000
3. Football	400,000
4. Soccer	185,000
5. Trampolines	100,000
6. Skateboards	87,000
7. Swimming pools	80,000
8. Swings, swing sets	76,000
9. Fishing	73,000
10. Weight lifting	68,000

Source: National Electronic Injury Surveillance System, Consumer Product Safety Commission
<http://www.cpsc.gov/CPSCPUB/PUBS/REPORTS/1999rpt.pdf>

swimming pools, while others have opted to retain this equipment and manage the risks. Given that diving is one of the few activities which develops kinesthetic sense (once trampoline has been eliminated), the benefits may be judged to outweigh the risks. Specific risks include divers hitting the ceiling, sides, or bottom of the pool when diving, slipping on or falling from the board, and landing on swimmers in the water. Spinal column injuries and head concussions are the most serious injuries resulting from these accidents. Schools should consult their insurance provider for the status of coverage. However, the following guidelines should reduce the risks of diving boards to an acceptable level:

- Diving boards and steps should have self-draining, slip-resistant surfaces.
- Diving boards should be inspected routinely to insure they are in working order.
- Diving boards higher than one meter should have guard rails installed.
- The diving area of the pool should be delineated from the swimming area, with a lifeguard chair positioned at the point of delineation.
- Follow NFSHSA standards for vertical and horizontal clearance, and for water depth.
- Establish safety rules and closely supervise the diving area.
- Post emergency procedures and maintain a backboard on the premises.

Informed Consent, Releases, and Waivers

An informed consent form is an agreement between the signers, usually students or parents and the school. These forms are used in situations where activities involve unusual risks beyond those in the ordinary curriculum (for example, when a student wishes to participate in tackle football). By signing this form, parents are giving permission for their child to participate in the activity. Consent forms must include notification of the risk of injury; otherwise, they haven't obtained "informed" consent (Ball, 1998).

In conjunction with the consent form, many schools require students and parents to sign a separate document that purports to waive the participant's right to sue for injury, or to release the school from liability for injury. In the past this form has been called a "waiver," but most jurisdictions are now calling it a "release." The primary goal of the release is to obtain a "summary judgment" that the participant accepted responsibility for the injury. However, releases often face legal problems for the following reasons: (1) It is impossible to release a future tort because of the unpredictability of future events. (2) Parents cannot sign away the rights of a minor child. (3) Children, being minors, cannot release the school from possible tort claims. (4) The form is usually unspecific as to the rights being waived or to all risks involved in the activity.

The following considerations should be incorporated into a release form to enhance its legal standing (Clarke, 1998):

1. The form should stand alone on one page, separate from the consent form.
2. A phrase to the effect, "In consideration of being allowed to participate in . . ." should be inserted at the outset of the release.
3. An assumption of risk should precede the actual release language.
4. The form should be identified at the top as a "Release" in capital letters.
5. Students should sign along with the parent/guardian to acknowledge by signature the risks and their responsibilities.

Although informed consent and release forms (figures 9-1 and 9-2) are not always found to be legally binding, it is still recommended that schools use them. The completion of such forms brings to the attention of students and parents that risks are involved, and that the school wishes them to understand that they are making a decision as to whether the student should participate. Thus, both parents and child are aware of risks assumed by participation in the activity. In addition, parents may be less inclined to sue for damages if they have signed these forms.

On the consent form, it is a good idea to obtain parental consent for emergency treatment along with contact information for parents and preferred medical providers. The severity of a situation may be lessened if the parents' wishes as to medical attention are known and followed in the event of an injury to the student.

INSURANCE MANAGEMENT

Because school districts may be financially liable for property damage or injuries, strategies are used to manage financial exposure for risks. A school district may transfer legal or financial responsibility for risks to other entities, such as insurance companies, or may decide to retain risks. Self-insurance is a risk-retaining option in which schools set aside a fund for the payment of claims. Purchasing insurance from a commercial carrier is usually less expensive and is the most common type of risk transfer. Some states allow schools to pay for insurance out of tax funds, while others do not (Gaustad, 1994). Schools require three types of insurance to manage risk: (1) liability insurance, which pays claims resulting from damages to persons or their property; (2) accident insurance, which pays medical expenses for injured students; and (3) property insurance, which covers school facilities against natural disasters, theft, vandalism, and other events.

Insurance for Teachers and Coaches

Teachers and coaches need to have professional liability insurance in addition to health insurance. If a student were injured during a supervised activity, the liability policy would cover any claim up to the amount of the policy limits (minus the deductible). Unless covered by a district policy or indemnified by the state, teachers and coaches should purchase a liability policy. Group policies available through professional organizations are the best option. The National Federation of Coaches Association (NFCA) provides its members with an insurance package that contains general liability, accident medical, and accidental death and dismemberment, at no additional fee. The insurance covers losses suffered while coaching any activities recognized by the state activities associations. Coaches also may contract privately with insurance carriers to obtain adequate coverage. However, group policies generally are more affordable.

Teachers can purchase group policies through professional associations or unions. The National Education Association offers its members professional liability and other insurance, as does the American Federation of Teachers. The American Association of University Professors also offers liability and other insurance for college teachers. Occasionally, homeowner or renter policies will include professional liability coverage for teachers in their professional duties.

Athletics Consent

HOOD ATHLETICS
INFORMED CONSENT FORM

The student-athlete and a parent/guardian MUST read this form carefully and SIGN it.

_____   _____
Name of student-athlete (Please print)   Sport(s)

The undersigned herewith,

A. Is aware that participating or training to participate in any sport can be a dangerous activity involving MANY RISKS OF INJURY.

B. Understands that the dangers and risks of participating or training to participate in the above sport(s) include, but are not limited to, <u>death</u>, <u>serious neck and spinal injuries</u> which may result in <u>complete or partial paralysis</u>, <u>brain damage</u>, serious injury to virtually all ligaments, muscles, tendons and other aspects of the muscular skeletal system and serious injury or impairment to other aspects of said student-athlete's body, general health and well-being.

C. Understands that the dangers and risks of participating or training to participate in the above sport(s) may result not only in serious injury, but in serious impairment of said student-athlete's future abilities to earn a living, to engage in other business, social and recreational activities and generally to enjoy life.

D. Comprehends the dangers of participating in the above sport(s) and recognizes the importance of following the instructions of the Athletics Staff regarding play/performance techniques, training and other team rules, etc. and agrees to obey such instructions.

E. Understands that if participating in CONTACT SPORT(S) (e.g., lacrosse, field hockey, soccer, or basketball) the risks of injury are even greater than for other sports.

In consideration of Hood College permitting me to try out for the _____team(s) and to engage in all activities related to the team, including, but not limited to, trying out, training for, practicing or playing/ participating. I agree to hold Hood College, its employees, agents, representatives, coaches, and volunteers harmless from any and all liability, actions, cause of action, debts, claims or demands of any kind and nature whatsoever which may arise by or in the connection with my participation in any activities related to the above sport team(s). The terms hereof shall serve as a release against Hood College, its employees, agents, representatives, coaches and volunteers by myself, my heirs, estate, executor, administrator, assignees and all members of my family.

_____   _____
Signature (student-athlete)   Date

_____   _____
Signature (parent/guardian)   Date

2002-2003

Figure 9–1.   Informed consent form (courtesy of Hood College, Sports Medicine Center)

SOUTH HADLEY HIGH SCHOOL
South Hadley, Massachusetts

PARENT/STUDENT ACKNOWLEDGEMENT/RELEASE FORM

TO: Athletic Director

This is to verify that I/we as the parents/guardians of _____
have received and read the Student Athletic Handbook describing the policies, rules and
regulations of the South Hadley School Department. I understand that my child may be
involved in dangerous activity that may result in serious injuries.

I give my permission for my child _____ to participate on the

_____ team.

_____
Parent(s)/Guardian(s) Signature   Date

This is to verify that I have received and read the Student Athletic Handbook regarding
describing the policies, rules, and regulations of the South Hadley School Department. I
understand that I may be involved in dangerous activity that may result in serious
injuries.

_____
Student's Signature   Date

In the event of a serious or life threatening injury, I give my permission for the coach of
my child's team (_____) to act as in loco parentis
(as if he/she were the child's parent).

_____
Parent's Signature   Date

**Figure 9–2.   Athletics permission and waiver form (courtesy of South
Hadley High School)**

Insurance for Students

Many schools pay the premiums on accident insurance for students, while others offer optional policies to parents. Accident policies either cover students on a so-
called door-to-door basis or provide twenty-four-hour coverage. Comprehensive accident policies cover students during athletics contests and practices, and during travel
to and from events. Rates for accident insurance are quite reasonable, providing
school districts with a cost-effective means of transferring risk.

Insurance for Athletes

Some states require school districts to document that student athletes have health insurance in order to participate in school sports or other extracurricular activities. In practice, schools often do not verify health insurance information before students participate in activities, nor do they follow up during the school year. It is not unusual for students to lose their insurance coverage at some point due to a change in their parents' job status. For this reason, schools are well advised not to rely solely on individual health insurance coverage.

An athletics insurance rider usually can be added at a nominal cost to the basic student insurance policy offered through the school. These riders cover accidents that occur during authorized games or formal practices. Virtually all school districts currently offer low-cost accident insurance to student athletes. In addition, some school districts provide free catastrophic accident insurance for athletes regardless of their personal insurance status. Catastrophic injury insurance is meant to cover the medical costs for serious injuries such as a fractured vertebra or paralysis. Catastrophic insurance is cost-effective, as the premiums are relatively low.

In practice, the parents, the school, or the booster club might pay basic insurance premiums, but athletes should be covered. Some school districts require all student athletes to purchase a policy through the district regardless of whether the athlete has private insurance. In cases where a policy is purchased through the district, typically the cost to the student is one-half of the total cost to the district. For instance, if the cost to the district for football insurance is $90, the student pays $45. When parents can't afford insurance, schools are meeting the needs of uninsured students through accident trust funds established at the district level and through community involvement (Shaul, 2000).

Property Insurance

School facilities and their fixtures represent a large investment. Public events such as athletics contests expose facilities to numerous risks. Consequently, school property needs to be protected through insurance coverage. There are two basic types of property insurance policies to consider. "Named Perils" insurance covers only certain specifically mentioned events (for example, fire or windstorms). An "all-risk" policy, which provides comprehensive coverage, offers a better choice, although these policies may contain exclusions. Two options are available regarding items covered by property insurance. "Blanket" coverage provides for all structures and their contents, while "scheduled" insurance allows the insured to specify items covered. Regardless of the type of coverage, administrators should maintain an inventory of all insured property (Farmer, Mulrooney, & Ammon, 1996).

Schools provide access to insurance for both ethical and legal reasons. The primary consideration is that students receive adequate medical care in cases of activity-related injuries. Insurance coverage also deflects potential lawsuits from parents or guardians of students who otherwise would face overwhelming medical and rehabilitation bills. However, administrators should be aware that insured status doesn't abrogate parents' or students' right to sue.

ACCIDENTS AND EMERGENCY MANAGEMENT

Anyone who teaches physical education and/or coaches for very long will be confronted with injury-causing incidents. The inevitability of injuries should motivate educators to develop an accident management plan. Basic considerations in managing accidents and other emergencies include developing and implementing an emergency response plan, timely and accurate assessment of damage and injuries, and performance of specific emergency procedures. The response plan should address personnel and their responsibilities (the extent of their duty to provide emergency care), availability of emergency equipment, and critical contact information for outside emergency services (e.g., ambulances, physicians, fire and police departments). Emergency response plans, once in place, should be reviewed and rehearsed periodically (Hawkins, 1998).

School administrators should make certain that all personnel who have teaching and coaching responsibilities are competent in CPR and first aid procedures. States vary in their requirements that school personnel hold current certificates in these areas. Depending on the seriousness of the injury, the duty of care can reach beyond first aid measures to obtaining timely emergency medical service (EMS). This duty may extend to participants, staff, and spectators. Regarding athletics, many schools cannot afford to retain a physician at all events. At a minimum, physicians should be present at contests involving contact sports. When team physicians are not on the sidelines, emergency-care decisions fall to athletics trainers, coaches, or student trainers. Coaches should be aware that liability risks increase when student trainers act without proper supervision. States also may require certain equipment such as stretchers and oxygen tanks to be present at team practices and at games (Hall & Kanoy, 1996).

Athletics trainers, and coaches who act in their stead, should keep the following points in mind: (1) Obtain medical advice and approval for any medical treatment of injuries beyond first aid. (2) Permit athletes to return to practice and competition following serious illness or injury only after securing medical release. (3) Be adequately concerned with the provision and use of properly fitted (and certified, where applicable) protective equipment for each athlete. (4) Be especially cautious about assuming responsibilities that officially or logically belong to the team physician.

Reports and Record Keeping

As mentioned above, consent for emergency care of students should be obtained from parents prior to participation in school activities. Administrators need to maintain consent forms for regular review and immediate retrieval. These records should be stored until the statute of limitations expires (between one and four years, depending on the state). Coaches or trainers should take copies of consent forms on road trips.

An accurately written accident report should be prepared as soon as possible after an accident occurs, to serve as a record of the facts. Every school district ought to have a well-designed accident report form (figure 9-3), along with the requirement that the form be completed and signed by the supervisor and the victim, or a witness in the place of the victim. If the student who had an accident violated a rule, document this on the report. Maintain the lesson plan and record of the student's attendance in class or at practice that day. Discuss the accident only with the parents, official medical personnel, school officials, and school legal counsel.

Reports should be filed in the central office so they can be retrieved readily in case of litigation. In addition to accident reports and treatment consent forms, most schools maintain the following student records: (1) medical history, (2) preseason physical examination forms, (3) physician referral and prescribed treatment records, (4) trainers' daily treatment logs, and (5) trainers' daily notifications to coaches (Hall & Kanoy, 1996).

PULASKI COUNTY SCHOOL ACCIDENT REPORT   129003

(SEE INSTRUCTIONS ON REVERSE SIDE BEFORE COMPLETING REPORT)

(Check One)   (Check One)
☐ School Jurisdictional   ☐ Recordable
☐ Non-School Jurisdictional   ☐ Reportable only

Report Accident Within 3 Days

GENERAL

1. Name   2. Address   3. Home Telephone

4. School   5. Sex (check one) ☐ Male ☐ Female   6. Age   7. Grade/Special School

8. The accident occurred:
   Date: AM: PM: Day of Week:

9. Nature of Injury

10. Part of Body Injured

INJURY

11. Degree of Injury (check one)   Death ☐   Permanent Disability ☐   Temporary Disability ☐   Non-Disabling ☐
                                                                      (Lost Time)   (No Lost Time)

12. Total Number Days   Total Number Days Lost From Activities   Total Time Lost
    Lost From School   (Other Than School Time)   (School and Non-School)

13. Cause of Injury

14. Accident Jurisdiction: (Check One)   School Building ☐   School Grounds ☐   To and From School ☐
                                         Other School Activities
                                         Not on School Property ☐   Bicycle ☐   Vehicle ☐   Home ☐   All Other ☐

15. Location of Accident   (Be Specific)

16. Activity of Person

17. Status of Activity

18. Supervision – If Supervised, Give Title and Name of Supervisor

19. Agency Involved

ACCIDENT

20. Unsafe Act

21. Unsafe Mechanical/Physical Condition

22. Unsafe Personal Factor

23. Corrective Action Taken or Recommended

24. Property Damage: School Property $ Non-School Property $ Total Property Damage $

25. Description. (Briefly give a word picture of the accident, explaining the who, what, where, when, and why of the accident.)

SIGNATURE

26. Date of Report

27. Report Prepared By: (Signature and Title) 28. Principal's Signature

White – Business Affairs (Insurance Reporting) Canary – Support Services   Pink – Pupil Personnel   Gold - School

Reprinted with permission of the Pulaski County (AR) Special School District.

Figure 9–3. Accident report form

CONTRACTS

A contract is a legally binding agreement representing two parties and describing duties and conditions. It is assumed that contracts are entered into voluntarily, in a spirit of agreement, and with a legal purpose in mind. Legal liability can arise under contract law as well as tort law. Disputes over contracts, although they fall within the category of civil rather than criminal wrongs, are treated separately from torts. A person entering into a contract voluntarily accepts certain duties or responsibilities, the violation of which might make the individual liable for legal action.

Teachers, coaches, and administrators enter into various kinds of contracts. Among these are employment contracts, supplemental contracts for extra duties, contracts with suppliers and construction personnel, and contracts with other institutions covering athletic events (the latter is addressed in chapter 16).

When an institution offers an individual an employment contract, the offer remains valid for a reasonable time, until a specified date, or until the offer is withdrawn. If the potential employee signs the contract, the individual and the institution become legally bound to the conditions of the contract, and neither party has the legal right to violate or change them unless the explicit permission of the other party is obtained. A similar situation exists with a contract involving an institution and a supplier or construction company or between two educational institutions, such as a contract covering an athletics contest.

Any contractual agreement involves the risk of not being enforceable under law. Administrators need to understand how to prevent this eventuality. Most states have a statute of frauds or equivalent statute that determines enforceability of a contract. These statutes vary; thus, administrators need to be familiar with state laws. Easter (2002) offers the following guidelines to insure that contracts are legally enforceable:

- Use written contracts for all agreements related to purchases over $500.
- Use written contracts for all agreements that will be performed for more than a year.
- Use written contracts for all agreements related to land purchase, lease, mortgage, and other similar matters.
- Establish written policies and procedures covering contracts.
- Engage in careful negotiations before an offer is made and accepted.

Although teacher contracts are usually signed annually, the contract has a different meaning if the teacher is tenured. Tenure is the right to hold a position indefinitely unless just cause exists for termination (see chapter 6). Most public schools handle coaching assignments with a separate contract, which is supplemental to the basic teaching contract. This permits flexibility in assigning coaches and changing their assignments without altering the basic contract. A few coaches in public schools are given multi-year contracts, but this is the exception, not the rule.

CIVIL RIGHTS ISSUES

Prayer at School Events

The first amendment of the U.S. Constitution states, "Congress shall make no law respecting an establishment of religion, or prohibiting the free exercise thereof." This constitutional right impinges upon laws in individual states and extends to their public schools. School administrators must ensure that their programs and activities violate neither the "free exercise clause" nor the "establishment clause" of the First Amendment. Errant decisions by schools are subject to lawsuits and review by the courts.

The establishment clause was interpreted by the Supreme Court in *Lemon v. Kurtzman* (1971). The so-called "Lemon test" states that in order to be constitutional under the establishment clause, a practice must have a secular purpose, must neither advance nor inhibit religion, and must not result in an excessive entanglement between government and religion. The *Lemon* ruling followed the landmark 1962 U.S. Supreme Court decision that struck down teacher-led prayer in schools as a violation of the establishment clause of the First Amendment. The issues seemed fairly straightforward until the 1990s, when the courts began addressing the issue of prayer at school events with less clarity. While some courts have allowed student prayers from the podium at graduation exercises, a federal appeals court in Houston ruled in 1999 that student-led prayer in a huddle before a football game is unconstitutional. Much of the recent controversy has revolved around prayer at school athletics events. Guidance was provided by the Supreme Court in *Santa Fe v. Doe* (2000), when it upheld a lower court ruling invalidating prayers conducted over the public address system prior to high school games at public school facilities before a school-gathered audience.

Regarding the free exercise clause of the First Amendment, the courts have consistently ruled that students' expressions of religious views through prayer or otherwise cannot be abridged unless they can be shown to cause substantial disruption in the school.

The following guidelines will help school administrators to meet the requirements of the First Amendment's establishment clause without violating the free exercise rights of students (Essex, 2001).

- Don't develop policies endorsing student-led prayer at school events.
- Follow state statutes that apply to student-led voluntary prayer. (Practices vary from state to state.)
- Don't rely on local community customs, standards, or mores that support religious activities in schools.
- Avoid ceremonies at school events that create a devotional atmosphere.
- Respond quickly to inquiries or complaints about religious practices in your school.
- Don't censor private speech rights of students.
- Consult the school district's attorney when in doubt about implementing new policies or changing policies regarding religious practices in your district.

Athletics: A Right or Privilege?

Amateur sport encompasses both restricted and unrestricted competition. School sports are examples of restricted competition. They are controlled by athletics conferences, associations, and leagues connected to schools and colleges. Athletes must be eligible to play as determined by the schools and governing organizations. Generally, athletes do not have an absolute right to participate in school sports. Although in certain states athletics have been established and maintained by statute, in most cases participation in athletics by a student has been held by the courts to be a privilege. Students and parents have challenged this doctrine increasingly over the last few years, and schools are finding themselves involved in litigation when denying eligibility or suspending athletes. In determining whether an athlete is eligible to participate, a court must first decide whether the individual has a right to play, as opposed to a mere privilege to play. Privileges can be revoked by the grantor of the privilege. If the individual has a right to participate, the court examines the individual's relationship with the school.

The rights of student athletes can be infringed by reasonable measures that are implemented for sound policy reasons. Eligibility criteria can vary from school to school, and even from sport to sport. "No pass, no play" rules or similar academic standards, which keep students off school teams because of low grades, have been ruled permissible in light of the school's overriding interest in education. Schools may also control the number of student athletes, allowing students to be cut from popular sports to keep athlete-to-coach ratios at manageable levels. Schools may enact other limitations, such as rules limiting the number of sports a student can play at one time, and rules authorizing students to be suspended or expelled from athletics for consuming alcohol or other drugs (West, "Sports Law," 1998).

School athletics are subject to the requirements of the Fourteenth Amendment to the U.S. Constitution and to federal civil rights statutes. To be in compliance with the law, a school may not discriminate unfairly among students and may be required to provide due process for certain actions regarding student athletes. Although the rightful suspension of a student from participation in athletics generally has been held to be valid, state and federal courts recently have disagreed in ruling on this issue. In some cases, when the student athlete might suffer career losses by disciplinary expulsion from varsity sports, the courts have required hearings prior to suspension or expulsion from athletics. In addition, unique requirements may apply to disabled or "at risk" students. Here also, the court rulings have not been consistent. Administrators should consult the school district's attorney before acting in these matters.

Gender Equity

The requirements for providing equitable programs for both sexes in physical education and athletics are discussed in chapter 12 and chapter 15.

Student Records

The Family Education Rights and Privacy Act of 1974 (FERPA)—often referred to as the Buckley Amendment—is the federal regulation that governs the release of educational records. FERPA addresses the methods of administration rather than any actual requirements concerning content of student records. This federal law creates the right of eligible students (over 18 years of age), and parents in the case of minor students, to see, inspect, reproduce, and challenge the accuracy of their education records. It also requires the student's (or parent's) written permission before allowing the disclosure of personally identifiable information to nonprivileged parties. This final provision has implications for teachers and coaches. When college coaches or prospective employers request specific information, athletics departments must be aware of all the restrictions on the release of such information.

Statutes Governing the Disabled

The Rehabilitation Act of 1973 prohibits discrimination against the disabled in all federally aided programs. The Education for All Handicapped Children Act of 1975

(PL-94-142) provides federal funding to aid state efforts in providing appropriate education to the disabled. In addition, each state has specific statutes dealing with the disabled with regard to participation in physical education, intramurals, and interscholastic or intercollegiate sports. (This issue as it relates to physical education instruction is covered at length in chapter 12, and in chapter 15 as it relates to athletics.)

References and Recommended Readings

Ball, Richard. 1998. "Warnings, Waivers and Informed Consent." In Herb Appenzeller (Ed.), *Risk Management in Sport: Issues and Strategies*. Durham: Carolina Academic Press, pp. 49–65.

Clarke, Kenneth. 1998. "On Issues and Strategies." In Herb Appenzeller (Ed.), *Risk Management in Sport: Issues and Strategies*. Durham: Carolina Academic Press, pp. 19–20

Cotten, Doyice. 1996a. "Defenses for Negligence." In Doyice Cotten & T. Jesse Wilde (Eds.), *Sport Law for Sport Managers*. Dubuque: Kendall/Hunt, pp. 42–53.

Cotten, Doyice.1996b. "Which Parties are Liable for Negligence?" In Doyice Cotten & T. Jesse Wilde (Eds.), *Sport Law for Sport Managers*. Dubuque: Kendall/Hunt, pp. 33–41.

Easter, Beth A. 2002. "Contract Law and Athletics Departments." *Journal of Physical Education, Recreation & Dance*, 73(4): 6–7.

Essex, Nathan. 2001. "Handling Student-Led Prayer at School Events." *School Administrator*, 58(5). Online article: <http://www.aasa.org/publications/sa/2001_05/2001_focrelig.htm>.

Farmer, Peter, Aaron Mulrooney, & Rob Ammon, Jr. 1996. *Sport Facility Planning and Management*. Morgantown, WV: Fitness Information Technology, pp. 86–87.

Fried, Gil. 2002. "Sitting Ducks." *Athletic Business* (April): 18, 20, 22.

Gaskin, Lynne. 1996. "Supervision." In Doyice Cotten & T. Jesse Wilde (Eds.), *Sport Law for Sport Managers*. Dubuque: Kendall/Hunt, pp. 84–93.

Gaustad, Joan. 1994. "Risk Management." *ERIC Digest,* 86. Online article: <http://eric.uoregon.edu/publications/digests/digest086.html>.

Hall, Ralph, & Ron Kanoy. 1996. "Emergency Care." In Doyice Cotten & T. Jesse Wilde (Eds.), *Sport Law for Sport Managers*. Dubuque: Kendall/Hunt, pp. 77–83.

Hawkins, Jerald. 1998. "Emergency Medical Preparedness." In Herb Appenzeller (Ed.), *Risk Management in Sport: Issues and Strategies*. Durham: Carolina Academic Press, pp. 209–13.

Lahey Clinic. 2000. "Trampolines." *Pediatric Health Tips*. Aug. 21. Online article: <http://www.lahey.org/Media/healthtips/PediTips/pedi_htip.asp?ID=18>.

Maloy, Bernard. 1996. "Immunity." In Doyice Cotten & T. Jesse Wilde (Eds.), *Sport Law for Sport Managers*. Dubuque: Kendall/Hunt, pp. 54–62.

Maloy, Bernard. 1996. "Safe Environment." In Doyice Cotten & T. Jesse Wilde (Eds.), *Sport Law for Sport Managers*. Dubuque: Kendall/Hunt, pp. 103–12.

Sawyer, Thomas. 2002. "Battery or Discipline?" *Journal of Physical Education, Recreation & Dance*, 73(3): 8–9.

Shaul, Marnie S. 2000. "Interscholastic Athletics—School Districts Provided Some Assistance to Uninsured Student Athletes." *FDCH Government Account Reports,* Sept. 12. Online article: <http://www.gao.gov> (report # GAO/HEHS-00-148, 22 pp).

Van der Smissen, Betty. 1996. "Elements of Negligence." In Doyice Cotten & T. Jesse Wilde (Eds.), *Sport Law for Sport Managers*. Dubuque: Kendall/Hunt, pp. 26–32.

West Legal Directory. 1998. "Entertainment: Sports Law." Online article: <http://www.wld.com/wsports1.htm>.

West Legal Directory. 1998. "Personal Injury and Torts." Online article: <http://www.wld.com/Conbus/weal/wtorts.htm>.

Online Sources

FindLaw Sports
http://sports.findlaw.com/sports_law/
This site presents case summaries of recent decisions in sports law from the National Sports Law Institute of Marquette University Law School, and provides sports law resources via topical hyperlinks.

Legal Information Institute
http://www4.law.cornell.edu/cgi-bin/empower
Click on the hyperlink "Law about . . . Sports." This Cornell University site covers antitrust, contract, and tort law in amateur and professional sports.

The National Center for Higher Education Risk Management
http://www.ncherm.org/ncherm/homepage.cfm
This center provides information on campus security, sexual misconduct and harassment, drug-related issues, and hazing.

10 Facilities and Equipment

"We shape our buildings and thereafter they shape us." This comment made by British Prime Minister Winston Churchill can serve as a useful guideline for planning school facilities. Educators shape schools, and schools shape children. It follows that schools should be attractive, safe, and conducive to learning. They should uplift and inspire their occupants. These standards apply in particular to physical education and athletics, as schools provide more facilities for these programs than for any other areas of the curriculum or extracurriculum. Furthermore, athletics facilities routinely are open to the public and shape general impressions of the school.

Meeting the above standards requires a firm commitment because school facilities are quite expensive to design, build, and maintain. Administrators are routinely faced with spiraling costs and tight revenue; therefore, it is paramount that facilities provide for maximum utilization at the lowest cost. Clearly, such facilities don't just happen; they have to be carefully planned, constructed, and then managed with this concept in mind.

FACILITY PLANNING AND CONSTRUCTION

Facilities should be carefully planned, for once constructed they remain in use for decades. Mistakes built into facilities are mistakes with which the occupants often have to live, because correcting design or construction flaws can be prohibitively expensive. The following guidelines for constructing physical education facilities can assist in the planning process:

1. Each school or school system should have an overall campus plan. Individuals selected to be involved with this plan should be competent and experienced, not only in planning but also in working with architects and contractors.

2. The functional capacity of a facility is of prime importance. Practically all other aspects rank below this important consideration.

3. Initially, plans should reflect the optimal or ideal facility to meet the particular circumstances. The reality of the restrictions will take over soon enough.

4. All planning and construction should conform to state and local regulations and acceptable facility standards. (Twenty-six states have OSHA-type regulations.)

5. Facilities should be designed so that they accommodate the needs of men and women (staff and students) equitably.

6. Facilities should be designed so that the different components can be used independently without opening and supervising the entire building or complex.

7. Adequate consideration should be given to possible future expansion of a particular facility, and this might influence design.

8. Offices should be designed for individuals, because having more than one person per office disrupts privacy and efficiency. Designing an oversized office encourages the housing two or more faculty members together.

9. The construction of an undersized teaching station is expensive because it forces class size to be small for the lifetime of the facility.

10. The special needs of disabled persons should be taken into consideration in facility planning, as stipulated by the Americans with Disabilities Act.

The following principles of facility construction are of further help in conceptualizing the program needs and requirements of a new building.

- *Validity.* Can the facility be used for what it was designed?
- *Utility.* Is the facility adaptable to different types of activities?
- *Accessibility.* Are areas easy to access by all groups who will utilize them?
- *Isolation.* Are quiet areas isolated from noisy areas?
- *Departmentalization.* Are functionally related areas adjacent?
- *Safety and sanitation.* Are professional and regulatory standards met?
- *Supervision.* Are areas accommodating students adequately visible?
- *Durability.* Are building materials long lasting and impervious to abuse?
- *Aesthetics.* Are the exterior and interior pleasing to look at and to inhabit?
- *Flexibility.* Can activity and instructional areas be expanded or modified?
- *Economy.* Is the facility the best possible value for the money spent?

Some applications of the above principles are worth illustrating. Regarding validity, many areas that accommodate interscholastic competition must be built to regulations. For example, volleyball courts require a minimum overhead clearance of seven meters (23 feet). The principle of utility explains why a gymnasium is the primary activity area in the building plan: many activities can be conducted in a gym. (Compare this to the limited utility of a handball court.) Accessibility means thinking about traffic patterns, inside and outside the facility. This is a crucial principle when designing areas to accommodate spectators. The principle of isolation dictates that noisy activity areas like weight rooms are removed in distance from classrooms, offices, and study areas. Equipment storerooms and locker rooms are located near activity areas because they are functionally related. High-risk areas such as natatoriums are designed to be highly visible. Flexible walls, partitions, and curtains on overhead tracks allow small teaching stations to be expanded into larger areas and vice versa.

Careful determination of facility needs for both the near term and long term is a primary responsibility of administrators. The desired program is the basis for projecting specific needs. In designing a facility to accommodate physical education and athletics, the following questions should be addressed:

- What are the future enrollment projections?
- Is the school likely to experience either consolidation or division?
- How many semesters of instruction in physical education will be required of all students, and what is the scope of activities?
- What will be the extent of participation in physical fitness and recreational activities for faculty and staff, in addition to participation by students?
- Will facilities be available for nonschool use, and if so, what kinds of use and how much?
- What will be the scope of the interschool athletics programs for men and women?
- What will be the scope of the intramural recreation program?
- Will the program include dance, aquatics, gymnastics, weight training, or other activities that require specialized facilities?
- If the facility is for an institution of higher learning, will the program include professional preparation curricula with research needs?
- What are the environmental and geographic factors that will affect the content of the program and the need for facilities?

Taken together, the above guidelines, principles, and questions provide a conceptual framework within which to plan the construction of a new facility or the renovation of an existing facility. Deciding whether to build or to renovate (and retrofit) a facility is an issue that must be addressed early in the process. *Renovation* refers to the rehabilitation of the physical features of a building, including rearrangement of spaces. *Retrofitting* implies the addition of new systems, features, materials, and/or equipment to a facility. A corollary question is whether the increased cost of maintaining an existing building will justify renovation rather the construction of a new building (Sawyer, 2002). A rule of thumb is that if the cost of renovation is more than 50% of new construction, the decision should be to build a new facility. Administrators also must take into account that part or all of the existing facility may be unusable during renovation or retrofitting.

Who Should be Involved?

Careful planning and timely construction of facilities, including the advance acquisition of land, is a complex and exacting responsibility. These tasks can be accomplished only by involvement of the right people at various levels within the educational system, and those representing the design and construction teams. It helps for an administrator to have clear insight into who needs to be involved and in what ways.

Following is a list of personnel who are ordinarily involved with planning and construction of facilities. Each situation differs in the amount of involvement of the personnel at each level.

College/University
Board of trustees or regents (including staff)

Institutional administrators
Campus planning committee
Office of New Construction
Office of Building Maintenance
Architect
Consultant (optional)
Administrators of the college/school and department where the facility will be used
Program specialists
Members of the planning committee, including representatives for the contractor

Public Schools
School board
School superintendent and principal
State specialists in physical education and athletics (optional)
District facility coordinating committee (optional)
Office of Facility Planning and Construction
Office of Building Maintenance
Architect
Representatives for the contractor
Consultant (optional)
Administrators and program specialists in physical education and athletics

A *program specialist* is one who is actively engaged in regular use of the facilities. This person is knowledgeable about the practical needs, issues, and problems of use. Examples of program specialists would be physical education instructors, intramural directors, and coaches. A typical example of input from a program specialist would be assuring that indoor walls in activity areas are flat to accommodate rebound drills and wall volley tests, and that ceilings in these areas don't trap balls hit too high. One or more program specialists should be involved in every aspect of planning from beginning to end. This will help to ensure that the facilities will function as intended.

A *consultant* is a recognized specialist in the area for which the facility is being planned. This resource person is familiar with recently constructed facilities of the same type, and with the latest innovations in design and materials. Sports facility consultants work with schools in planning and designing physical education, recreation, and athletics facilities. A consultant can be especially helpful in suggesting to the planning committee the names of architects who are well qualified for the particular project (many architects have no previous experience with this type of facility); assisting the committee in analyzing alternatives and determining priorities; translating program needs into structural design and square footage requirements; advising in connection with the public-relations effort; and serving as a coordinator between the program specialists, the architect, and the contractor during the planning and construction stages.

The *architect* is a key participant in the early planning of public buildings. This specialist will provide design proposals for review and help to ensure cost-effectiveness in selecting materials, implementing building procedures, and monitoring progress of the project. The architect often serves as the planning committee's liaison with the construction firm and assists the committee in revising plans when necessary. School administrators would be well advised to always work with licensed, reputable architects, preferably those with experience in designing schools or sports facilities (Olson, 1997).

A *construction manager* is hired in some instances to oversee the project. This person recommends sequencing of the work, suggests the most cost-effective means of packaging various components, and oversees the progress of construction (Farmer, Mulrooney, & Ammon, 1996).

The *contractor* works closely with the planning committee and architect in the construction phase of the project. The contractor creates detailed drawings from the architect's plans, which are the basis for construction, and then builds according to their specifications. A general contractor works with and oversees subcontractors such as a masonry contractor and electrical contractor.

Developing the Idea

The idea of a new or renovated facility develops through three important stages: *conception, presentation,* and *selling.*

Conceiving the need for a new or improved facility is the first step toward its accomplishment. The concept usually involves much discussion among those who have a particular interest in the project or who understand the circumstances that would justify it. Those personnel directly concerned should thoroughly discuss the matter, carefully think about where the need fits in terms of priority within a timetable, and determine what justification exists for the project. They should prepare themselves to defend the idea in terms of cost versus benefit.

The planning group may want to carry out a formal feasibility study. Such a study provides a comprehensive analysis of legal, usage, design, site, and administrative feasibility, as well as the cost of the project. In addition to initial construction costs, it is important to discuss the feasibility of a dedicated source of income or revenue stream for the completed project; otherwise, it will compete with other programs for existing funds (Farmer, Mulrooney, & Ammon, 1996).

After the idea has been thoroughly discussed and judged feasible by the initial planners, a proposal should be prepared for presentation to those in higher authority. Prior to or in conjunction with the formal proposal, informal approaches can be taken toward generating interest and support among administrative personnel and interested members of the community. Usually some discussion and exchange of information takes place over a period of time. This can provide valuable insight into what the formal proposal should contain once the idea develops to that point. Furthermore, it can pave the way for acceptance of the proposal.

For some projects, such as a practice field to be built on school property, little need may exist for public involvement. For other projects, such as an expensive indoor swimming complex to be used by both school and community, extensive public involvement and support are essential. At this stage, a project steering committee should be organized, with one of its prime responsibilities being the public-relations effort. The committee should apply the appropriate public-relations principles and strategies explained in chapter 8. This step is crucial because the public-relations effort, if not handled effectively, may result in rejection of the project.

Information Sources

In addition to the ideas gathered from consultants and program specialists, other information can be obtained via several strategies. One example is *field trips*, in which selected individuals travel to locations where similar facilities recently have been constructed. An on-site review of such facilities, combined with direct conversation with those who helped plan the facilities and those who use them, has tremendous potential for helping to sort out strengths and weaknesses of design. Careful pre-trip planning should make sure that the right people are included and that the desired information is clearly identified in the form of a checklist. Photographs of the facility are particularly helpful. (Many schools now provide a virtual tour of new facilities online.) After completion of the trip, the information gathered should be adequately discussed and properly applied.

Sometimes architectural drawings of a similar facility at another location can be reviewed. Borrowing a set of plans enables several people on the local scene to review them without the expense and time needed to travel to other sites. Good ideas can be obtained from a review of plans that would not be evident from an on-site inspection of a facility. Textbooks and articles on facility planning and construction are other useful sources of information. They include effective procedures to follow, desirable building features, and the advantages and disadvantages of various materials and products. These sources of information are accessible and inexpensive.

Two information sources are worth noting. The American Alliance for Health, Physical Education, Recreation and Dance (AAHPERD), in conjunction with Sagamore Publishing, produces a comprehensive, detailed guide to planning, construction, evaluation, and maintenance of facilities for physical activity. *Facilities Planning for Health, Fitness, Recreation and Sports*, a book edited by Thomas Sawyer (2002), is now in its 10th edition. In addition, *Athletic Business*, a trade journal, publishes an annual "Architectural Showcase" issue, which is a good source for previewing state-of-the-art facilities. This journal also publishes annually a list of architects and consultants who work in the sport facilities area.

Preparing a Facility Planning Document

A description of program needs should be written well in advance of the initiation of architectural planning to communicate the needs of the programs to the architect. The purpose should be to put into writing the current and anticipated school programs to be offered and to determine the facilities required to house these programs at the optimum level. Here are some guidelines for preparing the facility-planning document:

1. The information should be both concise and explicit as to program needs.

2. It should state the optimal as well as the minimal program.

3. It should be realistic in terms of potential financing.

4. It should be both factual and insightful in terms of present and future programming.

5. All aspects of indoor and outdoor facility needs should be considered.

6. A rough draft of the document should be reviewed by selected individuals, and their ideas incorporated into the final copy.

7. It should take into account trends or requirements that have an element of permanency, such as: (a) access for the handicapped, (b) application of the community-school concept, (c) equal opportunities for men and women, and (d) emphasis on physical fitness and lifetime sports.

8. Ample consideration should be given to the use of new products, materials, and equipment.

Even the most competent architect should not be expected to know how to adequately plan a building without considerable assistance from the specialists who will utilize the facilities. Thus, one challenge of writing a program is how to transmit a large number of messages from the school staff to the architect, conveying the staff's wishes, needs, and interests in such a way that these items eventually become transformed satisfactorily into wood, concrete, and steel. Efficiently communicating this information to the architect will increase the likelihood that the building will serve its purposes well and that the users will be satisfied.

Often, the program is written before the architect is selected. The written program then should not dwell on shape, form, or other specific aspects, but more on kinds of use, approximate size, special requirements, and particular features that are desired. The following table of contents of an actual program document gives the reader an overview of what should be included.

Introduction
Certification
Space requirements
Estimated costs
Location of building
Air conditioning and ventilation
Windows
Acoustical requirements
Interior surface materials
Ceiling heights
Location of dressing rooms
Location of laundry and issue rooms
Elevator
Service dock
Clocks and class bells
Audio and video equipment
Bulletin boards and display cases
Room or facility descriptions

Figure 10-1 reproduces a single page found in the *Space Requirements* section of the above outline. Each area within the building is given a facility number. Figure 10-2 is a description of facility 49, "Men's Locker and Dressing Area," taken from the plan.

Selecting and Utilizing an Architect

Architects often have been criticized for their shortcomings in designing school buildings. As noted above, too much responsibility is left with the architect without enough involvement by school administrators and other staff. Even though the architect may be highly competent in architectural design and construction, he or she cannot be expected to understand all of the functional characteristics needed in a physical education or athletics building. Architects have not had firsthand experience with utilization of such facilities and certainly are not experts on the programs that the facilities are expected to accommodate.

Some architects do truly wonderful things in both design and function, but even the best of architects needs all of the useful information available from the program administrators and subject-matter experts. The school should get what it needs to

	Facility Number	See Page	Total Sq. Feet
DRESSING ROOMS			
Women's Physical Education and Intramural			
Locker and Dressing Room	42	61	10,000
Shower and Toweling Area	43	62	4,000
Lavatory and Toilet Area	44	63	2,000
Hair Drying and Make-up Room Area	45	64	500
Towel and Uniform Issue Room	46	65	600
Women's Faculty Locker, Dressing & Shower Area	47	66	1,500
Women's Swimming Instructor Shower and Locker	48	67	160
Men's Physical Education and Intramural			
Locker and Dressing Area	49	68	5,600
Shower	50	69	1,050
Toweling Area	51	70	1,050
Lavatory and Toilet Area	52	71	240
Towel and Uniform Issue Room	53	72	450
Men's Faculty Locker, Dressing, Shower, and Toweling area	54	73	1,970
Men's Swimming Instructor Shower-Locker Room	55	74	160
Intercollegiate Athletics			
Football Locker, Dressing & Shower Area	56	75	3,500
Other Sports Locker, Dressing-Shower Area	57	76	2,846
Coaches' Locker, Dressing & Shower Area	58	77	456
OTHER ROOMS (36 Total)			
First Aid Room	59	78	150
Swimming Pool Equipment Storage	60	79	576
Dance Property Storage	61	80	840
Costume Construction and Storage	62	81	600
Playfield Cart and Equipment Storage	63	82	250
Canoe and Trailer Storage	64	83	500
Ski Storage	65	84	180
Garage for Driver Training Car and Equipment	66	85	500
Health Department Storage-Room A	67	86	480
Health Department Storage-Room B	68	87	300
Women's Physical Education Department Storage, Room A	69	88	120
Women's Physical Education Department Storage, Room B	70	89	120
Recreation Department Storage	71	90	150
Intramural Equipment Storage Room	72	91	150
Gymnasium Equipment Storage	73	92	2,200
Gymnasium Supervision and Equipment Issue Room	74	93	200
Gallery, Swimming Pool A (40 seats @ 7 sq. ft. ea.)	75	94	280
Gallery, Swimming Pool B (1,000 seats @ 5 sq. ft. ea.)	76	95	5,000
Gallery, Gymnasium A (200 seats @ 6 sq. ft. ea.)	77	96	1,200
Gallery, Gymnasium B (200 seats @ 6 sq. ft. ea.)	78	97	1,200

Figure 10–1. Sample page from the "Space Requirements Section" of a written program for a new facility (courtesy of Brigham Young University)

Facility 49
MEN'S LOCKER AND DRESSING AREA

Floor Area: 5,600 sq. ft.

Ceiling Height: Ten feet

Floor Material: Quarry tile

Wall Material: Ceramic tile

Ceiling Material: Moisture resistant and sound absorbent

General: Located adjacent to F. 50, F. 51, and F. 52;

Lockers arranged on raised pedestal, no legs, solid base, with 106 inches width for the aisle between the lockers;

8 in. benches, 30 in. in front of lockers;

360 dressing lockers, 12 in. by 15 in. by 60 in.; 50 of these to be designated as intercollegiate team lockers, 310 as lockers for physical education and intramurals;

1,800 storage lockers, 12 in. by 15 in. by 12 in.

Lockers to be arranged on a 5 to 1 basis. All lockers to have louvered doors, perforated backs, and tiers; tops of lockers are to be slanted for cleanliness;

Warm air is to be discharged under storage lockers section, forced upward through each locker (passing through perforations), and carried to the outside through central vents constructed in each locker section;

Mirrors to be arranged on ends of rows of lockers;

Two cuspidors and two drinking fountains, one set near toweling room, one set near entry.

Equipment needed in room:

Benches in front of all dressing lockers;

360 dressing lockers, 12 in. by 15 in. by 60 in., Worley Steel Lockers;

1,800 storage lockers, 12 in. by 15 in. by 12 in., Worley Steel Lockers;

Mirrors;

2 cuspidors;

2 drinking fountains.

Space utilization and importance to curriculum:

Dressing rooms and lockers for all men's activities which will be held in this building.

Figure 10–2.   A page from a written plan for a new facility. The information pertains to one area of the total facility (courtesy of Brigham Young University)

accommodate its programs, not only what the architect wants. This places a joint responsibility on the architect and school personnel to engage in cooperative, intelligent planning and to communicate thoroughly with each other throughout the course of the project.

Selecting the right architect is not a simple matter. One must consider knowledge and experience as well as personality and ability to establish rapport. The client should interview prospective architects, view their work, inspect buildings they have designed, and confer with other clients they have served. An architect's ability to provide good service should be investigated from every standpoint. The planning team for the project has the responsibility of working with the architect on details as the plan develops. Decisions must be made expediently in consultation with the architect.

The Design Process

The architect's work progresses through various stages of progressively detailed design. At the initial or *conceptual stage,* the architect provides drawings to assist in conceptualizing and "selling" the project. This is followed by the *schematic stage,* which consists of sketches and drawings scaled in sufficient detail to define functional areas and estimates of cost. In the *design phase,* drawings and documents describe the entire project, including structural, mechanical and electrical systems, safety, maintenance, materials, and equipment. The final *construction document* forms the basis for bidding by contractors (although schools may utilize a "design and build" option, in which the architect and contractor work for the same company). Documents at this stage should detail the previously mentioned features as they are affected by local building codes and should incorporate landscaping, schedule changes, and special requirements (Farmer, Mulrooney, & Ammon, 1996).

Planning Checklist

At this stage of the process, the planning committee may wish to employ a detailed checklist. The Facility Planning Checklist presented in Appendix C is intended to provide insight into the considerations that deserve attention in the planning process. It also serves as a model for a checklist that could be utilized in an actual planning situation.

Working with the Contractor

Once the plans are complete and the financial and other arrangements have been made to move ahead with the project, it is time to enter into an agreement with a contractor. Generally, a project should be put out to bid. All contractors who prepare bids must have access to exactly the same information about the project, so that the bids will be prepared on the same basis and can be logically compared against each other. When a project is put out to bid, the client accepts the best bid (which may not always be the lowest bid). Sometimes, the bidding procedure includes an understanding that the client has the right to approve the subcontractors whom the general contractor plans to engage. This assures that the client feels satisfied with all of the contractors who will work on the project.

During the construction phase and beyond, the architect has a continuing responsibility to the client to make sure that the construction is done exactly according to the approved plan. To do so, the architect works hand-in-hand with the contractor and performs periodic inspections to be sure that the conditions of the contract are met.

In addition, the school should assign an in-house inspector for each major construction project. The inspector should be a school technician whose main responsibilities in connection with the project are to make sure the materials and quality of work are in accordance with the contractual agreement. The inspector also serves as a liaison between the contractor and client and thereby communicates any clarifications and decisions needed along the way. When the project is nearly completed, the inspector should be involved with the architect and the contractor in semifinal and final inspections. These inspections should yield a list of all of the items that are uncompleted or unacceptable. These items must be completed to satisfaction before making the final settlement with the contractor and the architect.

On any large project, a number of changes are made during construction due to additional information that comes to light or the development of unforeseen circumstances. It is an advantage to both the client and the contractor to keep an open door for making such changes when they can be clearly justified. Whenever such a change is agreed upon, a *change form* should be completed and a copy kept by the client, the architect, and the contractor so that there is a written agreement explaining the change. A change is viewed as a modification of the original contract.

Site Considerations

Siting a school facility on adequate acreage with parking and hard- and soft-activity surfaces is a major planning challenge. A large high school site may encompass some 500,000 square feet of turf (equivalent to approximately 700' × 700') plus over 400,000 square feet of asphalt, not including tennis courts and a surfaced track. A middle school campus typically includes more than 200,000 square feet of turf and 200,000 square feet of asphalt, including outdoor courts. A typical elementary school needs 150,000 square feet of turf and 100,000 square feet of asphalt play area, including outdoor courts.

Site selection must take into consideration the following factors: (1) finance, (2) availability of utilities, (3) climate and other environmental factors, (4) contour of land and soil conditions, (5) zoning requirements, (6) accessibility, and (7) demographics of populations served (Walker, 1997).

Base materials are an important consideration and can provide a real challenge. Synthetic playing surfaces require a stable base to minimize cracking, and a smooth pavement. (Most of us have seen tennis courts with uneven surfaces or net posts leaning inward.) Paying attention to detail during construction can save headaches later on. Sawing joints in the asphalt helps to control cracking in locations that won't interfere with play. Base materials must be designed for local soil conditions, and groundwater can complicate construction. Adequate drainage must be considered.

Choices of surface materials must consider climate, safety, and requirements of the activities. Asphalt mix can be critical, especially in locations where intense sun and extreme dry climate make for short asphalt life. Surface material has become a critical issue for school playgrounds. Safety concerns, along with sanitary and upkeep issues, have led to increased use of a rubberized "fall surface." The recommended thickness of the surface varies in accordance to the height of the play equipment (Scheideman, 2000).

Parking is often given short shrift. An adequate number of parking spaces should be provided and located in functionally adjacent areas. For sports facilities that accommodate spectators, the rule of thumb is to figure three people to an automobile and divide by the number of spectator seats. For example, a basketball arena with 4,500 seats should have a parking lot with 1,500 spaces. Handicapped parking spaces must meet state and federal requirements.

Finally, the orientation of outdoor courts and fields needs to be considered. The direction of the sun is a factor in orienting football fields, baseball and softball diamonds, and tennis courts. Tennis courts, for example, are built on a North/South azimuth.

Choosing the Type of Structure

In view of accelerating construction costs, the concept of maximum useful space for minimum cost is crucial—keeping in mind the projected life span of the facility and the cost of maintenance. These decisions require careful analysis of the basic kinds of structure, the particular design features, and the materials used. The three most common construction materials for schools and sports facilities are steel, reinforced concrete, and masonry (brick and concrete block). Conventional brick construction has been by far the most prevalent material used for schools. Factors that influence choice of construction materials include: (1) costs, (2) environmental factors (e.g., climate, soil conditions), (3) location, (4) longevity, and (5) maintenance considerations (Farmer, Mulrooney, & Ammon, 1996).

Over the past several decades, considerable experimentation has been done and some practical use has been made of other forms of construction, such as hard-shell domes and air-support structures (figures 10-3 and 10-4). There is no question that such structures can provide large areas of space at less cost per square foot than conventional construction; therefore, these new forms of construction are sometimes worth serious consideration. The advantages and disadvantages of air-support structures must be considered, since there has been a mild trend toward the use of these structures in recent years.

Pure air-support structures and cable reinforced air-support structures have certain advantages when compared with each other or with conventional construction. The cable-reinforced structures must have solid walls or some other kind of structure to which the cables can be attached. The advantages of the various kinds of air-support structures as compared with conventional construction are: (1) The cost of air-support structures per square foot is only 25 to 35% that of conventional structures. (2) They are capable of providing a large span of unobstructed space. (3) They are rela-

Figure 10–3. Exterior view of an air-support structure at Dartmouth College (reprinted with permission of Sportexe)

Figure 10–4.   Interior view of an air-support indoor golf driving range (reprinted with permission of Air Structures American Technologies, Inc.)

tively easy and fast to build, and the smaller structures can be put up and taken down during different seasons of the year. (4) The maintenance costs are relatively low. (5) They are easy to light because of the penetration of natural light during the day and their ability to diffuse artificial light at night. (6) They are relatively fireproof and highly earthquake resistant.

Some of the disadvantages are: (1) The life of the air-support fabric is about 20 years, much less than that of conventional construction. This fact partially offsets the advantage of much lower initial construction costs. (2) In some environments, the architecture of air-support structures is considered unacceptable. (3) Temperature control has been a problem, especially in warm climates. Heat gain is a larger problem than heat loss. For this reason, air-support structures have proven to be more acceptable in cold than in warm climates. (4) In terms of acoustics, air-support structures are less than optimal but in particular instances are acceptable. (5) The potential for serious vandalism exists, but this fear has not been substantiated.

Indoor Activity Areas

Because of escalating construction costs, multipurpose areas are becoming more prevalent in schools. The exception is the combined gymnasium and auditorium, as it has been found to provide a poor-quality auditorium and compromises the integrity of the gym as an activity area. The following five indoor activity areas very in utility. Floor surfaces often are a major determinant in allowable use.

Court Areas

Large court areas can accommodate a variety of activities. Overlapping courts with colored-coded lines are a necessity for most programs, although overlap needs to be minimized to allow for traffic flow between courts. Motorized panels or curtains that provide separation for activity/teaching stations are utilized in large court areas found in gymnasiums. The solid panel system is more expensive than the curtain but provides better acoustical isolation and offers a rebound surface. Ceilings in court areas should be a minimum of 23–24 feet high and of a color that contrasts with balls in the air. Net standards for courts should mount into self-standing, socketed floor plates. Any spectator seating should be collapsible (preferably on a motorized system) (Malpass, Turner, & Waggoner, 1997).

Weight Training Areas

Because of the popularity of weight training as a fitness activity, weight rooms (or areas) are becoming more necessary to support athletics programs. A separate weight room is more secure and easier to supervise than an open area. A minimum space of 3,500 square feet (e.g., 70′ × 50′) is recommended to provide for weight machines and free-lifting areas. This facility should be located on the ground level near the periphery of the building away from quiet areas. Floors may be covered with either a rubberized surface or indoor-outdoor carpeting. Rubber floors are more expensive but last longer. Wall mirrors should be attached at least 18 inches off the floor to avoid rolling weights. Adequate ventilation is crucial to disperse odors and body heat (Malpass, Turner, & Waggoner, 1997).

Dance and Exercise Areas

Artistic dance, aerobic dance, and group exercise can be accommodated in one multipurpose area. A 3,000-square-foot room can accommodate 80 for exercise or about 30 dancers. The room should have two doors and be located near a large corridor or the lobby to handle large numbers of participants. Water fountains should be located near (but not in) the area. Large dance areas can be divided by a folding wall for multipurpose use. Adjacent storage is needed to accommodate sound equipment, as well as an area to store gear bags of participants. Speakers, mounted off the floor, should be of sufficient number to fill the room with sound. All interior surfaces should have acoustical properties sufficient to isolate noise carrying from the room.

Full-length mirrors are placed on at least two walls, 12 inches from the floor. Ceiling height ranges from 16 to 24 feet and may include ceiling fans. Temperature and humidity controls are crucial because of the number of active participants. No one type of floor has been found to be best, but the newer foam floors with interlocking sections are less costly than wood. Suspended wood floors are preferred by many dance instructors. Carpets are an option but may cause skin abrasions and, occasionally, allergic reactions. Experts agree that exercise/dance floors should have the four qualities of shock absorption, stability, traction, and resiliency. Hard surfaces require the additional purchase of floor mats for high-impact activities (Malpass, Turner, & Waggoner, 1997).

Mat Rooms

These multipurpose rooms can be designed to accommodate wrestling, gymnastics, and martial arts, with minimal compromises. If possible, the primary activity should be identified to determine the basic design. A 100-foot length is minimal for a gymnastic vaulting area. A wrestling room needs to accommodate two full-sized (42' × 42') wrestling mats. A 24-foot ceiling would serve all three activities, as would a hardwood floor. Acoustical block walls will isolate noise. Adjacent storage must be large enough for gymnastic apparatus, as well as to provide an area to accommodate rolled-up mats. If the room serves as a *dojo* for martial arts, additional storage space needs to be available for a judo mat—which differs from a wrestling mat. Anchored bars hanging from the ceiling can support heavy bags for karate (Malpass, Turner, & Waggoner, 1997).

Natatoriums

A natatorium is an indoor aquatic area that houses a swimming pool and may include a diving tank. (Few schools operate outdoor aquatic facilities.) Natatoriums are expensive (often in excess of $1 million), but more schools are building them for their obvious benefits. The type of pool constructed depends on its projected use: instruction, recreation, and/or competition. Pools for training people to swim have a depth of 3.5 feet or less. Competitive pools are built at lengths of 25 or 50 meters, with a depth varying from 4 to 6 feet (NFSHSA standards now stipulate a minimal depth of 4 feet for swim meets). Diving pools or areas, which should approximate 40 by 50 feet, have a depth of 13 feet or more depending on the height of springboards or platforms. L-shaped pools and rectangular pools with a diving tank at the side are the most common configurations. Moveable bulkheads allow dividing the pool into multiple use areas. Various materials are utilized in pool construction, including reinforced concrete, fiberglass, and vinyl liners (Olson, 1997).

Heating of water and air, along with ventilation to control humidity, are crucial design elements. Acoustics also are a major concern for indoor pools. Some natatoriums feature a retractable wall opening to an outside deck. As for fixtures, starting blocks are necessary for competition; lifeguard chairs are required for recreational use. Chemical and filtration systems should be housed in readily accessible areas near the pool. The natatorium may contain spectator seating. If so, seating should be located above the deck level with separate access. A natatorium office with reinforced glass walls on the deck level promotes supervision. A telephone outlet in the pool area is mandatory. (See the section on maintenance of natatoriums later in the chapter, and see chapter 9 for risk management issues.)

Floors

Floor surfaces in activity areas require serious consideration. A decision must be made as to whether to install hardwood floors or some form of synthetic surface in gymnasiums. Maple is the preferred choice for hardwood floors. The traditional subsurface options of a floating wood floor versus an anchored floor both have drawbacks. The floating floor tends to buckle in excess humidity; the anchored floor is a fatiguing surface on which to play. A relatively new hybrid system called a *restrained floating floor* seems to eliminate most of these problems. Water-based urethane finishes have replaced the old oil-based finishes, which were deemed a health risk by the Environmental Protection Agency (Steinbach, 2002).

Hardwood floors have become expensive, and some fairly suitable synthetic surfaces can be installed at less initial cost. It is generally agreed, however, that hardwood floors are still preferable in a gymnasium to accommodate a variety of sports. If the school gymnasium is used regularly for activities like meetings or social events, wood floors are not a good choice unless the school purchases protective floor covers. The popularity of roller hockey in the physical education curriculum has led to unprecedented damage to wood floors. A plastic tile system can be purchased to cover the floor; however, some damage still may occur depending on the types of wheels on skates. One option is to move roller hockey outdoors onto a blacktop area.

Today, several choices in synthetic indoor surfaces are available. Recycled rubber mat floors covered with urethane have become popular in Europe. Injection-molded polypropylene tiles are a popular choice for volleyball, and they stand up well for basketball and hockey. Vinyl and vulcanized rubber floors accommodate a wide range of performance specifications. Seams have virtually disappeared from synthetic surfaces due to improved sealing techniques (Steinbach, 2002).

The use of carpeting in schools has become controversial because of concerns about chemical emissions and carpet mold. These problems seem to be exacerbated by poor maintenance and cleaning. State health departments have issued cautionary statements, and some school districts have banned carpeting until the health concerns are resolved (Moffeit, 2000). Once safety concerns are resolved, the question of whether to use carpeting in activity areas should be analyzed in terms of relative initial cost, projected replacement and maintenance costs, acoustical advantages, and the positive emotional and behavioral influences it has on those who use the facility.

In large foyers, shower rooms, or other areas where a hard surface is desired, one must decide whether to use concrete, terrazzo, or some form of tile. Concrete is the least expensive in terms of initial cost. Terrazzo (chipped granite and cement mix) has been found to be highly satisfactory in terms of attractiveness, initial cost, and maintenance. Tile comes in many forms and with great variation in costs. The merits of various materials should be carefully assessed for each particular situation, based on current product information.

Indoor Lighting

The two basic options in lighting are natural light and energy-produced light. Reflected light from surfaces (walls, floors, ceilings) is also a factor in how much illumination is produced. Adequate brightness is the major consideration in activity areas. Glare, which is excessive brightness, should be minimized. Light intensity, measured in foot-candles, is a vital factor in seeing efficiency and eye comfort. Standards have been established for the amount of light striking the surface of various activity and instructional areas in schools, such as a gymnasium floor. These standards should be consulted during the design phase and then tested for compliance after lighting has been installed.

In addition to the amount of light on a surface, the quality of lighting and its costs must be factored into lighting decisions. Some types of lighting are more energy efficient than others, and this will impact fixed expenses. Current choices of lighting include incandescent lights, fluorescent lights, mercury vapor lights, and metal halide lights. Each has specific qualities, advantages, and disadvantages. Natural lighting can reduce utility expenses but also has some drawbacks. For safety reasons, and to comply with clearance regulations of various sports, lights in arenas, gymnasiums, and other activity areas should be a minimum of 24 feet above floor level. In short, providing the right amount and quality of illumination within budget is a complicated matter. Consequently, facility-planning committees may want to retain the services of an illumination engineer (Sawyer, 2002).

Outdoor Activity Areas

Playing Fields

Fields are the most prevalent outdoor activity areas. They should be laid out to allow for traffic between them and positioned to prevent balls and objects from hitting students on adjacent fields. (Ten-yard buffer zones between fields are recommended.) Light posts and other hard barriers in the proximity of playing areas should be well padded. The three choices for field surfaces are natural turf (grass), *prescription* natural turf (discussed below), and synthetic (artificial) turf. Natural grass provides a low-cost surface with simple maintenance requirements (watering, fertilizing, aerating). Optimal mowing heights vary by sport. Soccer and field hockey require closer mowing than football. Administrators should check with the state agriculture extension agent or local nurseries for advice on which variety of grass to plant (Steffen & Hall, 1997). The Sports Turf Managers Association is another resource. (See Online Sources at the end of the chapter.)

Natural turf's major drawback is damage from overuse. Maximum use for natural turf fields is estimated at thirty-six hours per week. The subsurface is a major factor. Most high schools have native-soil fields. Laying sand-based fields is an option, but it is prohibitive in cost for most public schools. High clay content of the soil leads to compaction and lack of percolation (root drainage). Mesh backing can help to stabilize and strengthen roots, but this too is an added construction expense. Drainage is crucial. The goal is to have water migrate to avoid puddles. Natural turf fields are built with a crown to assure drainage. Modern laser technology attached to tractors can assess the pitch of a field (Steinbach, 2002).

Prescription athletic turf (PAT®) employs a complex subsurface irrigation and gravity/vacuum drainage system. It is more expensive but also more durable than natural turf. The PAT system utilizes a subsurface polyethylene barrier and can recycle water, cutting the cost of irrigating fields. Prescription athletic turf allows a laser-leveled flat field and will accommodate a variety of grasses for warm or cool climates. The system includes software for computer-managed moisture control. PAT fields are more expensive to lay down than natural turf, and consequently are seen more commonly at the college level.

Synthetic turf is the most durable option but has a high initial installation cost. It is most cost-effective in high-use areas such as college intramural fields. Outdoor fields also require a crown for adequate drainage. Today's stitched seams are a big improvement over glued seams. Newly constructed fields also feature improved, longer fibers. Fiber systems with infill of crumb rubber or sand come the closest to resembling natural turf. Cushioning has been improved as well. However, despite improvements in subsurface padding, synthetic turf retains its reputation for causing more impact injuries than natural turf (Steinbach, 2002).

Track and Field Areas

It is recommended that school administrators work with a contractor who is a member of the U.S. Track Builders Association when designing an outdoor track to ensure that the facility meets all regulations. The track should include warm-up areas and multiple jump-and-throw areas, and should provide for running sprints in both directions. A common finish line for running events minimizes confusion at competitive meets. Safety fencing or barriers must be strategically located to prevent participants from being hit with misdirected implements. Protective cages are mandatory for discus and hammer events (Steffen & Hall, 1997). Most contemporary track surfaces are constructed with some form of granulated latex rubber, which can be colored, lined, and sealed. A variety of weather resistant and highly durable systems is available. Rubber surfaces can be patched if damaged.

Hard Surface Activity Areas

Blacktop or other synthetic hard surfaces facilitate a wide variety of outdoor activities, including low organized games, recreational basketball, and skating. These surfaces can be painted with court lines and other markings. The major drawback of hard surfaces is injuries caused by falling. Some schools utilize paved parking lots for after-school recreation. Basketball goals often are placed on the edge of parking lots. Students need to be protected from any residual automobile traffic. The use of portable barriers such as sawhorses can mark off recreational areas.

Outdoor Lighting

The main advantage of installing outdoor lighting is that it expands the number of hours during which activity areas can be utilized. Most schools build multipurpose sports fields, and lighting should be planned with this in mind. Sharing fields among several sports can save up to 40% on the cost of outdoor lighting. The height of lights and pole placement are two primary concerns. (All of us have had experiences playing ball on poorly lighted fields.) Lighting must be aimed and arranged to maximize visibility and minimize glare and spill.

The Illuminating Engineering Society of North America (http://www.iesna.org) publishes light-level guidelines for outdoor sports. Some sports require more lighting than others, and switching capacity should provide higher or lower levels of illumination. In addition, some consideration should be given to the location and accessibility of lighting controls (Sawyer, 2002).

Service Areas

Storage Space

Schools routinely are constructed with inadequate, poorly located or poorly designed storage space. Storage is an important element of the total operation, and it should receive adequate attention in planning and construction; however, storage space should be viewed practically. Consider, for example, that school building construction costs have risen to between $130 and $180 per square foot (depending on location). The question for administrators, then, is what kind of supplies and equipment are needed to occupy storage space at this initial cost, plus the cost of maintenance. Thus, efficient use of storage space is imperative. Storage areas for sports equipment should be equipped with metal double doors (for secure access) and no threshold, to allow large equipment to be moved easily.

Locker Rooms, Rest Rooms, and Shower Areas

The locker room should be large enough to seat informally an entire class or team. A general rule of thumb suggests a minimum of 20 square feet per person. Dressing areas should be immediately accessible from a corridor. A minimal number of grooming stations should be provided. Dry grooming stations are equipped with a mirror and a grounded outlet. Wet grooming stations include a sink (Sawyer, 2002).

Lockers should feature doors with open construction to facilitate drying of equipment. The tops of lockers should be slanted toward the front to discourage items from being left on them. The size of lockers depends on the nature of their use. Lockers need to be 75 inches in height by 18 inches in width to hold football equipment (Farmer, Mulrooney, & Ammon, 1996). Different locker arrangements can be considered, with each one having distinct advantages. Various arrangements for locker-room fixtures are illustrated in figures 10-5 and 10-6.

Shower rooms should be located close to dressing areas, with fixtures centrally located to minimize plumbing costs (figure 10-7). One showerhead to three students is the accepted standard. Open showers are the norm; however, some schools provide a few individual shower stalls for privacy. Shower facilities must meet ADA requirements (see Online Sources at end of the chapter).

Rest rooms and locker rooms receive more than their share of hard use and abuse. Materials and fixtures in these areas should be selected on the basis of hardiness and durability. Schools are installing warm-air hand dryers in locker rooms and rest rooms to cut the expense and nuisance of paper towels. Hand dryers reduce petty vandalism, such as toilets and sinks clogged with towels and the setting of fires in waste receptacles. Wall recessed, cast-iron dryers are virtually indestructible and pay for themselves over time. One dryer per two washbasins is recommended in a washroom with average traffic. Dryers should be positioned to create a traffic flow from washbasin to dryers to exit (Kilbryde & Ring, 2002). Schools are forgoing restroom doors and installing open zigzag entrances (common at airports) for easier access and monitoring of these areas.

Evaluation and Possession of New Facilities

The evaluation process is important, both during the building project and after it has been completed, to determine if conditions and specifications have been satisfied. While the owners take temporary possession of the completed facility, a checklist of problems, defects, and unsatisfactory quality of work is tabulated. These defects are the responsibility of the contractor to correct before possession is transferred permanently. The *facility guarantee* is usually in effect for a year against problems of work quality and installation of faulty equipment. At the end of this period a final evaluation is made. When the conditions are judged satisfactory, the owners take permanent possession of the facility. If major problems are detected subsequently, legal action may be necessary (Farmer, Mulrooney, & Ammon, 1996).

MANAGING FACILITIES

The initial concern of facility managers must be the maintenance of a safe environment. Most liability problems stem from the operation of a facility, not its design and construction. Notably, a majority of personal-injury lawsuits involving recreational facilities allege poor maintenance. For this reason, security, inspection, cleaning, and repair are critical areas of concern. The facility manager should develop a maintenance plan that incorporates these routine but necessary procedures (Maloy, 1996).

Once the safety of the environment has been assured, the focus shifts to the educational mission of the facility. Facility managers should have a clear concept of the objectives that determine appropriate use of the facility. The types of facility usage should be incorporated into a system of priorities that reflect program objectives. All potential users of the facilities should understand these priorities.

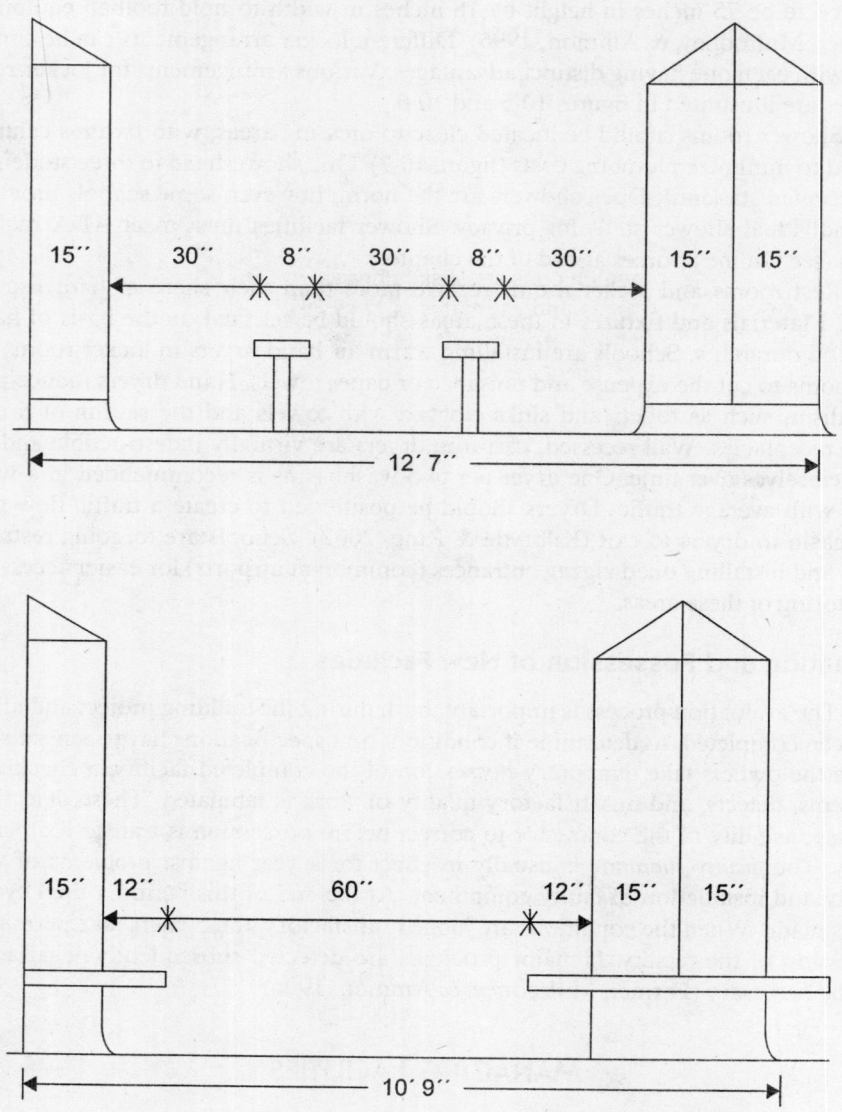

Figure 10–5. Two different arrangements for dressing room benches: isle mount (upper) and pedestal mount (lower). The pedestal mount has distinct advantages in terms of space utilization and traffic flow.

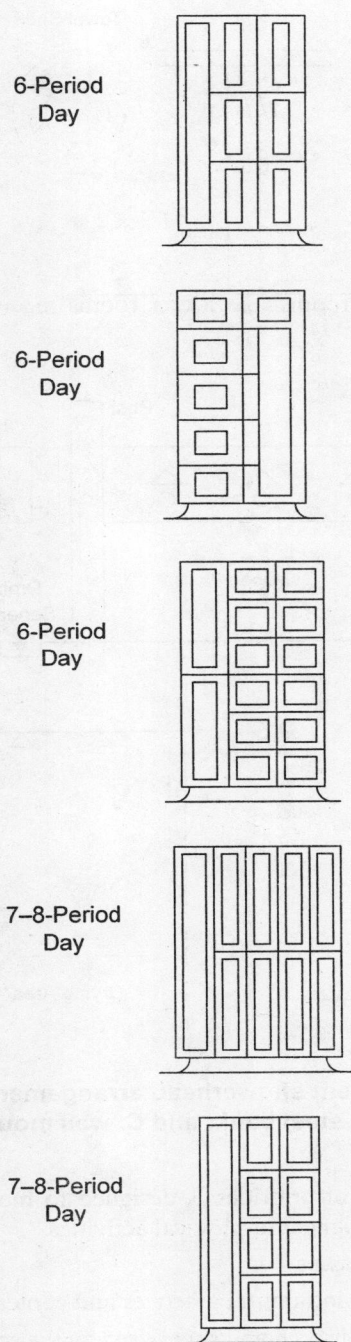

6-Period
Day

6-Period
Day

6-Period
Day

7–8-Period
Day

7–8-Period
Day

Figure 10–6. **Five different combinations of storage/dressing locker arrangements, with the small lockers for storage of uniforms between classes and the large lockers for clothes storage during class**

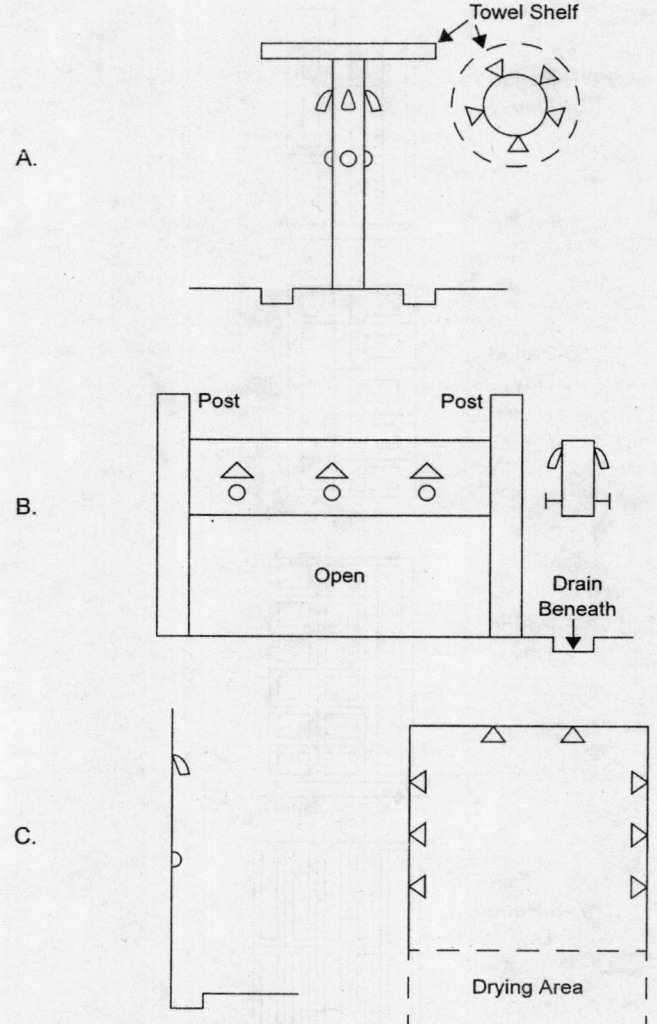

**Figure 10–7.   Three different showerhead arrangements: A. post style;
B. stainless steel bank; and C. wall mount**

The following ranking of priorities is designed to meet the various demands placed on school facilities designed for physical activities.

1st. Regular daytime classes

2nd. Interscholastic and intramural practices and contests.

3rd. Other activities sponsored and supervised by the school, such as rehearsals, evening classes, and free play.

4th. Activities sponsored by student groups or by faculty and staff groups.

5th. Approved activities sponsored by groups or agencies not affiliated with the school.

Facilities should be scheduled for optimal use in accordance with the established priority system to avoid conflict, misunderstandings, and disappointment due to errors. The person in charge of scheduling shouldn't be affiliated with any groups likely to come into conflict over use. Gymnasiums, in particular, can be areas of scheduling conflict due to their versatility and the variety of demands placed on them. Scheduling should be done through one scheduling office. On occasions when conflict or misunderstanding arises about the use of a facility, the information on the official schedule should apply.

It is recommended that a simple scheduling form be used to record facility use commitments, and a copy should be provided to the one initiating the request (see figure 10-8). In addition, an administrator should maintain a current master sheet for each teaching/activity station. In many schools, facility scheduling is now done on a computer. An instance where computerization would be a definite advantage is a college intramural program with a hundred or more teams and a limited number of courts and fields.

Maintenance must be coordinated with facility scheduling. For example, the sprinkling system for fields cannot be activated right before intramural flag football games are scheduled. Facility maintenance becomes more critical as the number of curricular and extracurricular activities increases. Excessive scheduling of activities on natural turf can

---

FACILITIES REQUEST FORM

Date _____

_____

Area Requested   Rental Fee

_____

Organization

_____

Date(s) Requested   Time

_____

Person in Charge

_____

Mailing Address   Phone

PERSONNEL NEEDED:
Building Supervisor   _____

                                 cost

Life Guard(s)   _____

                                 cost

Custodian   _____

                                 cost

TOTAL COST (rental fee
and personnel)   _____

REQUEST APPROVED BY: _____

Figure 10–8. Sample form for the request of facilities

cause enough wear and soil compaction to ruin the field. Sensible scheduling of use and maintenance keeps facilities attractive and protects them from deterioration.

Maintenance must be given consideration in facility planning and construction. If a facility is properly designed and well constructed, the maintenance costs can be minimized. (Annual maintenance cost of school facilities approximates $4 to $5 a square foot.) Unfortunately, facility maintenance often receives little consideration in the planning phase. Architects and contractors are most concerned about the initial cost of the building. Because maintenance and utility costs are often considered secondary in importance, maintenance staff should be involved in facility planning. They can anticipate built-in maintenance problems that have the potential of becoming extensive and expensive. Also, buildings should be designed to minimize utility costs, even if at increased initial cost. The savings on utilities over time may more than pay for the more expensive design features.

On the average, the lifelong maintenance costs of a building amount to about three times the original construction cost, excluding the effects of inflation. With buildings that are poorly constructed or that have poorly selected materials, the maintenance costs run much higher. Durable and easily maintained materials on the exposed areas—both inside and outside—are the greatest single factor in the solution to maintenance problems. Other important considerations include well-trained and conscientious maintenance personnel, adequate maintenance equipment located in strategic places for easy availability and use, and high expectations relative to maintenance among those associated with the use of the facility. Pride in the condition of the facility can make a major difference, and an administrator is in a good position to instill this kind of attitude.

Maintenance Plan

Many school districts require a *facility maintenance plan* to be developed and revised periodically. The maintenance plan should state policies and standard operating procedures for maintaining facilities and fixtures. It should include designated daily housekeeping tasks, provisions for supplies and equipment, and a schedule for completion of major repairs. Administrators would be wise to include maintenance personnel, instructors, program directors, and coaches in the formulation and updating of the maintenance plan. Table 10-1 illustrates a brief section of a maintenance plan for public school facilities.

Maintenance of Natatoriums

Natatoriums require special maintenance expertise due to health and safety concerns. Poorly managed swimming pools are an invitation to serious problems including public health violations and lawsuits.

School districts and colleges that operate natatoriums should employ a trained aquatic facility manager who has enrolled in courses or workshops in pool maintenance. The National Swimming Pool Foundation (NSPF) provides a Certified Pool Operator certification (CPO) course—routinely offered at colleges and community colleges. Many states and local communities require their public and commercial pool operators to be CPOs. Regardless of certification status, school aquatics managers, swimming instructors, and coaches should be familiar with basic sanitary and chemical maintenance procedures. State health department regulations typically address such concerns as onsite rescue equipment, water clarity and chemical balance, and

Area/ Items/Task	Statute Regulation	Guidelines/ Best Practice	Measure
Inspect fire alarm system	CFR* 29 1910.164 and 1910.165 http://www.osha.gov/	Fire alarm system inspected annually by a qualified technician.	__Exceeds the standard __Meets the standard __Partially meets standard __Does not meet standard
Inspect sprinkler system	NFPA** 101 http://www.nfpa.org/	Quarterly inspection by the state Fire Marshall.	__Exceeds the standard __Meets the standard __Partially meets standard __Does not meet standard
Test emergency and exit lighting	NFPA** 101 http://www.nfpa.org/	Lighting is tested monthly.	__Exceeds the standard __Meets the standard __Partially meets standard __Does not meet standard
Kitchen fire suppression system	NFPA** http://www.nfpa.org	Kitchen ranges and ovens have fire suppression systems.	__Exceeds the standard __Meets the standard __Partially meets standard __Does not meet standard
Kitchen fire suppression system tested annually	NFPA** http://www.nfpa.org		__Exceeds the standard __Meets the standard __Partially meets standard __Does not meet standard
Exits	29 CFR* 1910.36(b)(4) http://www.osha.gov/	All exits free and unobstructed.	__Exceeds the standard __Meets the standard __Partially meets standard __Does not meet standard
Exit doors	NFPA** 101 http://www.nfpa.org/	No wedges or devices holding exit doors open.	__Exceeds the standard __Meets the standard __Partially meets standard __Does not meet standard
Means of egress	29 CFR** 1910.36(d) http://www.osha.gov/	All means of egress free of obstructions.	__Exceeds the standard __Meets the standard __Partially meets standard __Does not meet standard
Exit signs	29 CFR** 1910.36(b)(5) and 29 CFR 1910.37(f)(4) http://www.osha.gov/	Every exit sign is conspicuously indicated and visible from any point.	__Exceeds the standard __Meets the standard __Partially meets standard __Does not meet standard
Lighted exit signs	29 CFR** 1910.36(b)(6) http://www.osha.gov/	All artificially illuminated exit paths have adequate, reliable illumination.	__Exceeds the standard __Meets the standard __Partially meets standard __Does not meet standard

*CFR = Code of Federal Regulations
**NFPA = National Fire Protection Association

Table 10–1
Maintenance Plan for Operations Service in Public School Facilities (courtesy of Jay Readinger, Maine State Dept. of Education)

recovery procedures for fecal matter incidents. These departments may also conduct chemical analysis of pools upon request. Technology for pool maintenance is constantly changing and improving, and pool operators need to keep current. For example, battery-operated photometers, which can assess chlorine, pH, bromine, and cyanuric acid, are replacing the traditional chemical test kits.

Community Use of Facilities

The community–school concept calls for an expansion of the traditional role of the school beyond the hours of the day, and days of the year, utilized for formal instruction. It also incorporates the idea of a more open and accessible physical campus. Although the community–school concept won initial acceptance, it has not been applied frequently enough. It should receive more consideration in every community and neighborhood. Promoters of this concept note that schools involve the largest single investment of tax dollars for most communities. Benefits include an economy in land ownership and costs of construction, operation, and maintenance through mutual use of areas and facilities. Putting this concept into action involves cooperative action between school and municipal authorities.

Opening up the school to community use does entail some additional risks. In order to protect the interests of the cooperating authorities and to assure the most effective application of a community–school plan, a formal agreement should be enacted by the parties involved. This agreement should stipulate the respective responsibilities for the purchase, development, operation, security, and maintenance of the community–school. In addition to agreements with local government entities, schools should develop written policies and procedures for renting facilities to outside groups, including private-sector users. Gymnasiums and other sports facilities are in high demand for use by groups in the community. Most states allow schools to rent facilities; however, local jurisdictions vary in permitting use of public school facilities on Sundays or renting to groups for religious ceremonies. Administrators need to be cognizant of limitations on use. Private-sector users may not be covered by school insurance and should have their own coverage for injury, loss, or damage (Olson, 1997).

BUILDING AND GROUNDS SECURITY

More stringent security measures have been implemented by schools following the increase of shooting incidents on campuses and the terrorist attacks in September of 2001. On a more mundane level, schools have long been the targets of vandalism, arson, and burglary. Many schools are installing high-tech security and alarm systems, including glass-break sensors, closed circuit television (CCTV) systems, motion and audio detectors, and magnetic contacts on exterior doors, along with metal detectors at entrances. Schools with a limited budget might begin with a basic alarm system in high-exposure areas. Portable alarm systems can be employed temporarily in areas where security breaches have occurred. In addition, it's important to train personnel in charge of facilities in the use of new security technology.

Schools should convene a panel to plan building and grounds security. The panel should include personnel from all levels: school security officials, fiscal managers, maintenance and operations personnel, program administrators, teachers, coaches, students, and district-level officials. Consulting an outside security expert in planning may save money and avoid mistakes. The four principal concerns are: (a) securing the campus and buildings, (b) maintaining a hazard-free environment, (c) minimizing theft and vandalism, and (d) creating a safe and inviting appearance for the buildings and grounds.

The security plan should encompass use of the campus for extracurriculur activities after hours. Because intramural and athletics directors must control access to school property outside of normal classroom hours, they need to be particularly sensitive to campus security issues. Parking lots and buildings with several entrances and exits increase the potential for unwelcome intruders, theft, vandalism, and defacement of vehicles and school property. An attempt should be made to minimize the number of entrances to campus. There are situations when access points to schools need to be supervised by individuals familiar with the student body. Campus traffic, both pedestrian and vehicular, should flow through areas that can be supervised. Perimeter fencing defines the campus and provides a means of restricting access. Direction signs should be placed in strategic, visible locations and be large enough to attract notice. Staff should be trained to courteously challenge suspicious persons. "May I direct you to your destination?" is a polite but firm way to approach visitors. Some schools now require students and/or staff to carry identification cards.

Athletics draw a wide variety of individuals to the campus during evenings and weekends. Public events should be organized so as to isolate spectator traffic within designated areas of the campus. Spectators should be kept apart from participants and game officials whenever possible. Access to spectator areas should not require transit through school hallways. If this cannot be avoided, wall-recessed security gates can be installed to block off-limit areas, or schools can purchase state-of-the-art access control systems that incorporate magnetic proximity cards allowing entrance into certain areas only at designated times. Convex mirrors at hallway corners provide a low-tech way to monitor access.

Activity areas are plagued with petty thefts and vandalism. Locker rooms, toilets, and shower areas seem particularly vulnerable. Locker rooms should not accommodate public restrooms and should be secured when not in use. Heavy-duty lockers with recessed locks can be purchased to discourage break-ins. Visibility in these areas should allow for adequate supervision. (Crowd control measures are discussed further in chapter 16.)

EQUIPMENT MANAGEMENT

Directors, teachers, and coaches are generally more involved on a continuing basis with the purchase and management of equipment and supplies than they are with facility planning and maintenance. Adequate attention to equipment needs is crucial, as the amount and quality of equipment has a significant impact on the instructional, intramural, and athletics programs.

Equipment managers need to understand the basic terminology utilized in this area of administration. *Equipment* refers to nonexpendable articles of tangible property having a useful life of more than one year, and not permanently affixed to the buildings or grounds (e.g., tumbling mats and volleyball nets). Some schools designate items costing over a minimum amount—often $500—as *capital equipment*. The term *attached equipment* is also used to refer to equipment attached to a structure, such as a basketball backboard. *Supplies* are inexpensive items expended in use or with a life span of less than a year (e.g., adhesive tape and tennis balls). The above terms aren't "etched in stone"; operational definitions vary among institutions.

Selecting and Purchasing Equipment

One of the most important objectives in the selection of equipment is to obtain the best value for the money. Many of the purchases made for school programs involve large expenditures; therefore, the one making the purchases must be thorough in searching for the right product from the vendor who will provide the best price and the best service. Compare prices and services offered by local vendors against those of national or regional distributors. While wholesale distributors typically offer lower prices, local vendors may be more responsive when emergency needs arise. Although some discretion in ordering is allowed, equipment must be purchased within the system of budgetary controls by which the district assures accountability. (Refer to chapter 7 for financial procedures involved in ordering equipment.)

Quality and quantity are the two basic considerations in ordering equipment. Because less expensive equipment often is substandard, performs unsatisfactorily, and wears out quickly, it is not prudent to purchase equipment based on price alone. A short life span is uneconomical, in terms of both money and time required for replacement. The purchaser should look for the optimal tradeoff between price and quality. A $30 badminton racket that lasts four years is a better buy than a $12 racket that lasts one year. Coordinate short-term and long-range purchase plans to save money.

Part of the task of selecting and purchasing equipment is knowing what is available. Researching products is a time-consuming but necessary responsibility. New types of sports equipment, new manufacturers, and new construction materials require a purchaser to keep up to date. Several strategies can be employed to stay current: (1) place yourself on the mailing list for equipment catalogs; (2) subscribe to periodicals that advertise sports and recreational equipment; (3) log on to Web sites of equipment manufacturers and distributors; (4) attend conventions and professional meetings where equipment is being exhibited; and (5) discuss new products with local vendors.

When ordering equipment, especially in large volume or through a bidding process, it is important to pay attention to detail. Precise specifications are necessary for such variables as color, size, weight, and composition. Failure to address specifications can result in unsatisfactory items that the vendor might substitute. An alternate approach is to order by catalog number and indicate "no substitutes authorized." Untested items or new models should be used on a trial basis before ordering in bulk. Consult with other teachers, coaches, and equipment managers to assess their experience with new items (Olson, 1997).

The quantity of equipment ordered is based on needs, which are determined by the nature of the program and the number of students in each instructional unit or team. Analyze how you are going to organize classes or team practices. It takes more balls for practice drills than for scrimmage. For items that must take into account "handedness," such as golf clubs and softball gloves, specify one left-handed model for each dozen ordered. Be sure to allow enough items to make up for breakage and loss. The amount of equipment to be ordered is the difference between program needs for the coming year or season and usable items remaining in inventory (discussed below).

Athletics directors or coaches order uniforms and footwear in addition to sports equipment and supplies. One must consider blends of materials, durability, sizes, styles, storage requirements, and ease of cleaning. Uniforms are expensive, and purchases must stay within the department budget. Outfitting one college football player can cost up to $1000. When purchasing clothing and protective equipment, proper fit and safety considerations should receive adequate attention. Football and hockey helmets must meet National Committee on Safety of Athletic Equipment (NOCSAE) certification standards and must be conducive to reconditioning and recertification.

Distributing Equipment

All equipment must be secured and accounted for with proper identification before being distributed. Well-defined policies and procedures should be in place for checking equipment in and out and for monitoring its use. Policies cover what can be checked out, when, for how long, and by whom. Many colleges require students to present identification cards, which may be held as collateral for equipment checked out. The individual is charged for items not returned. For athletics programs, long-term equipment checkout precludes the use of identification cards. Instead, the athlete establishes identity at the beginning of the season and signs an equipment checkout form. (Figures 10-9 and 10-10 show examples of such forms.) When equipment is

---

PHYSICAL EDUCATION ISSUE FORM

Name (print) _____

                     Last   First   Middle

Locker No. _____   Homeroom _____

I hereby accept full responsibility for the following uniforms and/or items of equipment checked out to me:

Uniform Items	No.	Equipment (list items)	No.
Blouse or shirt			
Shorts			
Socks			
Leotard			
Swim Suit			
Towel			
Other			

Address _____   Phone _____

Signature _____

*Note: Cross off items when returned.

Figure 10–9. A sample uniform and equipment checkout form for physical education participants

ATHLETIC EQUIPMENT ISSUE FORM

Name _____ Sport _____ School _____

Address _____ Telephone _____ Locker # _____

LIST OF EQUIPMENT

Practice		Game	
Belt _____	Pants _____	Coat, sideline _____	
Cap _____	Pants, warm-up _____	Headgear _____	
Guard, knee _____	Shirt, warm-up _____	Jersey _____	
Headgear _____	Shoes _____	Pants _____	
Jersey _____	Socks _____	Pants, warm-up _____	
Pads _____	Stockings _____	Shirt, warm-up _____	
Pads, hip _____	Tights, wrestling _____	Shoes _____	
Pads, shoulder _____	Undershirt _____	Socks _____	

I am responsible for the articles listed hereon, which are charged to me, and agree that they will not be used for any purpose but the school games or authorized practice while in my possession.

Signature: _____

Date _____

Figure 10–10. A sample form for the issue of athletic equipment

returned, the items are checked off the form, and both parties sign or initial it. The users of equipment must understand their responsibility for proper use and care, and their obligation for payment in the case of loss or damage through carelessness or abuse. The time or date on which items are to be returned and the check-in procedures must be clearly communicated during checkout.

When equipment is checked out to individual students and staff members for instructional or recreational use, it is important to design a workable and efficient system of distribution. A checkout window with a large countertop, adjacent to the storage area, facilitates the exchange of equipment and maintains security. Most colleges employ an equipment manager and/or student worker to check out equipment during hours of operation.

Inventory

Inventory involves keeping an amount of stock that is adequate to avoid frequent shortages and small reorders, yet not so large that it occupies an inordinate amount of storage space and expense. Keeping current and accurate inventory records is an important element in equipment accountability. This requires the use of an appropriate inventory form (see figure 10-11). As indicated, these forms should note

INVENTORY FORM				
Signature* _____ Title_____				
Date _____ Room of Location _____				
Description	Quantity	New or Used	Condition of Equipment	I.D. # or Label
*Signature of person accountable for the items				

Figure 10–11. A sample equipment inventory form

the number of and condition of items of equipment. It is also important to track the exact location of items in use and in storage. Schools routinely experience difficulty locating equipment during audits because of poor record keeping.

Computerized inventory procedures increasingly are being used by schools and colleges. A variety of inventory software is now available. Computerization greatly simplifies record keeping and makes the information more readily accessible. Computer inventory can easily interface with purchasing, preventive maintenance, and distribution. Before entering new equipment into the computer record, check all items received for compliance with the purchase order. If there are no discrepancies, items should be marked for ready identification (many school districts now use a bar code system for fixed assets). Only then should you enter the new items into the computer inventory.

Computers facilitate a *perpetual inventory* system. With this approach, items entering the inventory are recorded against items leaving it to obtain a current balance of items in stock and in use. The record is continuously updated as the inventory changes. When items are discarded as unusable, the computer can be programmed to signal their need for replacement and add them to the order list. The traditional system had been to carry out seasonal inventories, where items are accounted for at the beginning and the end of a sports season or an instructional period (semester, quarter, or school year). Perpetual inventory has proved more responsive to short-term needs.

Storage and Care of Equipment

Equipment represents money invested and must be properly stored in order to avoid loss. It should be reasonably protected from theft, vandalism, fire, dampness, exposure to the elements, and damage by rodents and insects. Therefore, it should be stored in areas that are clean, secure, properly ventilated, pest controlled, and protected from extreme temperatures and humidity. Proper maintenance of equipment can keep the items in usable condition and extend their life span. Equipment in poor repair results in disruptions and ineffectiveness in the program. In line with the above concerns, equipment should be inspected on a regular basis.

Equipment must be stored in a way to provide ready access without compromising security. The equipment-room manager has to determine what cabinets and other storage compartments to place under separate lock and key. Clear procedures should be established and followed relative to loss, theft, damage, or destruction. Equipment should be discarded in accordance with the prescribed procedures and by authorized persons.

Most equipment can be effectively maintained by simply using common sense. Specialized treatment or care instructions usually accompany equipment at the time of purchase. If additional information or clarification about care is needed, the distributor is usually the best source. Prudence and constancy in taking care of equipment is often more important than special instructions. Keeping equipment clean is fundamental, and maintenance should be carried out on a regular schedule. In-house staff can often make minor repairs, such as sewing and refinishing, to extend the life of equipment. A workbench should be built in the storage area to accommodate repairs.

Walker and Seidler (1993) offer the following guidelines for successful equipment-room management:

1. Locate the equipment storage and issue area near locker rooms and activity areas, and adjacent to corridors.

2. Provide double-wide, high doors for easy transportation of equipment into and out of storage areas.

3. Make certain that doors (including facings), hinges, and locks provide adequate security.

4. Windows, if any, should be small enough and located at a height to prevent unwarranted entry into the storage and issue area.

5. Keys to the area should be accounted for and issued on a limited basis.

6. The storage area should be designed to allow various types of storage compartments and the hanging of equipment from walls and ceilings.

In summary, a capable administrator should have positive answers to the following two fundamental questions: Do the employees consistently have all the supplies and equipment they need in order to be effective? Do the established procedures provide adequate protection against loss, breakage, and misuse, and do they encourage economy and efficiency?

References and Recommended Readings

Farmer, Peter, Aaron Mulrooney, & Rob Ammon. 1996. *Sport Facility Planning and Management.* Morgantown, WV: Fitness Information Technology, Inc.

Kilbryde, Linda, and David Ring, 2002. "Washrooms/Locker Rooms." *American School & University,* 74(6): 18.

Moffeit, Miles. 2000. "Invisible Invaders, Mold Discoveries, and Ailments on Texas Campuses Prompt a Renewed Push for Controls, but Obstacles Remain." *The Fort Worth Star-Telegram,* July 9. Online article: <http://www.cnn.com/2000/LOCAL/southwest/07/10/ftw.mold/>.

Olson, John. 1997. *Facility and Equipment Management for Sport Directors.* Champaign, IL: Human Kinetics.

Sawyer, Thomas (Ed.). 2002. *Facilities Planning for Health, Fitness, Physical Activity, Recreation and Sports: Concepts and Applications.* 10th ed. Champaign, IL: Sagamore Publishing.

Scheideman, Elton D. 2000. "Outdoor Physical Education Facilities." *School Planning and Management,* 39 (December): 50.

Steffen, Jeff, & Scott Hall. 1997. "Outdoor Sport and Adventure Activity Areas." In Marcia Walker & David Stotlar (Eds.), *Sport Facility Management.* Boston: Jones and Bartlett, pp. 183–90.

Steinbach, Paul. 2002. "Great Planes." *Athletic Business* (July): 79–86.

Walker, Marcia. 1997. "The Planning, Design, Construction, and Management Process." In Marcia Walker & David Stotlar (Eds.), *Sport Facility Management.* Boston: Jones and Bartlett, pp. 11–12.

Walker, Marcia, & T. Seidler. 1993. *Sports Equipment Management.* Boston: Jones and Bartlett.

Online Sources

ADA Home Page
http://www.usdoj.gov/crt/ada/adahom1.htm
This site provides information and technical assistance on requirements of the Americans with Disabilities Act for facility construction and renovation.

Athletic Business
http://www.athleticbusiness.com/facilitybuilder/specifications.asp
This site provides an online "Facility Builder" for subscribers. Select a sport, and court specifications are available as PDF files (requires Adobe Acrobat software).

Council on Facilities and Equipment

http://isu.indstate.edu/tsawyer/cfe.htm

This site focuses on concerns relating to facilities and equipment in relationship to physical activity. CFE develops policies, standards, guidelines, and innovations to ensure safe and effective health, physical education, recreation, dance, sport, and fitness facilities and equipment. CFE is affiliated with AAHPERD.

Occupational Safety and Health Administration (OSHA)

http://www.osha.gov/

OSHA provides information on how to construct and maintain facilities that are free from serious recognized hazards and comply with occupational safety and health standards.

Science Made Simple

http://www.sciencemadesimple.net/area.html

This site facilitates conversions among different units of measurement (e.g., from square feet to acres).

Sports Turf Managers Association

http://www.sportsturfmanager.com/my/shared/home.jsp

The association represents the sports turf industry. They provide research, training, education and other services.

11

The Evaluation Process

Schools cannot function effectively without evaluation. Everyone and everything connected with the educational process is evaluated at some point: students, teachers, coaches, administrators, buildings, basketball teams and basketballs, even the food service in the cafeteria. Evaluations are conducted for a variety of reasons including accreditation of programs, adequacy of facilities, efficiency of administrative procedures, allocation of resources, identifying instructional deficiencies, assessing student improvement, and demonstrating program effectiveness. Specific evaluations are made of program content, teacher effectiveness, and student performance and progress. This chapter covers four major areas of evaluation: *students, professional staff, programs*, and *facilities*.

Evaluation should be tied to educational goals and to how well those goals are being met. The primary purpose of evaluation is to identify strengths and weaknesses. However, evaluation that detects weaknesses serves little purpose unless the information is utilized toward positive change. The ultimate reason for evaluation is improvement. Without the intent to improve, why evaluate?

The term *evaluation* is inclusive of exact measurements, subjective assessments, and considered judgments. Evaluation can be based on test results, information secured from observations, interviews, questionnaires, scorecards, and other measures. It implies assessment against defined standards or goals by utilizing criteria and rubrics. (A *standard* represents a level of acceptance; a *criterion* determines the basis for judgment; *rubrics* [categories or classifications] are scoring guides utilized in subjective assessments.) Evaluation may be norm referenced, which incorporates comparison—with others schools, students, buildings, and so on—or may be criterion referenced, which implies meeting set standards. The latter approach is gaining in popularity and is typical of certification processes. Either way, the result is a determination of the status or worth of that which is being evaluated.

Evaluation should be approached with a constructive and positive attitude and with a sincere desire to accurately appraise and bring about constructive change. It is the responsibility of the administrator to justify and explain the purposes, goals, and uses of evaluation to stakeholders in the school.

ADMINISTRATIVE CONSIDERATIONS

Administrators should devote adequate time and energy to thinking about how to assess programs, students, staff, and facilities. Assessment choices are complex, and they entail tradeoffs in quality and feasibility. Many practical considerations constrain choices among assessment alternatives. The following factors must be taken into consideration before initiating an evaluation model.

Cost and Time. Two of the most precious resources in education are funds and time. A school needs to commit a considerable amount of staff time to produce new assessments or generate funds to purchase measurement instruments. To develop high-quality assessments, teachers may need specialized training in assessment design and use. Administering and scoring also demand a commitment of resources. Whether a school asks its teachers to administer and score the assessment tools or the job is contracted out, the costs in time or money have to be determined.

Acceptability. To have practical value, assessments must provide information that is useful and credible to those who apply the results: students, teachers, parents, and administrators. Assessments that require a great deal of time may meet resistance from teachers, who are often overburdened.

Purposes. Information gained from assessments is used by schools for various purposes: to meet accreditation standards, to justify school construction projects, to qualify for federal funds, to increase the tax base, to certify mastery of skills, and to evaluate program performance. The purposes to which assessment is employed can result in worrisome consequences, such as teachers "teaching to the test" if they feel undue pressure to ensure that students perform well.

Embedded versus stand-alone tasks. Traditional assessment methods such as objective tests involve testing students separately from learning activities. Many alternative assessments (see below) incorporate assessment directly into the learning activities as "embedded" assessment. The administrator must determine which approach is preferable.

Standardization versus flexibility. Most traditional assessments have a high degree of standardization in testing conditions, content, and scoring. Such standardization facilitates comparing the performance of different groups or programs. Alternative assessment allows a greater degree of flexibility in what is assessed, how it is assessed, and how it is scored, but it limits comparisons. Awareness of the impact of these issues is important in developing effective assessments (Rahn, Stecher, Goodman, & Alt, 1997).

Carefully considering all options and producing a clear statement of purpose will help optimize the evaluation model. Once these considerations have been resolved, the administrator can address the specific methods and procedures of evaluation. The three steps in evaluation are: (1) collection of useful information about the present status, (2) objective comparison of the information against norms or set standards, and (3) implementation of an effective plan to close the gap between the present status and that which is desired.

The two basic forms of evaluation are formative and summative. *Formative evaluation* is done "along the way" to identify weaknesses and problems so that needed adjustments can be determined and implemented. *Summative evaluation* is for used the purpose of making a final decision as to whether the person or program being evaluated meets the standards. Formative evaluation carries less risk to personnel and programs because its end is simply to bring about needed change. Summative evaluation, on the other hand, carries with it the risk of being judged a success or a failure. Its purpose may include a determination of whether to accept or continue a program, services of an employee, or other factors under consideration.

EVALUATION OF STUDENTS

Given the purpose for which schools exist, one aspect of evaluation transcends all others: the evaluation of students. Because student achievement is the ultimate goal of the educational effort, educators are vitally concerned with the kind and amount of student development resulting from participation in instructional programs. Student performance incorporates all the domains of learning: cognitive, psychomotor, and affective. (In addition, physical education evaluates organic development.)

Schools should adopt a workable evaluation model. A general model for instruction is: pretest, instruction, posttest, and feedback. Yonkers Public Schools in New York utilizes what it labels the Diagnostic/Prescriptive Process as their evaluation model (Yonkers Public Schools, 1999). The process includes the following steps:

- Assess student's level of knowledge and/or skill development.
- Analyze assessment data.
- Prescribe specific instruction based on the data.
- Provide targeted teacher support and assistance.
- Reassess student achievement to determine mastery.

The more information a teacher has about students' interests, abilities, and needs, the better equipped the teacher will be to design an effective program. Much of the necessary information can be gathered with the use of tests and measurements. Some of the ways that measurements are used include the following (Jensen & Hirst, 1980).

Diagnosis. Measurement is necessary to diagnose the level of, and differences in, abilities, interests, and needs of students in order to plan and conduct effective programs. Objective knowledge about specific student deficiencies is essential to provide both remedial and accelerated programs.

Classification. Sometimes it is an advantage to classify students into homogeneous or heterogeneous groups for a particular type of instruction, for competition, or for recreational experiences. Such classification can be based on the results of appropriate measurements.

Achievement. Objective measures and accurate records of student achievement and progress form the basis for the selection of program content and for the assignment of grades or marks, as well as for the advancement of students.

Improvement. Information obtained from measurements serves as a basis for guiding students into appropriate experiences to foster higher levels of achievement. Accurate measurements help to determine the success of students, and to determine whether students are ready to progress to the next higher level.

Selection of Tests and Measurements

The selection of tests and measuring instruments is based on established criteria. One criterion is *objectivity*. A measure is objective to the extent that it will produce the same results when conducted by different people under similar conditions (i.e., when the *subjectivity* of the test administrator is eliminated). Subjective measures are influenced by judgment, opinion, or bias. Teachers should strive for a high degree of objectivity in measurement but still should realize that not all important traits can be measured objectively. For example, a teacher's evaluation of a student's character includes some degree of objectivity but also involves a large amount of subjective judgment.

Tests and measuring instruments should also demonstrate an acceptable degree of validity and reliability. *Validity* is the extent to which the test actually measures what it claims to measure. While some measures have face validity (e.g., height), others (e.g., wall volley tests) must be judged against an external criterion (e.g., game performance). *Reliability* is the degree to which a test is internally stable and produces consistent scores over time. For example, if a student receives a similar score on a test given on consecutive days, this implies reliability. Tests and testing procedures also should be *standardized*. This means that test directions and conditions under which the test is conducted are always the same. Without standardization test results cannot be compared. (See "What Factors May Influence Test Results?" in the section later in the chapter on evaluation of athletes.)

Another administrative consideration is whether to employ group tests or individual tests. Group tests dominate, as they are more time efficient and cost-effective. They are justified with roughly homogeneous groups of students in mainstream classroom settings. Individual tests are often necessary in diagnosing or evaluating disabled students in line with individual education programs (IEPs). Teachers also should consider the educational applications of the evaluation experience. The test experience itself should be instructive and motivational.

Finally, the methods of evaluation should be harmonious with and complementary to course content. This is what is meant by *content validity*. More broadly, evaluation should reflect the major objectives of physical education. The following four objectives have been widely accepted:

1. Organic development (health-related fitness)

2. Motor skills (psychomotor development)

3. Knowledge and understandings (cognitive development)

4. Personal, social, emotional development—values, attitudes, social skills (affective development)

Several kinds of tests and measurements apply when evaluating these objectives. Specific examples appear in the measurement books listed at the end of this chapter. The cognitive objective implies the use of written tests and quizzes. A physical education teaching station should be designed to accommodate this type of testing.

Administration in an Instructional Setting

The following factors should be considered in approving forms of evaluation in an instructional setting (Miller, 2002):

- *Cost*—The expense in administering.
- *Time*—Requires no more than 10% of instructional time.
- *Ease of administration*—Instructions that are easy to follow; least amount of training required of test administrators.
- *Scoring*—Self-scored versus externally scored or computer scored.

- *Confidentiality*—Ethical and legal requirement.
- *Norms*—Availability of district, state, national norms.
- *Feasibility*—Facilities, equipment, and personnel available to accommodate testing.
- *Appropriateness*—Maturity of test takers, relevance to course objectives.
- *Safety*—Relative to testing area, testing personnel, capacity of test takers.

Careful attention should be given to the administrative procedures, as poorly administered tests can produce results that are misleading and useless to both the teacher and the students and can waste valuable class time. The following guidelines for test administrators will insure reliable results (Tritschler, 2000):

Preparation
1. Become familiar with the assessment instrument.
2. Plan the physical layout for the testing site.
3. Plan how you are going to organize the test takers for testing.
4. Plan for using assistants.
5. Plan the scoring and record keeping procedures. Utilize computers.
6. Collect and inspect all testing equipment (e.g., pencils, score sheets, timers).
7. Make arrangements for make-up tests.
8. Rehearse the explanation, demonstration, and testing procedures.

Data Collection
1. Explain the reasons for assessment to the test takers.
2. Administer the test in a way that is consistent with the intended use of results.
3. Spot check for accuracy in administration and scoring.

Posttest Procedures
1. Dismantle the test area.
2. Check recorded scores.
3. Tabulate results and perform statistical analysis.
4. Report results to students, parents, and other stakeholders.
5. File all pertinent information.

MARKING (GRADING) STUDENTS

Typically, the system of marking is determined at the school district level. Districts may allow variable marking systems across subject areas. In some situations, physical education is marked differently than other subjects. Regardless of the marking system in use, every teacher should develop effective methods of determining students' marks or grades. Marks may determine whether the student is admitted into college, wins a scholarship, is eligible for lower automobile insurance rates, or will secure a good position of employment in the future. To the students and their parents, marks represent an indication of success in school. Because of the importance placed on them, an accurate system for determining and reporting marks is imperative.

Purposes of Marking

Marks serve different purposes for students, teachers and guidance counselors, parents, and school administrators. A mark should indicate whether the student has met the standards on which the mark is based and has performed according to the requirements and expectations. A mark must be accurate and justifiable and have a strong basis, or it loses its value. It may even destroy the student's interest and incentive and result in a negative attitude toward the subject and the teacher.

Both teachers and guidance personnel may use marks for guidance and counseling. If accurately determined, marks can help the teacher predict how successful a student might be in certain pursuits and may also help the teacher evaluate the effectiveness of the teaching. Parents are usually eager to learn of their children's success in school and desire to know the reason for the lack of success. Marks that are considered accurate and reliable often result in guidance from home that helps overcome deficiencies.

Marks also are essential to administrators, who interpret them as symbols of progress and indicators of accomplishment. Since marks become permanent records of achievement, they are also used as a basis for promotion and scholastic honor awards. It is essential, then, that marks be established on a solid and fair basis, and that they be accurately reported and interpreted.

Criteria for Useful Marks

- Marks should be valid, reliable, objective, and based on standards or norms.

- Marks should be capable of *clear interpretation* with regard to what they signify. If the significance of an A is not evident in terms of quality and quantity of achievement, then the grade lacks meaning.

- Marking should involve an element of economy of the teacher's time and effort. Since marking is just one of the teacher's many tasks, only a reasonable amount of time can be justified for it.

- Marks should be *timely*. Final marks are recorded only at the end of semesters; but for purposes of incentive, students should be frequently informed of their progress and present standing.

Methods of Reporting Marks

Some of the several methods of reporting student achievement are as follows:

Five-letter. This is the most frequently used method. The teacher determines the criteria for each grade: A, B, C, D, F. Each student is given a letter mark based on performance in the class. The mark is reported, and sometimes an interpretation accompanies the mark, telling what it means in terms of level of performance.

Pass or fail. In this system, a standard of acceptability is established. If the performance is up to the standard, the student passes; if performance is below the standard, the student fails. Sometimes the terms *satisfactory* or *unsatisfactory* are used in place of pass or fail.

Percent. With this method the student is given a percentage score as a mark, with 100 percent representing excellent performance. Percentages can be converted to letter marks in order to make them more meaningful. One conventional set of standards is that 90 to 100% = A: 80 to 89% = B; 70 to 79% = C; 60 to 69% = D; below 60% = F.

Accumulative points. In this method, the student is awarded points for performing various activities throughout the unit of instruction. Each student's class standing is determined and reported on the basis of accumulated points. Points may be converted to marks.

Contract. Through agreement between the student and the teacher it is decided which projects the student will complete, and this agreement becomes a contract. The particular conditions of the contract define a predetermined grade that will be received if the student completes all the conditions. The completed contract should reflect what was learned.

Descriptive statement. This method consists of a short paragraph explaining the student's status and stating the student's strengths and weaknesses in the subject. The statement is reviewed by the teacher, student, and parent(s) or guardian(s) in reaching an understanding of the student's progress. Some schools include with the regular report card a sheet showing comparative physical fitness test scores.

Application of Results

The results of tests and measurement should be interpreted and used to the advantage of the students and the program. Teachers should be aware that the exactness and correctness with which they administer assessments influence their reliability. Test results that might have been valuable become useless when they are improperly administered and incorrectly interpreted.

Measurement results are used to motivate students toward achievement, evaluate the effectiveness of the educational program, evaluate student progress, and determine student grades.

Students generally want to be more knowledgeable, more fit, and more skillful. They are motivated by goals and want to demonstrate their achievements and be recognized for them. Furthermore, they want to see evidence that they are progressing, and they become stimulated by such evidence. Both written and motor-performance tests can assess achievement levels. The results of such tests provide students with evidence of progress or lack of progress in attaining these goals. Likewise, the lack of this kind of information can stifle motivation.

Student progress is the best indication of a successful program. To determine progress, the goals toward which the students should strive must be clearly defined, and sound methods for evaluating progress should be employed. One means a teacher has for determining progress is periodic measurement of student achievement. Measures of achievement in gymnastics, dance, and sports help to assess the success of these phases of the program. When achievement is low and progress is slow, the teacher should look seriously at the program content and the methods of instruction.

Assessing Students' Fitness

Fitness tests have moved away from an orientation to athletic fitness toward health-related fitness. In 1993, four national youth fitness tests were in use: AAHPERD Physical Best, Prudential FITNESSGRAM,® President's Challenge, and Chrysler Fund—AAU. The first two tests were nearly identical. Subsequently, in 1994 AAHPERD officially adopted the Prudential FITNESSGRAM in an effort to create a single youth fitness program for the nation's schools (figure 11-1). FITNESSGRAM incorporated the educational materials of AAHPERD's Physical Best program. A FITNESSGRAM Test Kit is available for purchase by schools through Human Kinetics Publishers <http://www.humankinetics.com>. The kit consists of software, a test manual, and accessories required to do fitness assessments. An individual FITNESS-GRAM and accompanying ACTIVITYGRAM can be printed out for each student, and interactive resources allow students to assess their activity regimen. Test administrators can refer to current measurement textbooks in physical education for critical comparisons of the available tests. (See "References and Recommended Readings" at the end of this chapter.)

Skinfold measurement to assess lean body mass is an important but sensitive item on health-related fitness batteries. Some parents or students may object to physical education or health education teachers taking these measurements. The Body Mass Index (BMI) can be substituted in this instance (however, skinfold assessments are the most valid measure of body composition). In either case, the results of the tests should be disseminated to students and parents, especially if the results suggest a student is overweight. Instructors should not be allowed to administer this procedure unless they have received adequate training. When administering skinfold assessments, the following guidelines should be followed to avoid controversy (Baumgartner & Jackson, 1999):

1. The triceps, subscapula, and calf are the recommended measurement sites for school children. The subscapular measurement requires raising the shirt or blouse. Sensitivity should be employed in this procedure to avoid embarrassing female and overweight students.

2. It is preferable that women instructors do skinfold measurements on female students, or act as observers if male instructors are the only qualified assessors.

3. Skinfold assessments are best administered in a private area. Partitions or curtains should be utilized in a gymnasium.

4. Verbally inform each student of the nature and purpose of the procedure before taking measurements.

Some fitness tests, such as the FITNESSGRAM, have adopted computerized evaluation and reporting systems, which makes them more convenient and economical to administer to large numbers of students. A printout sheet is available for each student tested. A copy of the sheet can be furnished to the parents to serve as a basis for discussion with the student or the instructor. The computerized information can be conveniently stored for long-term use for making comparisons, observing trends, and establishing norms. The computer can list the recipients of FITNESSGRAM awards. It can also identify those who score low and would benefit form remedial activities.

Alternate Assessment Options

The 1990s witnessed a new approach to student evaluation under the label of *authentic assessment*. In response, the National Association for Sport and Physical Education published *Moving in the Future: National Standards for Physical Education* (1995), which discusses alternate assessment options such as student projects, logs and journals, teacher and peer observations, self-assessment, group projects, and portfolios. This alternate approach was, in part, a response to criticism of evaluation techniques that relied heavily on traditional tests. Authentic assessment appears to have some definite advantages and also several drawbacks. The nature of physical education facilitates teacher observations of learning. However, large classes create logistical problems in adopting this approach. Authentic assessment should not be an "either/or" deci-

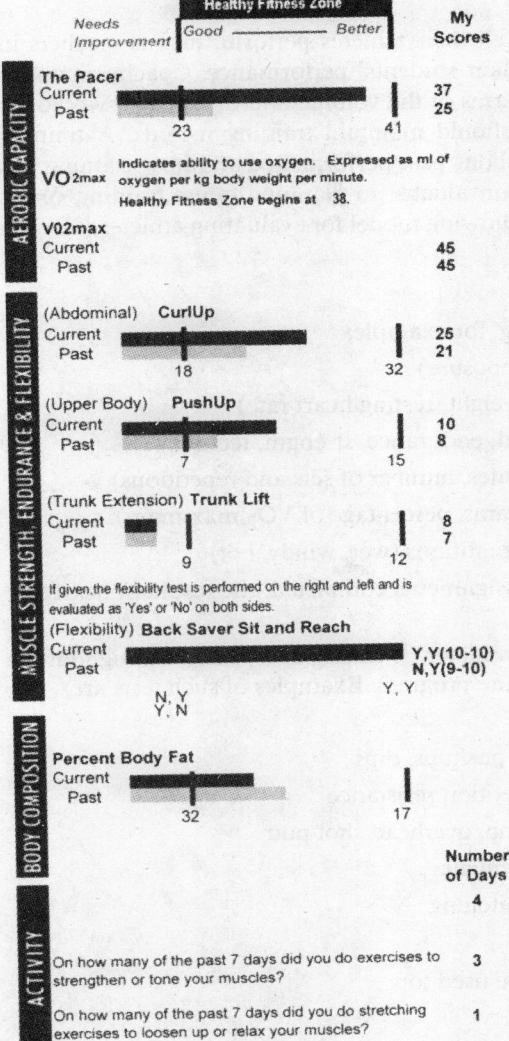

FITNESSGRAM®

Janice Jogger
Grade: 6   Age: 12
Franklin Middle School
Instructor: Marie Merritt

	Test Date	Height	Weight
Current	05/15/02	5'05"	122
Past	05/14/01	5'04"	118

Healthy Fitness Zone
Needs Improvement | Good ——— Better | My Scores

MESSAGES

AEROBIC CAPACITY

The Pacer
Current 37
Past 25
23 · 41

VO2max Indicates ability to use oxygen. Expressed as ml of oxygen per kg body weight per minute. Healthy Fitness Zone begins at 38.

VO2max
Current 45
Past 45

Janice, your scores on 5 of 6 test items were in or above the Healthy Fitness Zone. You are also doing strength exercises. However, you need to play active games, sports or other activities at least five days a week and do some flexibility exercises.

Although your aerobic capacity score is in the Healthy Fitness Zone now, you are not doing enough physical activity. You should try to play very actively at least 60 minutes at least five days each week to look and feel good.

MUSCLE STRENGTH, ENDURANCE & FLEXIBILITY

(Abdominal) CurlUp
Current 25
Past 21
18 · 32

(Upper Body) PushUp
Current 10
Past 8
7 · 15

(Trunk Extension) Trunk Lift
Current 8
Past 7
9 · 12

If given, the flexibility test is performed on the right and left and is evaluated as 'Yes' or 'No' on both sides.

(Flexibility) Back Saver Sit and Reach
Current Y,Y(10-10)
Past N,Y(9-10)
N, Y
Y, N   Y, Y

Your trunk extension may be improved by including trunk lifts in your strength activities. You may need to do more trunk lifts or be sure that you do them 3 to 5 days each week.

Your abdominal and upper body strength are both in the Healthy Fitness Zone. To maintain your fitness, be sure that your strength training activities include exercises for each of these areas. Abdominal exercises should be done 3 to 5 days each week. Strength activities for other parts of your body should be done 2 to 3 days each week.

Janice, your flexibility is in the Healthy Fitness Zone. To maintain your flexibility you should begin stretching slowly 3 or 4 days each week, holding the stretch 20 - 30 seconds. Don't forget that you need to stretch all areas of the body.

BODY COMPOSITION

Percent Body Fat
Current 17
Past 32

Janice, your body composition is in the Healthy Fitness Zone. If you will be active most days each week, it may help to maintain your level of body composition. You should also eat a healthy diet including more fruits and vegetables and fewer fats and sugars.

ACTIVITY

Number of Days 4

On how many of the past 7 days did you do exercises to strengthen or tone your muscles? 3

On how many of the past 7 days did you do stretching exercises to loosen up or relax your muscles? 1

To be healthy and fit it is important to do some physical activity almost every day. Aerobic exercise is good for your heart and body composition. Strength and flexibility exercises are good for your muscles and joints.

Good job, you are doing some physical activity in each of these areas. Additional vigorous activity would help to promote higher levels of fitness.

©The Cooper Institute for Aerobics Research

Figure 11–1.   Illustration of the FITNESSGRAM© (reprinted with permission from the Cooper Institute For Aerobics Research, Dallas, Texas)

sion. Administrators should allow teachers some discretion in deciding whether to use this approach. However, accreditation associations are prone to ask for evidence of authentic assessment, such as student portfolios (Baumgartner & Jackson, 1999).

Evaluation of Athletes

Coaches are interested in how well their athletes perform, just as teachers in other areas of study are interested in their students' performance. Coaches monitor practice sessions to assess progress in terms of the volume and intensity of workouts. Both coaches and individual athletes should maintain training records. A training diary can provide useful information about past performance and how training has progressed. Information of this kind is invaluable in planning future training. *Sports Coach*, an online magazine, offers the following model for evaluating athletes ("Evaluation," 2002).

What Should be Recorded?
A. Day-to-day information from training, for example:
- State of the athlete (health, composure)
- Physiological data (e.g., body weight, resting heart rate)
- Focus of the training unit (speed, endurance, strength, technique)
- The training load (number of miles, number of sets and repetitions)
- The training intensity (in kilograms, percentage of VO_2 maximum)
- The prevailing environmental conditions (wet, windy, hot)
- The response to training (the assignments completed, the resultant heart rate recovery, fatigue)

B. Information that measures status of athletes. If testing is repeated throughout the program, it can then be used to measure progress. Examples of such tests are:
- Time trials—speed, endurance
- Muscular endurance—chinups, pushups, dips
- Maximum strength—single repetition resistance
- Explosive strength—vertical jump, overhead shot putt
- Mobility—agility, range of movement
- Event-specific skills—passing, catching

What are the Benefits of Evaluation?
The results from performance tests can be used to:
- Predict future performance
- Indicate weaknesses
- Measure improvement
- Enable the coach to assess the success of the training program
- Place the athlete in appropriate training group
- Motivate the athlete

What Factors may Influence Test Results?
The following factors may have an impact on the results of a physical test:
- The ambient temperature and humidity

- The amount of sleep the athlete had prior to testing
- The athlete's emotional state
- Medication the athlete may be taking
- The time of day
- The athlete's caffeine intake
- The amount of time elapsed since the athlete's last meal
- The test environment

Testing athletes offers additional benefits in that it breaks up and adds variety to practice sessions. Tests can be used to satisfy the athlete's sense of competitiveness. Since they demand maximum effort, they are useful as a training device in their own right.

Postcompetition Evaluation

Following competition it is important that the coaches and athletes get together as soon as possible in order to evaluate the athletes' and/or team's performance. Many coaches use game films for this purpose. Performance may be compared against the precompetition strategy or game plan. Evaluations are provided during a team meeting or individually. Evaluation forms can be useful to the athlete and the coach when conducting an individual performance review. It is important that the coach discusses athletes' evaluations with them in a constructive and nonpunitive manner, with the goal of improved performance.

EVALUATION OF PROFESSIONAL STAFF

Staff accountability is as much a part of administration as financial accountability. School administrators should conduct evaluation of both teaching and nonteaching personnel.

Evaluation of Teachers

The evaluation of teaching effectiveness is a very important responsibility, considering the mission of schools. The most direct measure of good teaching is effective learning; however, some important aspects of learning are difficult to identify and even more difficult to measure. Because of these limitations, schools usually resort to the subjective evaluations by those who have made firsthand observations of teaching. Because this approach lacks objectivity, additional methods should be utilized. Methods to consider are evaluation by students, self-evaluation, evaluation by the supervisor or administrator, and peer evaluation. It is important for the teachers to have confidence in the methods used.

In evaluating teachers, one should take into account the source of information; its validity, objectivity, and reliability; and how well the information applies. Following are five basic principles that should be applied in the evaluation of teachers: (1) it should enhance the growth and development of the teacher as an individual; (2) its main focus should be on teaching effectiveness; (3) evaluation should include a discussion of goals for the future and how they can be achieved; (4) it should involve the person in penetrating self-analysis; (5) some aspects of evaluation should be formalized with pre- and post-observation conferences (see the section on supervision in chapter 6), and these should be carried out in a nonthreatening manner.

Evaluations by Students

Students should play a major role in evaluating teachers, as they are the group most directly affected by the quality of teaching and are firsthand observers in the classroom on a daily basis. They see teachers at their best, their worst, and at their typical level of performance. Students can be involved in both formative and summative evaluation of teachers. Formative evaluation should take place around mid-semester so that the teachers can improve based on feedback, and students can benefit from these changes. Summative evaluations are administered near the end of the semester, but not on days when students are being tested. Evaluation methods range from formal to informal. An example of an informal method would be a student journal. Formal methods most often consist of rating scales. (Figures 11-2 and 11-3 are sample evaluation forms that allow students to rate their instructors.)

Rating instruments should be standardized, valid, and reliable. Because internally developed (in-house) instruments usually haven't been subjected to these psychometric procedures, most experts recommend that schools adopt one of the standardized, published instruments available (see Arreola, 2000). Schools also should standardize the procedures for administering the instruments. Finally, both teachers and students should be apprised of the intended uses of the evaluations (Lacy, 1999).

Evaluation of teachers by students is a method used more frequently at the college level than in the public schools. If conducted properly, evaluations by students can be useful at the secondary level as well. Student evaluations of teachers should not be relied on exclusively, as they do have well-established limitations (e.g., students enrolled in elective courses rate teachers higher than those taking required courses).

Self-Evaluation

Self-evaluation by teachers should also be used. This approach is valuable to determine whether the instructor's perceptions are consistent with those of other observers. Differences in perceptions can be used to target blind spots the instructor has about his or her teaching. Many teachers' perceptions tend to confirm their own theories of instruction, which are sometimes anecdotal and faulty. Self-evaluation that results in improvement in the classroom requires reflection, usually indicating the assessment of one's own work and passing judgment on it. This approach makes a great deal of sense in educational settings, as teachers are generally motivated by feelings of self-determined competence.

Self-evaluation should be a part of formative (but not summative) evaluation of teachers. A variety of techniques are recommended for self-evaluation: videotapes of teaching, standard forms, journals, and portfolios are some that have been used with success. Figure 11-4 is a sample form that can be used in self-evaluation. It can also be used by a colleague or administrator. The administrator's role is to help facilitate the self-evaluation process and to make resources available. This method should be conducted only with the cooperation of the instructor, for obvious reasons. Some teachers find techniques like videotaping to be threatening. Instructors should be apprised of the uses that will be made of self-evaluations (Lacy, 1999).

Evaluation by Supervisors or Administrators

The supervisor or administrator should approach the evaluation process as a resource person and a facilitator, as well as an evaluator. This approach will help to reduce the apprehension of the teacher being evaluated. The process should provide the teacher with a setting in which to talk freely and seek help and solutions. Evaluation by administrators is inherently an anxiety-producing experience for teachers, as

TEACHER'S NAME _____ DATE _____

COURSE NUMBER AND SECTION _____

INSTRUCTOR EVALUATION

	1	2	3	4	5	6	7
1. Has excellent knowledge of the subject matter.							
2. Is enthusiastic about the subject.							
3. Is well prepared for each class.							
4. Makes good use of class time.							
5. Gives clear examples and explanations.							
6. Makes helpful evaluations of my work (e.g., papers, exams).							
7. Clearly explains difficult concepts, ideas, or theories.							
8. Responds respectfully to student questions and viewpoints.							
9. Is genuinely interested in helping me understand the subject.							
10. Is available to students during regular and reasonable office hours.							
11. Motivates me by his/her example to want to learn about the subject.							
12. Has produced new knowledge, skills, and awareness in me.							
13. Starts/dismisses classes at scheduled times.							
14. Seldom misses class.							

15. Overall rating: ___Poor ___Fair ___Good ___Excellent

Code: (Place an X in the appropriate square)

1 = Strongly disagree	5 = Agree
2 = Disagree	6 = Strongly agree
3 = Somewhat disagree	7 = Very strongly agree
4 = Somewhat agree	

Figure 11–2.   A concise but well designed teacher evaluation form to be completed by students (courtesy of Brigham Young University)

TEACHER'S NAME _____ DATE _____

COURSE NUMBER AND SECTION _____

STUDENT EVALUATION OF TEACHER*

Do you feel that the course objectives were clearly stated?   Yes___ No___

Do you feel that the course objectives were within your abilities?   Yes___ No___

Were the tests and evaluations directed toward the course objectives?   Yes__ No__

How well do you feel the instructor demonstrates those skills he or she teaches?

Very Well					Needs Improvement				
1	2	3	4	5	6	7	8	9	10

How well were you able to learn those skills which the instructor taught?

Very Well					Needs Improvement				
1	2	3	4	5	6	7	8	9	10

How well do you feel the instructor evaluates your skills?

Very Well					Needs Improvement				
1	2	3	4	5	6	7	8	9	10

Do you have the necessary equipment to do an effective job of learning?

(comment)

Which of the following is expected of you?

___Memorization and recall.

___Making a judgment about something, using given information.

___Understanding the meaning of ideas or concepts.

___Putting information together in an unusual way to solve problems.

___Examining a complexity and breaking it down into parts.

___Using or applying information in a new situation.

Does the instructor show an interest in you?   ___Always ___Usually ___Seldom

Does the instructor encourage you to ask questions and to state your views?

___Always ___Usually ___Seldom

Which of the following, if any, needs improvement?

___Speech   ___Sense of humor

___Planning for class sessions   ___Tolerance for students' ideas & views

___Willingness to help students   ___Ability to inspire students

___Enthusiasm   Other: _____

Figure 11–3.   An extensive form with which students can evaluate teachers (courtesy of Laramie County Community College, Cheyenne, WY)

STUDENT EVALUATION *(Continued)*

Does the instructor encourage new ideas and concepts? Yes___ No___

How would you rate your instructor's communications skills—questioning, answering, discussing?

___Outstanding ___Very Good ___Good ___Adequate ___Poor

	Exciting									Boring
Rank the presentations	1	2	3	4	5	6	7	8	9	10

	Outstanding									Poor
Rank the instructional planning and organization	1	2	3	4	5	6	7	8	9	10

	Outstanding									Poor
Rank the effectiveness of the instructional methods	1	2	3	4	5	6	7	8	9	10

Does the instructor make appropriate use of instructional materials such as textbook, films, magazines, manuals, etc.? Yes___ No___
(comment)

Does the instructor possess an adequate knowledge of course content? Yes___ No___
(comment)

Overall, do you feel the course was good? Yes___ No___
(comment)

What aspects of the course do you feel were of most value to you?

_____

_____

	Super									Rotten
Rank the instructor	1	2	3	4	5	6	7	8	9	10

Is the instructor available to help you? ___Always ___Usually ___Seldom

What do other students enrolled in this course think of it?

___Like   ___Don't Talk About It

___Dislike   ___Afraid to Say

___Don't Care   ___Don't Care Either Way

Figure 11–3 *(continued)*

their salaries, promotions, and/or job security may depend on it. Where merit pay has been instituted, the salary increases awarded must be tied to objective evaluations. Moreover, if a teacher challenges reasons for his or her dismissal, the school must be able to show that they have evaluated the teacher regularly and fairly. Teachers have a right to be evaluated and apprised of their status. Therefore, evaluation of teachers by

PERSONNEL EVALUATION

Name _____ School _____ Date _____

Years of Teaching Experience _____

Years in Alpine School District _____

Grade or Subject Area _____

Conference Held: Date _____   Date _____

This checklist is used for the purpose of improving instruction and/or evaluation for future employment. This basic set of criteria is the official reference for observation and evaluation. Teachers are encouraged to use this form as a self-evaluation and to discuss these criteria with their administrators.

Final evaluation is to be completed by the principal and reviewed by the teacher. Teacher's signature is only an acknowledgement of having reviewed the final report.

I. TEACHING TECHNIQUES	Superior	Satisfactory	Needs Impr.
Is fair and impartial			
Demonstrates creativity in planning, preparation and implementation			
Varies materials and techniques			
Demonstrates a knowledge of subject matter			
Meets the individual needs of students			
Develops critical thinking			
Uses good English oral and written			
Demonstrates skill in motivating students			

III. PERSONAL ATTRIBUTES	Superior	Satisfactory	Needs Impr.
Shows poise, self-control, tact			
Demonstrates a wholesome sense of humor			
Maintains a professional appearance			
Maintains regular attendance, health, and vitality			
Demonstrates positive attitude toward teaching			
Demonstrates leadership			
Has a good self-concept			
Shows dependability			

Figure 11–4.   Example of a self-evaluation form to be used by a teacher. The form can also be used by another person, peer, supervisor, or administrator (courtesy of Alpine School District) (Continued on next page)

Helps students understand an appraise their own work							
II. SCHOOL MANAGEMENT							
Maintains classroom control							
Shows concern for care and appearance of room and equipment							
Completes all school assignments							
Uses and maintains appropriate school records							
Establishes a positive learning environment							
Shows evidence of daily planning and development of long-range objectives							
IV. PROFESSIONAL AND SOCIAL ATTITUDES							
Observes professional ethics							
Adheres to school and district regulations							
Accepts suggestions for improvement							
Shows evidence of professional growth							
Demonstrates genuine liking for teaching and students							
Views own assignment in relation to total school program							
Develops positive parental relationships							
V. TOTAL FINAL EVALUATION							

Teacher's Summary Comment: _____

_____

_____

Teacher's Signature _____

Principal's Summary Comment: _____

_____

_____

Principal's Signature _____

Figure 11–4. *(Continued)*

their superiors must be viewed as an important responsibility. Administrators must act in an impartial manner and carry out evaluations in a way that promotes improvement in instruction.

The department chair may play a limited role in firsthand observation of teachers. Evaluation of classroom instruction is more normally carried out by supervisors (described at some length in the section on supervision in chapter 6). Evaluation of teachers by department chairs or school principals covers more ground than what goes on in the classroom. Department chairs have a primary role in linking evaluations to professional development and enhancement. Some departments utilize personnel committees to assist in conducting evaluations. Standardized forms have been developed for evaluating teachers, and such forms are utilized by most schools. Figure 11-5 provides one example of a form that can be used or adapted for use by the administrator or supervisor in evaluating a teacher's classroom effectiveness.

Peer Evaluation

Teachers' colleagues have a different perspective than administrators of how well an individual is functioning on the job. Indeed, studies indicate that a fairly low correlation exists between supervisor evaluations and peer evaluations. This implies that the two forms of evaluation are not redundant but provide complementary information. Peer evaluation has both advantages and disadvantages. The advantages are that colleagues are generally mature and responsible individuals who are familiar with the teaching process, the subject matter (if they are from the same discipline), and departmental goals. The disadvantages result from familiarity and peer pressure. In some cases, the relationship between two colleagues is too positive or too negative to allow the evaluator to be objective. The evaluator may consciously or unconsciously desire to do the other person undue good or harm. These biases should be taken into account.

Peers can be employed in both formative and summative evaluations. Procedures include classroom observations, review of instructional materials, and critiques of course design or course syllabi. Classroom visitations are not recommended for summative evaluations (because of poor interrater reliability) but can be a valuable type of formative assessment of new teachers to assist in their professional development. Reliability improves with the number of visits by a peer evaluator, but this can be expensive in terms of faculty time and effort. Sometimes, schools employ observation teams made up of colleagues.

Finally, peer evaluation often causes apprehension among both evaluators and those being evaluated, because colleagues usually do not like to evaluate each other. When the circumstances are favorable, however, evaluation by colleagues provides an additional valuable dimension to the total evaluation process (Lacy, 1999). Figure 11-5 provides an illustration of a form that can be used or adapted for evaluation by a colleague.

Evaluation Based on Student Progress

A sound argument can be made that student progress is the only meaningful measure of teacher effectiveness. If all forms of student achievement could be measured with sufficient accuracy and consistency, perhaps there would be little use for other forms of teacher evaluation. The most valid and appropriate way to measure learning is by analyzing scores on standardized tests. The scores of students taught by one teacher can be compared with those taught by others. In addition, the record of students' scores over several years can give an indication of improved teaching. However, a caveat is in order. The students in one teacher's class may be significantly different from those in another class prior to and during instruction. Thus, the difference in test scores could be attributed to factors other than teaching quality. The same warning applies when comparing standardized test scores of students from different schools.

Unsolicited Information

People who work closely with each other have a good feel for how well the other person is performing. The day-to-day observations and impressions mount up to a large amount of information and clear indications as to a person's effectiveness. The unsolicited comments that are passed on from students, parents, and colleagues provide additional information about the effectiveness of each member of the teaching force. These sources of information can serve a useful purpose when combined with the information obtained through more structured methods. Here again, a caveat is in order. Recognize that this type of information is impressionistic, highly subjective, and may not be representative. It should supplement, not substitute for, more objective evaluations.

Evaluation of Administrators

Many administrators fail to recognize the importance of evaluating their own leadership. They recognize the need to evaluate everyone else in the organization, yet from the standpoint of overall success there is no one whose performance is more important. Therefore, it is vital that administrative effectiveness be accurately appraised and the results applied toward improvement. Some of the same sources of information for the evaluation of teachers can be used in evaluating the administrator. Figure 11-6 illustrates a structured procedure that can be used for evaluation of those in administrative positions.

Evaluation of Athletics Coaches

Some of the methods of evaluation discussed on the previous pages are appropriate for coaches, as well. Regardless of methods used, the director of athletics should evaluate coaches each school year. Criteria need to be broader than won-loss record or measures of on-field performance. Often coaches are led to believe that "winning is the only thing" when it comes to being evaluated. Other coaches complain that they receive little feedback—positive or negative—from their superiors. Schools need to implement a more systematic approach to evaluation. Leland (1988) suggests an eight-step process for athletics directors to evaluate coaches:

Step 1 (preseason). Develop a written job description for each coach which is clear, comprehensive, detailed, and personalized.

Step 2 (preseason). Develop performance review criteria (PRC) so that coaches know how they are going to be evaluated before they enter the performance period.

Step 3 (preseason). Discuss PRC with coaches and reach a consensus in meetings with each coach.

Step 4 (preseason). Set goals for each of the performance criteria. The goals do not need to be totally objective; leave room for professional judgment.

Step 5 (in-season). Observe contests and team practices. Perceptions of the evaluator gathered through direct observation determine the salience of feedback.

Step 6 (postseason). Conduct a written evaluation, elaborating on positives and negatives vis-à-vis the criteria. Consider a rating system (e.g., 1 = excellent; 2 = needs improvement, etc.) for each criterion.

TEACHER EVALUATION FORM

Please evaluate the teacher by placing an 'X' in the appropriate square for each item. It is hoped that your comments will be frank, realistic and to the point.

Teacher _____ School _____ Date _____

Period of employment from _____ 19___ to _____ 20___

Secondary–Elementary: _____

Subject or Grade _____

Major and Minor _____

Description of Ratings: P–Poor; F–Fair; G–Good; VG–Very Good; O–Outstanding. Example: P F G VG O

PERSONAL TRAITS:	P	F	G	VG	O
1. Dependability					
2. Responsiveness to suggestions					
3. Judgment and tact					
4. Industry					
5. Sense of humor					
6. Voice quality and articulation					
7. Health					
8. Personal appearance					
9. Initiative					
10. Poise					
11. Ethical qualities					
12. Genuine liking for boys and girls					

Analytical Statement:

SCHOLARSHIP:	P	F	G	VG	O
1. Mastery of subject					
2. Knowledge of source materials					
3. General educational background					
4. Correct and fluent English					

Analytical Statement:

CLASS MANAGEMENT:	P	F	G	VG	O
1. Organization of routine details					
2. Control of classroom situation					
3. Maint. of good "emotional climate"					
4. Adaptability					
5. Attention to physical conditions and appearance of classroom					

Analytical Statement:

OUT-OF-CLASSROOM ACTIVITIES:	P	F	G	VG	O
1. Participation in sponsorship of student activities and in the supervision of out-of-classroom situations					
2. Supervision of hallways or playgrounds as required					
3. Prompt and accurate in filing reports					
4. Prompt in arrival at school and classes					

Analytical Statement:

Figure 11–5. Evaluation form for an administrator, supervisor, or colleague to rate a teacher's effectiveness. This form could also be used by a teacher in the self-evaluation process. (Courtesy of the Provo School District.)

TEACHING METHODS:	P	F	G	VG	O
1. General skill in planning					
2. Motivation of student interest					
3. Provision for individual differences					
4. Creativity					
5. Awareness and use of life situations					
6. Stimulation of critical thinking					
7. Use of resources and instruction aids					
8. Evaluation of student progress					
9. Use of pupil personnel information					

Analytical Statement:

PROFESSIONAL ATTITUDE:	P	F	G	VG	O
1. Cooperation with staff					
2. Attitude toward teaching profession					
3. Parent relationships					
4. Membership in professional organizations:					

Local: Yes ☐ No ☐
State: Yes ☐ No ☐
National: Yes ☐ No ☐

Analytical Statement:

Signature _____

Title _____

Figure 11–5. (Continued)

ADMINISTRATOR EVALUATION FORM

Please evaluate the administrator by placing an 'X' in the appropriate square for each item. It is hoped that your comments will be frank, realistic, and to the point.

Description of ratings: P–Poor; F–Fair; G–Good; E–Excellent; O–Outstanding

In terms of support, the administrator

	P	F	G	E	O
1. has been successful in gaining a high level of respect from other members of the organization.					
2. has been successful in securing a reasonable level of budget and other kinds of support for the programs.					
3. is instrumental in maintaining a high level of morale among the faculty and staff.					
4. help the organization toward a high level of respect within the larger educational system.					

In Personnel Matters, the administrator

	P	F	G	E	O
1. is fair and impartial in dealing with other employees.					
2. communicates effectively with other employees.					
3. is sensitive and responsive to the feelings, needs, and interests of others.					
4. presents a model and expectations of others which stimulates them to higher performance.					
5. demonstrates desirable ethical and professional behavior and encourages others to do the same.					
6. causes employees to feel both comfortable and challenged in their positions.					
7. keeps the faculty and staff free of unnecessary difficulty, stumbling blocks, and disruptions that might detract from their productivity.					
8. adequately prevents conflicts, disputes, controversies, and frustrating circumstances from diluting the efforts of the faculty and staff.					
9. relates well to other people and is respected by them.					

Figure 11–6.   An example of an administrator evaluation form

With respect to management, the administrator	P	F	G	E	O
1. provides clear direction and adequate insight toward well-established goals.					
2. is energetic and productive and utilizes time and energy in a desirable manner for the benefit of the organization.					
3. permits and encourages ample participation by others in decision making.					
4. makes decisions that are consistent, timely, and accurate.					
5. is efficient and effective in performing various administrative duties.					
6. follows correct administrative procedures in work with both the staff and higher administrative levels.					
7. is a recognized expert in the field of administration and contributes professionally to the overall good of that discipline.					
8. is an effective manager in all aspects of administration, especially in the management of people and financial resources.					
9. is consistent, predictable, and forthright in administrative duties.					
10. is both objective and fair in dealings with other people.					

Figure 11–6. *(continued)*

Step 7 (postseason). Hold a formal evaluation interview after the coach has received and reviewed the written evaluation. The interview should take place soon after the written evaluation so that any misunderstandings can be cleared up face to face.

Step 8 (postseason). The coach submits a response to the evaluation. The response can be in writing, if the coach so desires. Consider modifying the evaluation following the response, if facts or an altered point of view justify doing so.

Dartmouth College's Athletic Council established the following nine criteria for evaluating coaches (Leland, 1988):

Philosophical structure. Ability to articulate goals; general attitude toward the school, colleagues and athletes.

Administrative performance. Ability to organize an office, carry out administrative tasks, solve problems.

Interpersonal relations. Relationships with faculty, students, other coaches, parents.

Public relations. Ability to promote athletics, the department, and the school; relationship with fans, alumni, media.

Knowledge of sport. Knowledge of teaching/learning principles, of fundamental skills and technical aspects of the sport.

Practice organization. Ability to organize practices and motivate athletes, manage discipline, and promote safety.

Team performance. Demeanor before, during and after games; quality of game plans, ability to analyze game situations; team performance and improvement over season.

Recruiting. Knowledge of regulations; ability to analyze needs, implement effective strategy, and enroll admissible student athletes. (Not a suitable criterion for prep school or public school coaches.)

Professional relations. Involvement in professional associations, coaching groups; league or conference activities; contact with peers in coaching.

Some coaches feel threatened by formal evaluations. The athletics director must convey to the coaching staff that the purpose of evaluation is to increase understanding. Conferences should be held in a positive atmosphere, characterized by open communication. The director must be professional and straightforward, and must exercise good listening skills. Eliminate innuendo and informal evaluations. Formal evaluations can serve to protect the employer and employee. Concurrent salary discussions only cloud performance issues. While the evaluation may be the basis for merit increases and contract extensions, the evaluation process should precede any salary discussions (Leland, 1988). The director's evaluation can be supplemented by a self-evaluation by each coach, using a format provided by the department. Also, athletes should be interviewed as part of the process of evaluating coaches.

PROGRAM EVALUATION: CURRICULUM AND INSTRUCTION

Evaluating program effectiveness complements the evaluation of teaching. A school could have a superb curriculum and still not achieve superior results due to teacher ineffectiveness. Conversely, what can be accomplished through effective teaching is compromised if the curriculum is weak. The ideal is to develop a quality curriculum combined with quality instruction. Program evaluation can identify both strengths and weaknesses, as well as obvious omissions in the curriculum. The purpose of program evaluation is to determine how well curriculum and instruction contribute to educational goals, and how they could be modified to better realize these goals. In its broadest sense, program evaluation assesses not only outcomes based on student-attainment goals and levels of implementation, but also external factors such as budgetary restraints and community support.

Evaluation of instructional programs may be formative, summative, or *ex post facto* ("after the fact"). *Formative evaluation* is an internal function that feeds results back into the program to improve an existing unit of instruction. This type of evaluation is used frequently by teachers and administrators to compare outcomes with goals. *Summative evaluation* is carried out at the end of an instructional unit for the purposes of demonstration and documentation. Summative evaluations help programs analyze their unique characteristics and determine how to best achieve pedagogical goals. An example of summative evaluation is measuring the success of students who have emerged from the program. *Ex post facto evaluation* attempts to determine if programs are achieving the desired goals though continuous analysis of data over time (longitudinal) and, when possible, with data of similar programs (cross-sectional). Results provide the basis to recommend improvements or termination of programs (Beswick, 1990). Figure 11-7 shows a flow chart that illustrates the steps involved in curriculum construction and revision. The flow chart has built into it the impetus to ask penetrating questions about curriculum effectiveness and about how the curriculum could be improved.

The first and most important issue in program evaluation is how well students achieve mastery of skills. Often, this can be measured by standardized tests. However, standardized tests have their drawbacks, which has led some program evaluators to adopt other techniques to measure student attainment. Several alternative evaluation methods include: (1) standardized interviews that allow students' responses to be compared and summarized; (2) direct tests (such as skill demonstration), which enable teachers to gauge strengths and weaknesses and determine competency beyond answers on a test; and (3) students' notes, portfolios, and other material that can be

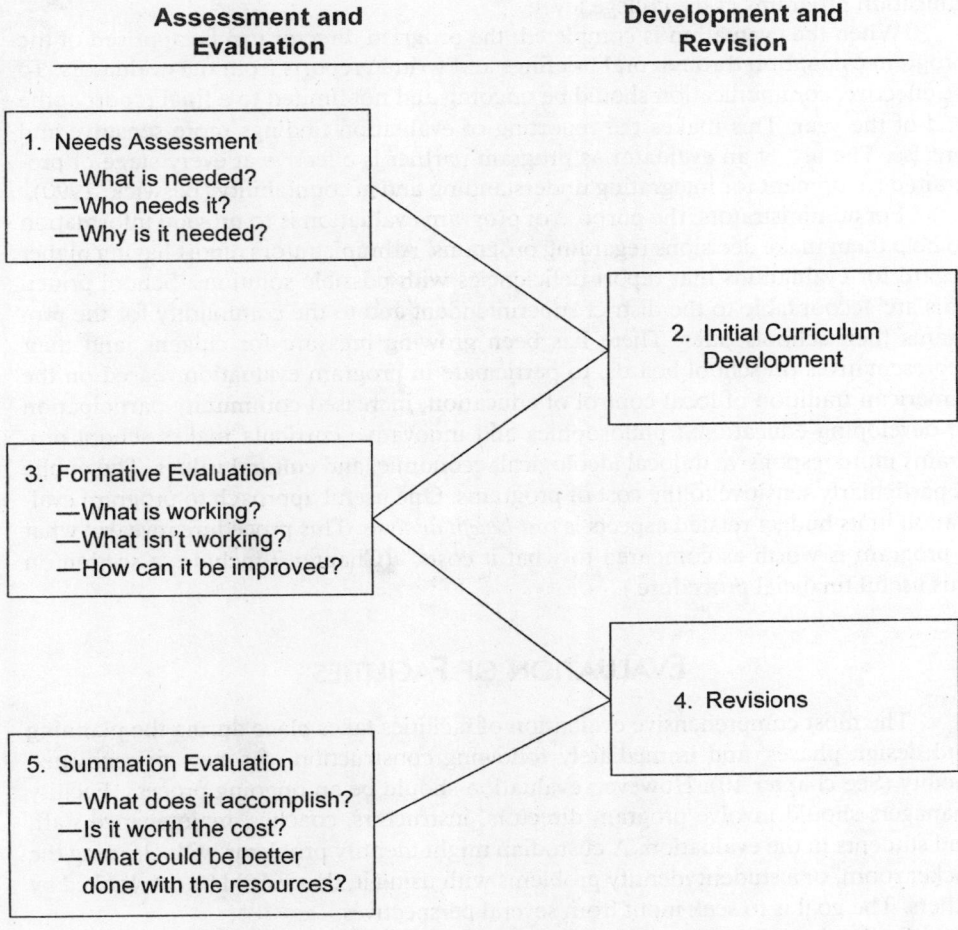

Assessment and Evaluation

Development and Revision

1. Needs Assessment

   —What is needed?
   —Who needs it?
   —Why is it needed?

2. Initial Curriculum Development

3. Formative Evaluation

   —What is working?
   —What isn't working?
   —How can it be improved?

4. Revisions

5. Summation Evaluation

   —What does it accomplish?
   —Is it worth the cost?
   —What could be better done with the resources?

Figure 11–7. A flow chart of the steps involved in curriculum evaluation and development

inspected for evidence of mastery (Beswick, 1990). An important aspect of program evaluation involves evaluating individual courses or units of instruction by the students in the class. Figure 11-8 shows a form prepared for such use.

Program/curriculum evaluation carried out by staff can include the utilization of standard forms, checklists, or scorecards with criterion and standards built in. Appendix D lists criteria for evaluating the physical education curriculum in grades 9–12.

Program evaluation should be conducted periodically as an ongoing procedure. It may be done annually or less often. Some schools form a Program Review Committee (PRC) composed of administrators, chairpersons, and a program specialist. Each year the PRC conducts a review of one or two programs, so that all programs in the school receive scrutiny once every few years. Program evaluation can be an internal procedure conducted by school staff, carried out by outside consultants, done in conjunction with an audit, by state accreditation, or by professional accreditation associations. As an example of the latter, NASPE works in conjunction with the National Council for the Accreditation of Teacher Education (NCATE) to evaluate teacher education programs at the college level.

When the evaluation is completed, the program director can be apprised of the program evaluation through oral briefings and written reports from the evaluators. To be effective, communication should be ongoing and not limited to a final report at the end of the year. This makes the reporting of evaluation findings more sensitive and precise. The use of an evaluator as program partner is effective at every stage of program development for integrating understanding and accountability (Beswick, 1990).

For administrators, the purpose of program evaluation is to provide information to help them make decisions regarding programs. Administrators report having higher regard for evaluations that report deficiencies with possible solutions. School principals are accountable to the district superintendent and to the community for the programs their schools offer. There has been growing pressure for citizens, and their representatives on school boards, to participate in program evaluation. Based on the American tradition of local control of education, increased community participation in developing educational philosophies and innovative curricula makes school programs more responsive to local ideological, economic, and cultural values. The public is particularly sensitive to the cost of programs. One useful approach to program evaluation in its budget-related aspects is *cost-benefit analysis*. This procedure reveals "what a program is worth as compared to what it costs." (Chapter 7 includes a section on this useful financial procedure.)

EVALUATION OF FACILITIES

The most comprehensive evaluation of facilities takes place during the planning and design phases, and immediately following construction of a new or renovated facility (See chapter 10). However, evaluation should be an ongoing process. Facility managers should involve program directors, instructors, coaches, maintenance staff, and students in the evaluation. A custodian might identify problems with cleaning the locker room, or a student identify problems with using it, that would be overlooked by others. The goal is to seek input from several perspectives.

Evaluation of facilities should be conducted systematically to include instructional, recreational, clerical, and service areas. All of the systems and features of the facility should be assessed: surfaces (floors, walls, ceilings), ventilation, electrical sys-

COURSE TITLE_____ DATE _____

PERIOD (TIME)_____ INSTRUCTOR _____

COURSE EVALUATION

	1	2	3	4	5	6	7
1. Course objectives are clear.							
2. Course is well organized.							
3. Student responsibilities are clearly defined.							
4. Course content is relevant and useful.							
5. Assigned workload is appropriate for credit hours.							
6. Assigned homework is *not* just busy-work.							
7. Text(s) and other materials have helped me understand the course topics.							
8. Exams concentrate on important points of the course.							
9. Exams are clearly worded.							
10. Exams are good measures of my knowledge, understanding, ability to perform.							
11. Grading procedure is fair and impartial.							
12. Assignments are appropriately distributed throughout the semester.							
13. Course as a whole has produced new knowledge, skills, and awareness in me.							

14. Overall rating: ___Poor ___Fair ___ Good ___Excellent

Code: (Place an X in the appropriate square)

    1 = Strongly disagree   5 = Agree

    2 = Disagree   6 = Strongly agree

    3 = Somewhat disagree   7 = Very strongly agree

    4 = Somewhat agree

Figure 11–8. Example of a simple but useful course evaluation form to be completed by the students

tems, lighting, and plumbing. Buildings also need to be evaluated periodically to determine if the support infrastructure remains current with changing technological standards. Instructional areas and offices may need to be refitted to accommodate the latest computer, Internet, and distance learning technologies. This also applies to stadium press boxes that accommodate the electronic news media.

Facilities must be evaluated on a regular basis for maintenance tasks such as floor resurfacing, painting, and repair of major fixtures. Frequent evaluations can reduce wear and damage caused by inattention. In schools, long-term planning is necessary following evaluation, as major maintenance must be scheduled during the summer months. Work has to be contracted in advance so as not to interrupt programs. For this reason, schools should develop time lines for facility evaluation (Olson, 1997).

Checklists provide a valuable tool for facility inspection and evaluation. Facility managers can develop their own checklists adapted to specific facilities or use published checklists. A comprehensive checklist for swimming pools and locker rooms is found in Appendix B of Marcia Walker and David Stotlar's *Sport Facility Management* (1997). Similar facility checklists may be obtained from professional associations representing architects and building contractors or government agencies (See Appendix C in John Olson's *Facility and Equipment Management for Sport Directors* [1997] for a list of these resources.)

References and Recommended Readings

Arreola, Raoul. 2000. *Developing a Comprehensive Faculty Evaluation System.* 2nd ed. Boston: Anker.

Baumgartner, Ted, & Andrew S. Jackson. 1999. *Measurement for Evaluation in Physical Education and Exercise Science.* Boston: McGraw-Hill.

Beswick, Richard. 1990. "Evaluating Educational Programs." *ERIC Digest*, Series No. EA 54, Report # EDO-EA-90-8. Online article: <http://ericae.net/db/edo/ED324766.htm>.

"Evaluation." 2002. *Sports Coach.* Online article: <http://www.brianmac.demon.co.uk/>.

Jensen, Clayne, & C. Hirst. 1980. *Measurement and Evaluation in Physical Education and Athletics.* New York: Macmillan.

Lacy, Alan. 1999. "Using Multiple Sources to Improve the Evaluation of Teaching." *The Chronicle of Physical Education in Higher Education*, 10(3): 11–13.

Leland, Ted. 1988. "Evaluating Coaches—Formalizing the Process." *Journal of Physical Education, Recreation & Dance* (Nov./Dec.): 21–23.

Miller, David K. 2002. *Measurement by the Physical Educator: Why and How.* 4th ed. Boston: McGraw-Hill.

Olson, John. 1997. *Facility and Equipment Management for Sport Directors.* Champaign, IL: Human Kinetics.

Rahn, Mikala L., Brian M. Stecher, Harvey Goodman, and Martha Naomi Alt. 1997. "Making Decisions on Assessment Methods: Weighing the Tradeoffs." *Preventing School Failure*, 41 (Winter): 85–89.

Tritschler, Kathleen. 2000. *Practical Measurement and Assessment.* 5th ed. Philadelphia: Lippincott, Williams & Wilkins.

Walker, Marcia, & David Stotlar. 1997. *Sport Facility Management.* Boston: Jones & Bartlett.

Yonkers Public Schools. 1999. *Annual Report, 1998–1999.* Yonkers, NY: Yonkers Public Schools. Online report: <http://www.yonkerspublicschools.org/pdf/ypsannual9899.pdf>.

Online Sources

ERIC Clearinghouse on Educational Management. "Trends and Issues: Teacher Evaluation."
http://eric.uoregon.edu/trends_issues/instpers/selected_abstracts/teacher_evaluation.html
This article describes a holistic, integrated approach to teacher evaluation and retention, collegial teaching and professional support, resource allocation, teacher professional growth, and opportunities to learn within a state and district standards-driven context. It includes a case study of school reform.

NASPE: Program Accreditation
http://www.aahperd.org/naspe/template.cfm?template=programs-accreditation.html
This site provides links to NASPE/NCATE, NASPE/NASSM and NASPE/NCACE professional program accreditation.

Part III

Program
Administration

12

The Instructional Program

THE INSTRUCTIONAL PROGRAM

An old maxim in physical education states, "It's not what the child does to the ball, but what the ball does to the child." Actually, these goals are reciprocal; both ends are realized in a good instructional program. Physical education should promote student achievement in sports, gymnastics, and dance. In addition, the program must be concerned with how these activities contribute to the development of the students. An effective program should be tailored for students' particular needs and characteristics. At the same time, take care to avoid a program that attempts to do some of everything with nothing done well, or repeats the same experiences year after year without regard for progression or expansion.

In the design and management of the instructional program, the following considerations should receive serious attention: (1) appropriate class size to enhance effective instruction; (2) appropriate length of class periods and adequate instructional time per week; (3) the use of competent teachers, assigned in accordance with their strengths; (4) provision of adequate facilities, equipment, and supplies; (5) adequate variety to meet the students' needs and interests; and (6) involvement of the right sequence of progressions to lead effectively toward educational goals.

A well-organized instructional program is the foundation for all that is encompassed within the broad area of physical education. Instruction accompanied by exercise provides the basis for motor fitness and skill competency, which promote the enjoyment of physical recreation and success in higher-level competition. Figure 12-1 illustrates the fundamental relationship among levels of a comprehensive physical education program.

CLASS SCHEDULING

The curriculum on paper cannot be implemented successfully without effective scheduling. Physical education should be scheduled as an integral part of the total

259

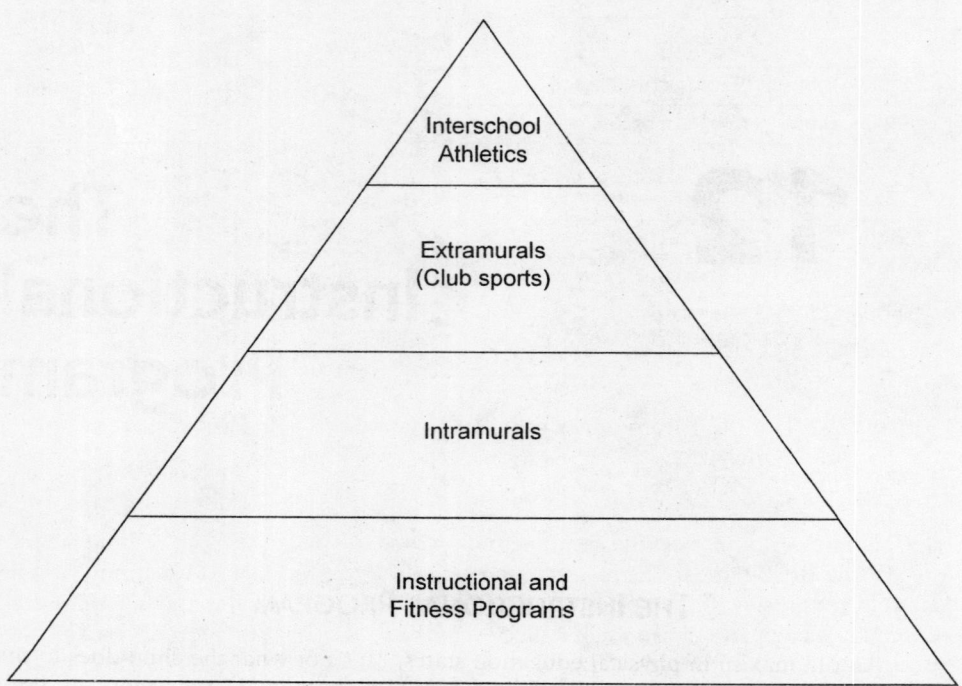

Figure 12–1.   Hierarchy of the elements in a comprehensive physical education program

school program. In large programs, scheduling can be tedious and time consuming; however, advances in computer technology have increased the potential for efficient and innovative scheduling. Three organizational patterns used for scheduling include traditional, modular, and block.

Traditional scheduling is the assignment of a given number of students to an instructor for a designated class period. Usually each class period occupies an equal amount of time. Typically, one teacher (occasionally assisted by a teacher's aide) instructs the class for the entire semester (Stillwell & Willgoose, 2002). This approach to scheduling provides widely acknowledged benefits. It enhances the opportunity for developing close teacher–student rapport, and it is the simplest method of scheduling. It also has disadvantages. The number of students in the class usually remains constant regardless of the instructional approach and changes in learning needs during the semester. Time per period remains constant when a longer or shorter period would be beneficial. Traditional scheduling seldom involves homogeneous groupings of students. Figure 12-2 illustrates a simple example of traditional scheduling.

Physical education can be scheduled less frequently (two or three times a week), with each session of instruction lasting two class periods. This results in proportionately more instructional time, because less time is utilized in dressing, showering, taking attendance, and other nonteaching activities. Within the framework of traditional scheduling, *team teaching* can capitalize on the expertise of different teachers.

With *modular* or *flexible scheduling*, the school day is divided into short periods of time called modules (or "mods"). Each module is a standard length, anywhere from 15

Time	Monday	Tuesday	Wednesday	Thursday	Friday
8:00– 8:55	Biological Science	- -			│
9:00– 9:55	Social Studies	- -			│
10:00–10:55	Study	- -			│
11:00–11:55	Industrial Arts	- -			│
12:00–12:55	Lunch	- -			│
1:00– 1:55	Physical Education	- -			│
2:00– 2:55	Geometry	- -			│
3:00– 3:55	Chemistry	- -			│

Figure 12–2. **A student's class schedule using the traditional scheduling method.**

to 30 minutes. Modules can be combined to provide the desired length of time for different instructional units. The instructional periods for a particular subject may vary in length from one day to the next, meaning that a student or a teacher may have a different schedule each day of the week. Once established, the cycle (if weekly) remains the same throughout the term. Some schools are on six-day schedules with two alternating cycles (labeled, for example, ACE and BDF). Physical education is scheduled on alternate days for students. Teachers meet half the students on each cycle.

In a highly flexible system, an instructor may request varying class lengths and frequency of meetings on a weekly basis. In such cases, students and teachers pick up new schedules prior to the beginning of each week (or cycle). This kind of scheduling requires the use of a computer, along with careful planning and coordination among teachers. It is not a frequently used method, but it does function effectively in certain situations. Flexible scheduling also allows for variations in class size for different purposes and facilitates ability grouping and independent study. An example of a student's class schedule using a modular or flexible system appears in figure 12-3.

Practical applications of flexible scheduling in physical education might include the following. Instructors would schedule a forty-minute module to show a 35-minute film on soccer to all their classes in one sitting in the school auditorium. This would eliminate the necessity of showing the film repeatedly to each class and free up time for other instructional needs. Assigning lockers during the first week of the semester might be accomplished in a 20-minute module for each class. A longer-than-traditional time period also has advantages. A 100-minute module might allow time to play an entire game at the end of an instructional unit.

In *block scheduling*, the daily schedule is organized into large blocks of time of more than sixty minutes to allow flexibility in instructional activities. The variations might involve reconfiguring the lengths of terms as well as the daily schedule. Examples include:

- Four ninety-minute blocks per day with the school year divided into two semesters, allowing former year-long courses to be completed in one semester
- Alternate-day block schedule where six or eight courses are spread out over two days, with teachers meeting half of their students each day

- Two large blocks and three standard-sized blocks per day, with the year divided into sixty-day trimesters and a different subject taught in blocks each trimester

The advantages of block scheduling are similar to those of modular scheduling. Larger blocks of time allow for a more flexible and productive class environment with opportunities to use varied teaching methods. Other benefits include more effective use of school time, decreased class size, increased number of course offerings, and a reduced number of students with whom teachers have daily contact. The major challenges to block scheduling are the planning time required and the adjustment of the teaching staff. Block scheduling is best implemented gradually over a couple of years with appropriate staff training (Irmsher, 1996).

CLASS SIZE

Class size is influenced by several factors: regulations and policies issued by higher authority, the size of teaching stations, the nature of the subject or activity, and the number of students that must be accommodated with the facilities and teachers available. A tennis class with six available courts would have to be smaller (accommo-

TIME	M	T	W	Th	F
8:00	English	Indep. Study	English	Social Studies	English
8:30				Indep. Study	
9:00	Indep. Study	Math	Guidance	Math	Social Studies
9:30			Social Studies		
10:00	Math	English	Math	English	
10:30	Indust. Arts		Indust. Arts	Typing	
11:00	Boys' Chorus	Typing	Boys' Chorus	Physical Education	Math
11:30	Indep. Study		English		
12:00	L	U	N	C	H
12:30	Social Studies	Social Studies	Indep. Study	Indep. Study	Guidance
1:00	Physical Education				
1:30		Indust. Arts		Indust. Arts	Physical Education
2:00			Physical Education		
2:30					English
3:00					Boys' Chorus

Figure 12–3. A student's class schedule for one week, using flexible scheduling with 30-minute modules (reprinted by permission of the American Alliance for Health, Physical Education, Recreation and Dance, 1900 Association Drive, Reston, VA 22091)

dating 24 students) than a soccer class that had the use of two regulation soccer fields. A modern dance class in a studio 30 × 40 feet would have to be restricted in numbers as compared with one taught in a studio 40 × 60 feet. The nature of the activity also influences class size. For example, a class of 25 in modern dance or ballet would be large for one teacher to handle effectively, whereas in ballroom or folk dance one teacher might handle 40 or more students. In beginning swimming, an optimum ratio is one teacher for about 15 students, whereas in certain field games a teacher can manage as many as 30 students effectively.

A few small classes can be justified, provided there are oversized classes to compensate. The important point is to strive for an optimum average class size, with each particular class adjusted in accordance with the influencing circumstances. The average class size recommended by the AAHPERD and the President's Council on Physical Fitness is 35 students (or smaller if circumstances permit). Where students with disabilities are mainstreamed, the class size should be reduced.

School administrators occasionally treat physical education instruction differently from "academic" classes by assigning as many as 50 or 60 students to one class. In effect, they use physical education as a dumping ground for excess students. The practice is most common where school principals have witnessed a lack of instruction in deference to a "throw out the ball" approach. In such cases, physical education is little more than organized recess. Teachers must teach the material in the curriculum and persuade principals of the instructional nature of their classes. This approach promotes administrative support for a manageable class size.

TIME ALLOTMENT

Individual states and professional associations have prescribed minimum instructional time. (See table 12-1 for state physical education requirements.)

For *elementary schools*, the AAHPERD recommends that students should participate in an instructional program of physical education for at least 150 minutes per

---

Table 12–1
Physical Education Requirements in the States and Territories of the United States as of 2001

*States that require 4 units:**
Illinois**	New Jersey

States that require two units:
California	Nebraska
Nevada	New York
Virginia	Washington
District of Columbia	

States that require 1½ units:
Louisiana	Texas
Utah	Vermont
Wisconsin	

States that require one unit of physical education in high school (Grades 9–12):
Alabama	Alaska
Connecticut	Delaware

Table 12-1 *(continued)*

States that require one unit of physical education in high school (Grades 9–12):

Hawaii	Indiana
Iowa	Kansas
Maine	Missouri
Montana	New Hampshire
New Mexico	North Carolina
North Dakota	Oregon
South Carolina	West Virginia

States that require ½ unit:

Arkansas	Florida
Georgia	Kentucky
Maryland	Michigan
Ohio	

States with requirements set by the local school districts:

Arizona	Idaho
Massachusetts	Minnesota
Pennsylvania	Rhode Island
Tennessee	Wyoming

States with no requirement:

Colorado	Mississippi
South Dakota	

Oklahoma shifted from specific time requirements to a standards-based requirement.

Grading practices:

Twenty-nine states plus the District of Columbia give a grade for physical education and include it in the grade point average. Georgia, Kansas, and New Hampshire do not include the physical education grade in the grade point average. In the remainder of the states, local school districts determine whether grades are included in the students' grade point average.

Substitution practices:

Twenty-one states allow no substitutions for instructional physical education. Twenty-seven states allow substitutions for high school physical education. The justification of substitution may be due to medical or religious reasons; participation in varsity athletics, ROTC and marching band; or other special activities.

*A Carnegie unit represents one hour per day each academic year or 180 to 190 hours of classroom contact.
**Requires daily physical education K–12; however, a waiver program allows exemption from the mandate, and there are no time or content guidelines.

Adapted from NASPE, *Shape of the Nation*, 2001.

week, in addition to time allotted for free play during recess and noon hour. Instruction should occur on a daily basis. The Society of State Directors of HPER recommends a daily instructional period of at least 30 minutes, exclusive of time for dressing, showering, and free-play periods.

For *junior high/middle school* and *high school* students, the President's Council on Physical Fitness recommends physical education for one standard class period per day, 5 days per week. The Society of State Directors of HPER recommends at least 300 minutes per week spread over 3 or more days at this level.

Whatever the arrangement, the use of time is just as important as the amount. As little time as possible should be spent in dressing, showering, calling roll and other matters peripheral to instruction and exercise. The important question is what is accomplished toward the educational goals rather than how much time is spent. Teachers should exercise good classroom and locker room management techniques. Studies show that as much as 20 minutes of class time can be expended in noninstructional activities.

Taking attendance the traditional way by calling names from a grade book is inefficient. An alternate method involves assigning students numbers and having them respond verbally or cover their numbers (painted or stenciled on the floor, wall, or locker room bench) during roll call, with the uncovered numbers quickly recorded. Another strategy is to have squad leaders take roll. Squad leaders also can assist in leading calisthenics and distributing equipment to enhance efficient use of class time.

TEACHING STATIONS

Teaching stations are areas assigned to classes or individual students to accommodate instruction. In physical education, both indoor (gymnasium, natatorium) and outdoor (fields, running track) teaching stations are required. Elementary schools may provide a large multipurpose activity area rather than a gymnasium. Optimally, a conventional classroom near the activity area should be available for occasional use by physical education classes. Outdoor stations should include both turf and hard surfaces. The most effective instruction occurs when a variety of indoor and outdoor multipurpose facilities are available. In practice, physical education programs too often are limited by a lack of adequate teaching stations. Conducting activities in school hallways or empty classrooms should be a last resort.

The initial task in scheduling the curriculum is to determine the number of teaching stations required. One method of calculating the needed number of teaching stations for a *secondary school* is illustrated in table 12-2, based on a school of 1,000 pupils with each pupil attending 30 periods per week, five of which are for physical education (one period daily). The result of the calculation indicates seven teaching stations needed for physical education. The following formulas are useful:

$$\text{Number of sections to be offered} = \frac{\text{total number of students}}{\text{anticipated class size}}$$

$$\begin{array}{c}\text{Number of teaching stations needed}\\\text{(classes per class period)}\end{array} = \frac{\text{total number of students}}{\text{class size} \times \text{number of periods/day}}$$

$$\text{Number of teachers needed} = \frac{\text{number of sections}}{\begin{array}{c}\text{number of periods taught/day}\\\text{by each teacher}\end{array}}$$

Figure 12-4 is a chart that shows teacher assignments correlated with teaching stations and class periods. In order to assure efficient use of staff and facilities, a two-way chart such as this should be constructed on the computer and reproduced as a hard copy. In the chart, the teaching stations appear horizontally across the top, and the time periods are shown in the vertical left-hand column.

Table 12–2
Plan for Calculating the Number of Physical Education Teaching Stations Needed for 1,000 Pupils in a Secondary School

Plan	Example
1. Divide the anticipated enrollment by the desired average class size. This will give the number of sections needed.	$\dfrac{1000 \text{ students}}{30 \text{ pupils per class}} = 34$ sections
2. Multiply the number of sections by the number of periods per week physical education is offered.	34×5 periods = 170
3. Multiply the resulting number by 1.25 to allow for a utilization factor.	$170 \times 1.25 = 203$
4. Divide the resulting figure by the number of teaching periods available in the school week (30 periods).	$\dfrac{203}{30} = 6.7 = 7$ teaching stations

Source: Willgoose, 1979.

Figure 12–4. A scheduling form for coordinating periods with the assignment of teachers and facilities (teacher and station assignments)

CLASS PERIODS	GYMNAS- IUM #1	GYMNAS- IUM #2	WRESTLING ROOM	OUTSIDE FIELD	SWIM. POOL
1	Smith	Hansen	Shaw	Jones	Black
2	Smith	Black		Jones	Moss
3	Black	Hansen	Shaw	Jones	Moss
4	Smith	Black	Shaw		Moss
LUNCH 5		Hansen		Jones	
6	Smith	Hansen	Shaw	Black	Moss
7	Smith	Hansen	Shaw	Jones	Moss

To calculate the number of teaching stations required for an *elementary school*, proceed as illustrated in table 12-3. In this example, an elementary school of approximately 500 students would require two teachers of physical education. Each would teach 25 30-minute periods a week, aside from any intramural supervision or program planning.

STUDENT GROUPING

Traditionally, grouping for instruction in schools has been based on chronological age and grade level. However, placement of students into classes and groups within classes should reflect more than administrative expedience. Classes are learning groups, and students should be assigned to these groups according to their learning needs. With the advent of coeducational classes, physical education teachers have

---

Table 12–3
Plan for Calculating the Number of Physical Education Teaching Stations Needed in an Elementary School

Plan	*Example*
1. Estimate the enrollment by counting the number of grades and number of classes in each grade.	A school with 6 grades of 3 classes per grade = 18 classroom groups
2. Decide on length and number of class periods.	A total of ten 30-minute periods per day = 50 periods per week
3. Decide on the number of periods per week for pupils.	5 periods per week
4. Calculate number of teaching stations:	

$$\text{Teaching stations} = 18 \text{ classroom groups} \times \frac{5 \text{ periods per week}}{50 \text{ periods per week}} = \frac{90}{50} = 1.8$$

Source: Willgoose, 1979.

---

been confronted with a wider range of size and ability among students and a growing need to provide optimal environments for learning and competition. This situation suggests more sophisticated approaches to student grouping. Grouping students in physical education safeguards the health and safety of participants, facilitates effective learning, equalizes competition, and ensures progress.

Categorically, groups can be homogeneous (alike on a given characteristic) or heterogeneous (different on some characteristic). Criterion-referenced assessments can be utilized to sort students into homogeneous groups. Grouping students by ability in physical education generally has been associated with improved learning. Although such grouping seems to work better for highly skilled students, heterogeneous groups offer the advantage of exposing low-skilled students to models of success by the highly skilled, who also can act as peer teachers. Heterogeneous groups are more difficult to manage, however. When a teacher's aide is available, the two instructors can work individually with different groups. The teacher must consider all the above factors in deciding if and when to group students. The class can be grouped by ability for part of the instructional period and then brought together after students have mastered the basic skills (Stillwell & Willgoose, 2002). Grouping classes and teams for competition is often based on such factors as height and weight.

UTILIZATION OF FACULTY

The *traditional* method is to assign a class for which the teacher is responsible during the term. This is the simplest pattern and offers the teacher the advantage of establishing a good rapport and becoming well acquainted with the students. The main disadvantage is the lack of flexibility in utilizing the strongest attributes of several teachers.

The *team teaching* method is one in which two or more teachers share the responsibilities for a particular class. The purpose is to capitalize on the strengths of the teaching staff. The three basic forms of team teaching are the unit team approach, the differentiated approach, and the topical approach. The *unit team approach* features a master teacher and supporting teachers. The master teacher is "the expert" who directs the class, gives overall leadership to the learning process, and supervises the supporting teachers. The supporting teachers work with subgroups in specialized phases of instruction. The *differentiated approach* is usually based on skills to be taught. Members of the teaching team are assigned different skill units to teach, or students are grouped by ability and the team members teach different skill levels. With the *topical approach*, members of the teaching team provide instruction on a particular topic or activity in which they are highly specialized. For example, a swimming specialist might teach a four-week course on swimming, then a gymnastics specialist would teach gymnastics for four weeks, then a tennis specialist would teach a four-week course of tennis during the term.

Assignment of Teachers

The main element in a successful educational environment is a competent and concerned teacher. Teachers should be assigned to classes based on their qualifications and records, not on personal preference or convenience. Teachers of activities with inherent risks, such as swimming or gymnastics, should have the specialized knowledge necessary to conduct the activities at a reasonably safe level. In such cases, certification by national agencies associated with the activity is highly desirable.

Staffing methods differ by level of instruction. The methods of staffing physical education in *elementary schools* (addressing its advantages and disadvantages) are as follows:

Physical education specialists. One or more specialists in each school teach all of the physical education classes. The advantages are the specialist's expertise in physical education subject matter and methods, and knowledge of the stages of child development vis-à-vis motor learning. Disadvantages are that the specialist does not know the students as well as the classroom teacher, and teaching the same subject all day at this level can become monotonous.

Rotating specialists. A specialist rotates among the schools assisting classroom teachers by team teaching with them at least once a week and providing leadership and program ideas for the instructional periods when the specialist is not there. There are advantages: (1) a specialist is knowledgeable about physical education subject matter and methods; (2) a specialist understands developmental motor learning; and (3) through in-service workshops, master lessons, and coaching, a specialist can improve the classroom teacher's physical education techniques. However, there are also disadvantages: (1) specialists often must travel to two or more schools; (2) some classroom teachers do not work effectively with the specialist; and (3) the specialist is unable to become well acquainted with the pupils, due to limited contact.

Classroom teachers trading assignments. One teacher instructs physical education for another, who in turn teaches another subject for the first teacher. The main advantage is that teachers can teach in areas of preference and expertise. The main problem is finding teachers who are prepared and willing to do the reciprocal instruction needed to make this method work.

Classroom teachers teaching physical education. The advantages are as follows: (1) the teacher can integrate physical education with other subject matter; and (2) the teacher is familiar with the students and knows their personalities and abilities. Disadvantages include: (1) lack of expertise in teaching physical education among classroom teachers, and (2) no opportunity for a break from the routine of supervising the same students every period of the day.

Most curriculum consultants recommend that physical education specialists be employed in some capacity at the elementary-school level, so that the teaching of movement activities is not left solely to the classroom teacher.

Physical education classes at the secondary level should be staffed with teachers who have at least a bachelor's degree with a major or minor concentration in physical education/kinesiology. There are several important considerations in staffing classes in *secondary schools*:

1. Is it better for a physical education instructor to teach only physical education, or to teach some classroom subjects as well? A combination of classroom and physical education duties may provide some advantages.

2. Is it desirable for a teacher to instruct in the physical education program several hours during the school day and then have a heavy coaching assignment after school? It is generally agreed that this is too much. A better combination is some classroom teaching along with physical education instruction and coaching.

3. Is it appropriate to perpetuate the belief that a successful classroom teacher or a successful athletics coach can get by with an inferior effort in physical education instruction? The principle should be enforced that teaching 30 students in physical education is just as important as teaching 30 students in language arts and mathematics, or coaching 30 students on the soccer field. Any teacher/coach who views this differently ought to be prohibited from teaching physical education.

4. Should secondary-school physical educators be informed about the instructional programs in physical education that precede those on their grade level? The answer is yes, in order that the subject matter and teaching methods fall into sequence and build progressively onto what the students have already learned. Without this coordination, the idea of graduated levels of instruction fails.

Licensing, Certification, and Accreditation

Certification of teachers and specialists and accreditation of programs have increased over the last several decades as physical education became more specialized. Teacher certification within each state is well established. Public school programs may be accredited at the state level by departments of education, and/or by regional accreditation associations. Accreditation of various professional preparation programs has been gaining ground in colleges and universities. Teacher education programs with concentrations in physical education/kinesiology are accredited jointly by the National Council for the Accreditation of Teacher Education (NCATE) and NASPE.

Individual teachers can now apply for national board certification, a process established by the National Board for Professional Teaching Standards (NBPTS). This is a voluntary process consisting of a rigorous performance-based assessment over a year. By early 2002, some 16,000 teachers in the United States had been board certified, including over 180 physical education teachers. States or local school districts often reimburse the fees for undergoing this assessment and reward the attainment of certification with salary adjustments ("Physical Educators Earn . . .", 2002).

The National Consortium for Physical Education and Recreation for Individuals with Disabilities (NCPERID) has developed national standards and a national certification examination for the field of adapted physical education. (Most states have not defined the qualifications teachers need to provide adapted physical education.) The Adapted Physical Education National Standards (APENS) and the accompanying certification examination developed by the Consortium have been endorsed by AAHPERD. Teachers who pass the certification exam earn a seven-year certificate that endorses them as Certified Adapted Physical Educators and allows them to use the acronym C.A.P.E. after their names to designate their professional status.

The standards, which address fifteen areas of adapted physical education, can be found in the National Consortium's APENS manual (1995) and study guide. Standard 7 on Curriculum Theory and Development is divided into five components (Kelly, 1995):

1. Understanding the foundation of curriculum design.

2. Selecting content goals based on relevant and appropriate assessment.

3. Conducting learner analysis.

4. Designing learning objectives based on relevant and appropriate assessment.

5. Planning learning experiences.

Teaching Loads

In *colleges* and *universities*, a teaching load should not exceed 12 class hours per week of lecture courses, and no more than 9 hours for faculty teaching graduate courses. At community colleges, the teaching load is occasionally 15 hours a week; however, this practice is not recommended. Activity courses usually are scheduled at the ratio of two clock hours for each credit hour. A load of 18 clock hours should be the absolute maximum for activity course instructors.

In *secondary schools*, it is recommended that class instruction per teacher not exceed 5 clock hours per day, or 1,500 minutes per week. Six clock hours per day or 1,800 minutes per week should be considered an absolute maximum, and this should include any after-school responsibilities. A daily load of 200 students per teacher should be the maximum. Finally, each teacher should have at least one free period daily for consultation and conferences with students.

The number of hours per day or week is not an absolute measure of teaching load; other factors must be considered. If a person teaches different subjects, more preparation is required as compared with teaching multiple sections of the same class. The correct balance between the two factors is important because the necessity of too many preparations contributes to inefficient use of the teacher's time; whereas teaching too much of the same subject can make the teacher's approach routine and dull.

Another consideration is the length of the class periods. A large number of short class periods is more difficult to teach than a smaller number of longer class periods, because more classes require more preparations and contact with more students. For example, six 40-minute classes would be more demanding than four 60-minute classes, even though the total minutes of instruction are the same.

In addition to the time spent in class, a teacher needs time for planning and preparation, for student consultation, and for consultation with administrators and parents. Another consideration is that physical educators may have extra duties after hours: coaching athletics, conducting intramurals, coaching cheerleaders and marching groups, or serving as advisors for activity clubs. Two prevalent methods are used by schools to compensate teachers for performing extra duties. One method is *extra pay* for extra work, and the other is to give the teacher *released time* from the normal school day.

Whichever method is used, the arrangement should be fair and equitable for both the school and the individual. (Extra duties for coaches are discussed in chapter 15.)

INSTRUCTIONAL TECHNOLOGY

Administrators should stay current with developments in instructional technology and provide the technical support and infrastructure to meet program needs within the budget. Following are examples of current instructional technology utilized in physical education classes.

Pedometers that clip onto shoes or belts are used to motivate students toward fitness. Wrist-mounted heart rate monitors are utilized to allow students to assess their fitness levels, with data downloaded and stored on a personal computer. Personal Digital Assistants (PDAs) are palm-held mobile electronic organizers that allow the teacher to build a mobile databank. PDAs are generally compatible with word processing and spreadsheet programs. They can be used to take attendance and record grades, display lesson plans, disseminate course materials, keep track of students' performance and progress, organize students according to their abilities, and measure levels of fitness. PDA software programs (available on the Internet) are designed to support these functions as well as to integrate with digital cameras and voice recorders for multimedia presentations. Students can utilize PDAs to create a portfolio of their work (assignments, projects) or maintain a journal including written and verbal notes (Juniu, 2002).

Power Point software allows teachers to make visual presentations if the facilities are wired to accommodate the technology. Utilization of PCs or laptop computers, videotapes, recordings, and programmed learning modules allow instruction to occur at individual learning stations apart from the classroom. Workstations proximate to the physical education teaching station should be available to accommodate these learning aids.

DISABLED STUDENTS

The Rehabilitation Act of 1973 prohibits discrimination against "handicapped individuals" in any federally aided programs. This objective was enhanced by the Education for All Handicapped Children Act of 1975, which provided federal funding to aid state efforts in providing an appropriate education for the disabled. This second law, known as Public Law 94–142 (later amended), was intended to ensure an appropriate public education for all disabled students. In part, the law enables

. . . specifically designed instruction, at no cost to parents or guardians, to meet the unique needs of a handicapped child, including classroom instruction, *instruction in physical education* (emphasis added), home instruction and instruction in hospitals and institutions.

The provisions in sections 2, 3, and 4 of the law that apply directly to physical education are:

1. Search for disabled students.
2. Identify the population that needs service.
3. Provide the least restrictive educational environment.

4. Prepare an individualized education program (IEP) for each disabled child identified as having special needs.

5. Do an evaluation of the progress of each student at least annually.

6. Conduct a fair hearing for students, or parents on behalf of their children, to establish the needs of the student if there are any questions (grievances) regarding the individualized education program.

The All Handicapped Children Act was upgraded in 1990 by Public Law 101-476, the Individuals with Disabilities Education Act (IDEA). This law, referred to as IDEA–Part B, mandated that appropriate physical education services must be made available to every disabled child who receives a free education. The provisions of the above-mentioned legislation that address individualized education programs and least restrictive environment require further clarification.

Components of an Individualized Education Program (IEP)

In order to develop the individualized education program in compliance with the requirements of federal laws, the following components must be included for every disabled child requiring special education and related services:

- A statement of the child's present levels of educational performance, including academic achievement, social adaptation, prevocational and vocational skills, psychomotor skills, and self-help skills.
- A specific statement describing the child's learning style.
- A statement of annual goals that describes the educational performance to be achieved by the end of the school year under the child's individualized education program.
- A statement of short-term instructional objectives, which must be measurable intermediate steps between the present level of educational performance and the annual goals.
- A statement of specific educational services needed by the child (determined without regard to the availability of those services), including a description of:
  a) All special education and related services needed to meet the unique needs of the child, including the type of physical education program in which the child will participate.
  b) Any special instructional media and materials needed to implement the individualized education program.
- The date when those services will begin and length of time the services will be given.
- A description of the extent to which the child will participate in regular education programs.
- A justification for the type of educational placement that the child will have.
- A list of the individuals who are responsible for implementation of the individualized education program.
- Objective criteria, evaluation procedures, and schedules for determining, on at least an annual basis, whether the short-term instructional objectives are being achieved.

Least Restrictive Environment (LRE)

Inherent in the above legislation was the "mainstreaming" concept, the idea that disabled learners should be integrated into the least restrictive environment. The continuum of placement within physical education is represented by figure 12-5. As one of the basic principles (indicated on the diagram), students should be placed in the environment that will allow them to function best. The physical education teacher usually is the person best equipped to make the judgment as to the kind of physical education setting in which the student belongs. In making such judgments, teachers should remember that disabled students have the same basic needs, desires, and interests as the nondisabled. The disabled are much more like others than they are different.

The concept of least restrictive environment means that each disabled child should be placed in the most normal situation possible. The physical education administrator should be knowledgeable about each situation, so that if called on, the administrator can provide an accurate report on the status of the individual in question. The administrator must also be aware of how effectively teachers in the department are working with disabled students. As can be seen on the continuum of physical education placement, four of the placement levels involve the assistance of a staff member other than a physical education teacher in a regular class. If the law is applied correctly, it might enable a school to hire specialized staff to assist those disabled students requiring special attention.

Many schools are expanding LRE through the implementation of the "Regular Education Initiative," which became known as the "Full Inclusion Concept." This initiative, although not directly supported by any of the legislation, has extended the mainstreaming concept to move as many students as possible, even those with severe disabilities, into the regular classroom. Under this process of full inclusion, all students with disabilities have a regular class as their home base. In this environment, special support programs are brought into the regular class setting so the students with special needs can stay with their peers.

One concern is that the shift toward placing all students with disabilities in a regular setting may be detrimental or deprive some students of needed special physical education services. Indeed, services provided in segregated environments are beginning to disappear. Parents who have accused schools of "counterfeit mainstreaming" or "dumping" disabled students into regular classes have filed numerous lawsuits (Greenwood & French, 2000).

The Society of State Directors of Health, Physical Education and Recreation makes the following points (reprinted with permission) regarding the teaching of disabled students:

- Physical education should be adapted for those pupils who have special needs. To the extent feasible, such pupils should take part in regular classes. Schedule special classes for pupils who are severely disabled or who are otherwise unable to participate successfully in the regular program.

- A physician's recommendation in consultation with teachers, parents and the student should be the basis for assignment to special classes. The physician should work with the physical education teacher in determining the activities that should be prescribed for each child who requires adaptive physical education.

- Students may need adaptive physical education because of poor coordination, lack of strength, or similar deficiencies.

- The services of teachers with special preparation in adaptive physical education should be available to every school.

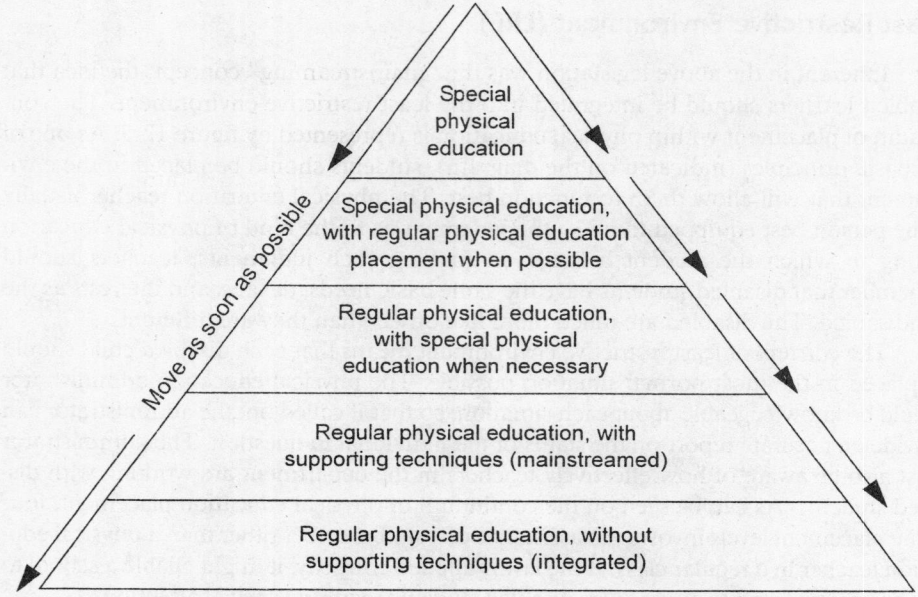

Figure 12–5. Continuum of physical education placement of disabled students

Administration of Physical Education for Disabled Learners

Organizational structures should be established in the school that accommodate students within a full range of disabilities and facilitate both heterogeneous and homogeneous grouping. Generally, additional funds will be needed to provide these modifications and accommodations. Some school districts have had to hire employee consultants to facilitate the planning, as well as have additional staff available to provide direct services. Preservice and in-service training of staff and paraprofessionals is recommended for those who work with disabled learners. Establish a record-keeping system so that all requested information and forms can be produced on demand (Greenwood & French, 2000).

Program coordinators must cultivate a close working relationship among all the individuals involved in meeting the federal requirements for physical education of the disabled. This includes special physical education instructors, school health personnel, special education teachers, family physicians, and parents. Establish a communication network as part of this effort. Cooperation is also essential with professionals outside the school. This may include medical and psychological service providers, as well as physical therapists, vocational counselors, and clinical social workers (Stillwell & Willgoose, 2002).

HEALTH AND SAFETY CONSIDERATIONS

The teacher or supervisor of a class or activity is fully responsible for seeing that the activity is conducted in a manner to insure optimum health and safety of students.

This involves providing a safe environment in terms of facilities and equipment, adequate supervision, and the proper selection of activities for the particular group or individual. A student should not be allowed to engage in activities against a physician's recommendation. Sometimes, because of health reasons or disability, a physician will prescribe modified activity or excused absence for a student. Such a recommendation should always be honored. Figure 12-6 shows an example of a physician's recommendation form.

CONFORMITY TO TITLE IX

Title IX of the Education Amendments Act of 1972 applies to all schools that receive federal financial assistance, and it is enforced by the Office of Civil Rights within the U.S. Department of Education. The legislation is based on the principle

Physician's Recommendation

This is to certify that I have examined _____

from _____ school and have found the following

abnormal condition _____

_____

Therefore, I recommend that participation in physical education be as checked below:

_____ Normal program—all activities

MODIFIED PROGRAM

_____ Relaxation and rest

_____ Mild exercises, done lying or sitting on mat

_____ Nonvigorous games or dance

_____ Body building exercises modified as to need

_____ Other (specify) _____

Duration: for_____ weeks, for _____months, until next examination

Date: _____ Signature _____ M.D.

Telephone _____ Address _____

COMMENTS:_____

_____

_____

Figure 12–6.   Example of a physician's recommendation for modified physical education or excused absence

that all educational activities have equal value for both sexes, and therefore school programs shouldn't discriminate on the basis of gender. Title IX doesn't specify curriculum content but stipulates that both sexes must have equal access to the curriculum. The regulations require that physical education classes be coeducational. (See also the section on Title IX in chapter 15, which explains its application to athletics.)

Section 86–34 of the law specifically states that an institution or agency may not:

> . . . provide any course or otherwise carry out any of its education program or activities separately on the basis of sex, or require or refuse participation therein by any of its students on such basis, including health, physical education, industrial business, vocational technical, home economics, music, and adult education courses.

With respect to *physical education*, appropriate modifications should be made in activities to accommodate optimal learning and equalize competition. In doing so, teachers and administrators should be aware of the following specific provisions:

1. The law does not prohibit grouping of students in physical education classes and activities by ability as assessed by objective standards of individual performance, developed and applied without regard to sex.

2. The law does not prohibit separation of students by sex within physical education classes or activities during participation in wrestling, boxing, rugby, ice hockey, football, basketball, and other sports if the purpose or major activity involves bodily contact.

3. When use of a single standard of measuring skill or progress in a physical education class has an adverse effect on members of one sex, the recipient shall use appropriate standards that do not have such effect.

4. Portions of classes in elementary and secondary schools that deal exclusively with human sexuality may be conducted in separate sessions for boys and girls.

POSITION STATEMENT OF THE SOCIETY OF STATE DIRECTORS

The Society of State Directors of Health, Physical Education and Recreation has taken a position on conditions that should exist in the physical education instructional program. Some of the major points in that position statement follow (reprinted with the permission of the Society):

- Credit in physical education should be required for graduation from high school. Standards should ensure satisfactory achievement based upon reasonable objectives for each individual student.

- All students should wear physical education clothing that permits freedom of movement and safe participation. Showering should be encouraged whenever class activities are sufficiently vigorous to cause perspiration. Appropriate instruction and supervision should be provided to ensure safe, healthful practices in dressing and showering.

- Physical educators should be alert to changing patterns of school organization, methodology, and educational technology so that the educational potential of new practices and innovations may enhance the physical education program.

- In addition to the necessary space, indoor and outdoor facilities, equipment and supplies for conducting the instructional program in physical education, good teaching requires an adequate collection and proper use of books, periodicals, audiovisual materials and other teaching aids. Innovative use of modern technology is encouraged whenever it can contribute to improved instruction.

- Evaluation of the curriculum by both students and teachers should be continuous. Evaluation should be used to assess student progress toward stated goals, to provide guidelines for adjusting current programs and planning new ones, and to indicate needs for instructional improvement. Evaluation is an important means of interpreting the program to students and adults in order to improve community understanding of the values and outcomes of physical education.

- Certification standards for physical education personnel, consistent with general certification requirements, should be established and periodically reviewed by each state department of education. Increased interest and involvement in certification on the part of professional associations and teacher preparation institutions should be encouraged and welcomed.

- Elementary classroom teachers should have a clear understanding of the contribution good physical education makes to the education of boys and girls, including the relationships of physical growth and development to the general educational progress of all pupils, and have sufficient preparation and laboratory experiences in the basic skills, methods, and content of physical education to enable them to conduct an effective program under the guidance of a physical education specialist or supervisor, and to relate the experiences in physical education to ongoing classroom instruction.

- Teachers who work with students who have special needs or characteristics, e.g., inner-city, disadvantaged, preschool, disabled, ethnic or non-English speaking, should be specially selected and prepared. The tremendous personal and social benefits of physical education for such students should be maximized through the program.

- All teachers of physical education should be provided with assistance from supervisors or resource persons who are qualified by education and experience to provide professional leadership and guidance in physical education. In addition, clinics, workshops, and other in-service opportunities should be provided for both classroom teachers and specialists.

- Teacher aides, when utilized, should work under the direct supervision of a certified physical education teacher and only in a supportive and supplementary capacity.

FINAL STATEMENT OF PRINCIPLE

Every student is a unique individual with differing physical, mental, emotional, and social needs. Students have varying abilities, aptitudes for performance, and rates at which motor skills are perfected. Regardless of differences in gender or ability, every student, including the disabled, has the need for and the right to benefit from well-planned physical education experiences.

Physical education is an important part of the development of the whole person. Even though the emphasis is on the development of physical fitness and motor characteristics, good physical education experiences also contribute to social, psychological, and mental development. Without an effective physical education program in conjunction with academic subjects, it is impossible for students to achieve the well-rounded development they need and deserve.

An effective instructional program: (1) enhances growth and development, including physical fitness, motor skills, and desirable body composition and postural characteristics; (2) contributes to one's health and wellness; (3) has potential for enhancing self-image, psychological adjustment, and social adeptness; and (4) develops desirable lifelong recreational interests and skills.

References and Recommended Readings

Camaione, D. N. 1979. "A Method for Determining Instructional Load Credit." *Journal of Physical Education, Recreation & Dance*, 50(5): 58.

Greenwood, Mike, & Ron French. 2000. "Inclusion into Regular Physical Education Classes: Background and Economic Impact." *The Physical Educator*, 57(4): 209–16.

Irmsher, Karen. 1996. "Block Scheduling." *ERIC Digest* #104 (March) Online article: <http://eric.uoregon.edu/publications/digests/digest104.html>.

Juniu, Susana. 2002. "Implementing Handheld Computing Technology in Physical Education." *Journal of Physical Education, Recreation, and* Dance, 73(3): 43–48.

Kelly, Luke E. 1995. *Adapted Physical Education National Standards.* The National Consortium on Physical Education and Recreation for Individuals with Disabilities. Champaign, IL: Human Kinetics Publishers.

"Physical Educators Earn Profession's Top Honor." 2002. *NASPE News* (winter): 12.

Stillwell, Jim, & Carl Willgoose. 2002. *The Physical Education Curriculum*, 5th Ed. Prospect Heights, IL: Waveland Press.

Willgoose, C. E., 1979. *The Curriculum in Physical Education*, 3rd ed. Englewood Cliffs, NJ: Prentice-Hall, 170.

Online Sources

Bonnie's Fitware Resources
http://www.pesoftware.com/Resources/resource.html
This site provides several links to products and resources related to teaching physical education.

PE Central
http://pe.central.vt.edu/
A Web site for health and physical education teachers, sponsored by Flaghouse Inc., PE Central offers instructional resources and products, best practices, lesson ideas, and applications of research.

Adapted Physical Education National Standards
http://teach.virginia.edu/go/apens
The APENS site, which is sponsored by Texas Women's University, is designed to inform adapted and general physical educators about national certification in adapted physical education.

Physical Education Lesson Plan Page
http://members.tripod.com/~pazz/lesson.html
This site provides a place for physical educators to share lesson plans for low-organized games. Focuses on K–6 level.

High School Sample Lesson Plans
http://www.stan-co.k12.ca.us/calpe/High.html
This California Physical Education Resources Web site lists numerous lesson plans along with unit standards and unit outlines by grade level for various activities.

Sites featuring educational software for PDAs:
Teacher's PET <http://www.coffeepotsoftware>
PrestoGRADE <http://www.aptustechnologies.com>
Bonnie's Fitware <http://www.pesoftware.com>

13 Curriculum Development and Management

THE NATURE OF EDUCATION

Education is a process of positive change. In a broad view, it is any act or experience that has a formative effect on the mind, character, or physical ability of an individual. In this sense education never ends, because we learn from experience throughout our lives. Its purpose should be to help people grow and succeed. *Formal education* is the process by which society, through schools, colleges, and other institutions, deliberately transmits the cultural heritage—its accumulated knowledge, values, and skills—from one generation to another. Such formalized education involves planned programs and distinct methods for accomplishing these objectives. Such programs and methods make up the curriculum.

Physical education is an element of formal education. It can be viewed in two ways: education *through* the physical, and education *of* the physical. Education *through* the physical implies changes in the total individual—knowledge, values, skills—as a result of participation in movement activities. Much can be learned about our environment, about others, and about ourselves through the educational medium of human movement. Furthermore, it can significantly influence the development of one's personality and ability to function effectively in society. At the same time, the primary benefit of physical education is the development *of* the physiological systems. It enhances the body's ability to function effectively and perform efficiently. The total thrust of the physical education program ought to emphasize all these aspects of learning in their proper balance. Physical activity is the *medium* for learning—not the end but the means. The ultimate end should be the development of human potential in all of its aspects.

The school curriculum must be based on sound values that are in harmony with the philosophy of a free society, the values of the local community, and the goals of

279

the school system. The following guidelines help frame the curriculum within these contexts. (A review of chapter 1 will help in understanding the importance of goals and working toward them, and of sound values on which to base the curriculum.)

CURRICULUM DEVELOPMENT

Framework of the Curriculum

Throughout the history of U.S. education, meaningful statements have been set forth that have served as guidelines for the direction of education and for the placement of emphases in the curriculum. As far back as the early 1900s, the Commission on the Reorganization of Secondary Education (1918) listed the following Seven Cardinal Principles of Education:

- Health,
- Command of the fundamental processes,
- Worthy home membership,
- Vocational preparation,
- Civic education,
- Worthy use of leisure, and
- Ethical character.

These principles still are widely accepted as guides to the nature of education in a free society.

Similarly, in 1966 the American Association of School Administrators listed nine imperatives in education. These are guidelines that should be at the forefront as school curricula are developed and modified to meet the needs of America's students (Cooper et al., 2002):

- To make urban life rewarding and satisfying,
- To prepare people for the world of work,
- To discover and nurture creative talent,
- To strengthen the moral fabric of society,
- To deal constructively with psychological tensions,
- To keep democracy working,
- To make intelligent use of natural resources,
- To make the best use of leisure time, and
- To work with other peoples of the world for human betterment.

In 1994, educator John Goodlad (*What Schools Are For*) re-examined the purposes of education and listed the following eight goals for schools:

- Mastery of basic skills or fundamental processes,
- Career or vocational education,
- Enculturation,
- Interpersonal relations,
- Autonomy,

- Citizenship,
- Moral and ethical character, and
- Self-realization.

Common themes run through these lists, indicating essential values and goals that have persevered over the years. The lists should help the reader identify the basic purposes of education and inform a rationale for including physical education in school programs.

The Nature of the Curriculum

A carefully planned curriculum incorporates activities that promote the afore-mentioned outcomes and objectives. Following are some commonly used definitions of what a curriculum is.

1. All of the planned experiences that a student undergoes through his or her participation in a school's instructional program or a particular area of it.

2. An orderly sequence of courses and/or experiences offered by an educational institution to create an environment in which the objectives of the program can be achieved.

3. A systematic set of educational experiences designed to efficiently produce the most important learning and behavior needed for successful living.

4. All of the instructional and extra-instructional activities sponsored by a school.

Steps in Curriculum Planning

The logical steps in curriculum development begin with a well-defined purpose and clear directions, which include the following:

1. Study the local conditions and attitudes that should influence the curriculum.

2. Study and clarify the philosophy and goals of education within the school system.

3. Become well informed about opportunities and limitations relative to resources—facilities, teachers, and equipment.

4. Thoroughly evaluate the present curriculum in terms of both strengths and weaknesses.

5. Recommend curriculum changes along with the justification for them.

6. Assist with the implementation of approved changes, including dissemination of information to teachers, students, and public, and in-service training of the staff as needed.

These steps are best carried out by a curriculum committee made up of knowl-edgeable individuals (see the following section). Those responsible for curriculum development must be both planners and doers, and they should be well aware of the expertise available through national, state, and local organizations and institutions. Curriculum managers should be diligent about keeping abreast of useful research and curricular trends and innovations.

In the area of physical education, sources of information include the AAHPERD (publications, staff specialists, position statements, conferences, conventions, and workshops), the state association of HPERD (conferences, conventions, and publications), the President's Council on Physical Fitness and Sports, specialists from the State Department of Education, and college professors who are curriculum specialists.

THE CURRICULUM COMMITTEE

The school curriculum committee typically designs and implements the instructional program on the local level. These committees take different forms in different situations. For example, a large high school might have a curriculum coordinating committee and several subcommittees representing different areas of the school program. In such a situation, each subcommittee would study the curriculum content in its particular area of responsibility and recommend revisions. The coordinating committee would consider recommendations from the subcommittees; assist the subcommittees in identifying weaknesses and possible solutions; watch for curricular conflict, duplications, and voids; and give broad leadership to curriculum development for the school.

In addition to the curriculum committee and/or subcommittees in a particular school, the school district often has a curriculum coordinating committee with two important functions: to serve as a resource body for the curriculum committees in the various schools of the district, and to coordinate the curricular content of the different schools.

The curriculum coordinating committee should have representation from the administration, faculty, student body, and the community (through the PTA or the school board). An administrator can provide insight into such matters as budget, facilities, faculty resources, and the like. Teachers can represent the grassroots professional level, because they work daily with students and are in close touch with what will and will not work. Students can provide information regarding their interests and the relevance of learning experiences. A representative of the public can help interpret the usefulness of curricular content in a practical sense as viewed from outside the school.

The local educational system must be compatible with the basic philosophic values that exist in the local community and comply with the needs and interests of the residents. The curriculum also must be consistent with government legislation and regulations. Currently, about four-fifths of the states determine the basic minimum content of physical education in their schools (Greenwood & Stillwell, 1999).

The Curriculum Guide

The purpose of a written curriculum guide is to give direction to the program. It translates the desired outcomes and objectives into action by describing what the instructional program purports to do and how it purports to do it. Curriculum guides may be written at the local, district, or state levels, and also are produced by professional associations. At the local level, the guide is usually developed by a curriculum committee or subcommittee composed of about five or six members.

Before proceeding, the committee may want to gather a representative sample of guides from other similar school districts to use as references. Curriculum guides tend to be consistent in content but have varied formats. The committee should choose the most effective curriculum model for their particular setting and students. Curriculum guides iterate the school's philosophy and general objectives, and they often translate these into performance objectives. Three main elements found in curriculum guides are: *standards* (national and state), *benchmarks* (targets for measuring student progress), and *sample tasks and activities* (including assessments). Program content can be organized by grade levels. The guide may also include instructional strategies and resources, methods for evaluation, and references. Upon completion, the guide is submitted to the proper authority for approval and adoption (Stillwell & Willgoose, 2002).

THE PHYSICAL EDUCATION CURRICULUM

If physical education is to justify its place in the school curriculum, then it must demonstrate that it contributes to the above principles and imperatives of education. The specific goals and objectives of physical education should be consonant with the general outcomes of education.

AAHPERD (1986) stated the goal of physical education as that integral part of total education which contributes to the development of the individual through the natural medium of physical activity—human movement. It is a carefully planned sequence of learning experiences designed to fulfill the growth, development, and behavior needs of each student.

The outcome of the curriculum should be the physically educated student. According to NASPE (1995), a person is physically educated when he/she:

- has learned skills necessary to perform a variety of physical activities,
- is physically fit,
- participates regularly in physical activity,
- knows the implications of and the benefits from involvement in physical activities, and
- values physical activity and its contribution to a healthful lifestyle.

This description of the physically educated student was translated into twenty outcome statements and sample benchmarks for selected grade levels. The outcome statements and benchmarks are available in NASPE's publication, *Moving Into the Future: National Physical Education Standards: A Guide to Content and Assessment* (1995).

Major Objectives of Physical Education

Outcomes and goals should be consistent with objectives, which facilitate their accomplishment and guide the development of the curriculum. Four major objectives of physical education have been widely accepted. They are listed below with their specific capacities/learnings:

1. Organic development (physical capacities)
     muscle strength and endurance
     cardiorespiratory endurance
     flexibility
     lean body composition
2. Neuromuscular development (psychomotor domain of learning)
     fundamental movement skills: running, jumping, throwing, catching, etc.
     kinesthetic sense
     agility
     eye-hand/eye-foot coordination
     balance
     sports skills

3. Development of knowledges and understandings (cognitive domain of learning)
   knowledge of terminology, etiquette, rules, safety, equipment, understanding of game strategies, physiological principles, and human movement

4. Personal/social/emotional development (affective domain of learning)
   attitudes
   values
   appreciations
   habits
   social skills

These objectives are to be realized through activities introduced into the physical education curriculum. For example, the teacher must consider what types of activities contribute to flexibility: Can this capacity be enhanced by dance, gymnastics, swimming? A good curriculum will contribute to most of the learnings and physical capacities listed under these objectives.

Curriculum Models

Educators in all disciplines should be aware of the various curriculum models that are common in the schools and understand the characteristics of each. These models provide a basic context within which instruction takes place. Three basic curriculum models follow (Stillwell & Willgoose, 2002).

The *Separate Subjects Model* has been the traditional pattern of curriculum organization. It constitutes the teaching of separate subjects—history, language arts, science, physical education—for a set amount of time each day. Little attempt is made to relate one subject with another. The tendency is for students to learn isolated facts and skills without perceiving them as being related. This approach works best on the elementary level when there is a physical education specialist.

The *Broad Fields Model* seeks to eliminate sharp lines between subjects by grouping subjects under a common umbrella. This approach helps students see relationships among specific subjects. It provides for fewer subject-matter areas and, therefore, longer time periods for each of the broader areas. Physical education can be combined with health, safety, and recreation under one umbrella, although health is often combined with the biologic sciences. The model requires administrative flexibility in scheduling.

The *Humanistic Model* focuses on what the students should *be* rather than what they should *know*. It is sensitive to student needs, and emphasizes methodology as much or more than curriculum content. This approach utilizes ability grouping and nontraditional evaluation. In physical education, the model suggests movement exploration activities, skill mastery, and activities that support objectives in the affective domain.

These curricular models can be employed across disciplines. Specific curriculum models for physical education, which derive from and are compatible with these general models, can be found in curriculum textbooks and in the periodical literature. The following ten curriculum models are listed by Stillwell and Willgoose (2002).

Developmental Model

Movement Education Model

Fitness Model

Academic Discipline Model

Personal-Social Development Model

Sport Education Model

Adventure Education Model

Multi-Activity Model

Games for understanding Model

Eclectic Model (composed of two or more of the above models)

Daryl Siedentop's *sport education model* provides an example of a curriculum model in physical education. In the sport education model, students form teams; take on roles such as coach, equipment manager, athletic trainer, and sports information director; and engage in league and tournament play. Special-duty teams officiate, keep score, and record statistics. Teams and roles are reassigned usually after about 20 lessons (a "season") to provide maximum interaction among class members. Each game category extends through a season, including league and tournament games in the various sports within that category. The "invasion game" category includes rounds of competition for games like basketball, soccer, flag football, and floor hockey. League play occurs each day, and win-loss records are kept to determine seeding for tournament play at the end of the "season." For individual sports like tennis and badminton, teams of four are formed, and team members are seeded. Team captains select the seeds, in consultation with team members. The number-one seeds from each team play each other in singles, and so on. The first and second seeds pair up to play doubles, as do the third and fourth seeds. Lower-seeded players can challenge their ranking within the team.

The sport education model emphasizes development of social responsibility. For example, when disputes or confrontations arise during competition, the offending students are quickly reminded that the "level" of social responsibility they are demonstrating falls below the expected level (Oslin, Collier, & Mitchell, 2001).

Selecting Curriculum Content

Determining the content of the curriculum from kindergarten through twelfth grade, and for the college level, is a complex task. The curriculum must have: (1) proper sequence of learning experiences; (2) adequate breadth and scope; (3) the right kind and amount of emphasis at the different stages of student development; and (4) practicality in terms of weather conditions, facilities, equipment, time, and faculty resources. Sources of information that can be useful in selecting curriculum content include:

- Teachers' opinions based on their own orientation, their professional preparation, and their evaluation of the students' needs and interests within the framework of what is feasible under the circumstances.

- Information from questionnaires completed by parents and/or students.

- Consultation, oral or written, with known experts in the field, whose information and judgment on curriculum matters are respected.

- Information produced and judgments made by curriculum committees.

- Curriculum content recommended in curriculum guides and textbooks on curriculum.

Since the amount of school time that can be allotted to any particular subject is limited, those individuals determining curricular content should be highly selective. Well-defined criteria such as the following can aid in the selection process.

- Is the activity consistent with the stated goals of physical education for the school? Will it contribute significantly and consistently to the goals?
- What are the strengths of the activity, be it physical fitness, lifetime sports, skills, social interaction, character development, or the development of useful knowledge, attitudes, and appreciations?
- Is the activity suitable for the particular student level (state of development), and does it fit properly at this point in the overall curriculum sequence?
- Is the activity relevant, meaningful, and interesting to students, and do students relate to it, have an interest in it, and see a need for it?
- Does the activity contribute to a logical sequence of curricular content?
- Is the activity safe, or under what conditions can it be made safe at an acceptable level?
- Does the school have the resources—faculty, facilities, equipment, budget— to sponsor the activity?
- Is the activity compatible with the local situation; is it within the limits of what the public would perceive as acceptable, useful, important, and appropriate?

Additional criteria may be listed to accommodate local circumstances, but proposed curricular content must measure up to these basic criteria. Don't assume that each criterion would the same value in the selection process. It is often a good idea to assign weights to criteria so that each one carries the influence that it logically deserves.

Curriculum Trends

At the opening of the twenty-first century, progressive physical education programs feature climbing walls, Skate in School® programs, and sophisticated conditioning programs that employ pedometers and heart monitors. The federal government has recognized the need to promote innovative physical education programs and has made available some $400 million in incentive grants over five years through the Physical Education for Progress (PEP) Act, passed in December of 2000.

Innovations and trends have always been part of the school curriculum. However, a curriculum must have a strong element of stability, as it is gradually improved and updated based on changing circumstances and new information. Rapid change for the sake of change and blind adherence to tradition are both extreme approaches that should be avoided. Some other recent trends in curriculum development include:

- Greater emphasis on physical education at the elementary school level, with concentration on perceptual motor learning and movement education.
- Increased attention to the individual needs and interests of students, resulting in: (a) more student involvement in learning methods, (b) students' choice of activities rather than school-imposed activities, (c) equal opportunities for both sexes, (d) emphasis on physical fitness activities and lifetime sports, and (e) increased use of audiovisual media for instruction.
- More flexibility in scheduling in terms of time, space, student grouping patterns, and staff made possible by the utilization of computers.
- Increased accountability for the effectiveness of the curriculum and instructional methods.

Many schools across the nation have adopted AAHPERD's "Physical Best" curriculum. Physical Best is the educational component of the comprehensive health-related fitness program that supports a curriculum to help students meet NASPE fitness standards. It is designed to raise students' awareness of their fitness level and motivate them to participate in physical activity to become their *physical best*. The program contains a health-related fitness assessment, an educational component, and an awards system. Curriculum guides and kits are available to assist the physical education teacher in implementing this program on both the elementary and secondary levels (Stillwell & Willgoose, 2002). A Physical Best CD-ROM, available for purchase from Human Kinetics Publishers at <http://www.humankinetics.com>, contains worksheets, posters, station cards, activity cards, forms, and charts for the program, ready to print out. FITNESSGRAM® is the testing component compatible with the Physical Best program.

Sequence of Content

The physical education curriculum should be planned so that students experience at least a minimum of exposure to those categories of activities listed in table 13-1. Although a minimal exposure may be required, students are encouraged to continue exploring activities in their areas of primary interest. In these areas, a progression of developmental experiences should be evident. This suggests a planned sequence.

Table 13–1
Upper Middle and Secondary School Physical Education Activities

Aquatics
springboard diving
SCUBA diving
lifesaving
beginning swimming
intermediate swimming
advanced swimming
synchronized swimming
water polo

Conditioning Activities
aerobic exercise
circuit training
free exercise
interval training
jogging
weight training

Gymnastics
balance beam
horizontal bar
parallel bars
uneven parallel bars

vaulting horse
pommel horse
rings
stunts and tumbling

Individual/ Dual Activities
archery
badminton
bowling
cycling
deck tennis
fencing
golf
racquetball
shuffleboard
ice skating
inline skating
roller hockey
table tennis
tennis
track and field
wrestling

Rhythms/Dance
ballroom/social dance
contra dance
folk dance
modern dance
square dance

Team Sports
basketball
field hockey
flag football
flickerball
floor hockey
ice hockey
lacrosse
soccer
softball
speedball
team handball
volleyball

Source: Adapted from Stillwell & Willgoose, 2002.

Curriculum sequence refers to the order in which the learning experiences occur. The sequence should be logical and interrelated. One of the weaknesses of physical educators has been a failure to provide a graduated sequence of instruction. (For example, too often the same basketball unit is taught repeatedly without offering new learning experiences.) The content of the physical education program should be organized into a continuous flow of experiences through a carefully planned, graduated sequence of skills, knowledge, and fitness levels from preschool through high school and into college. The sequence should be developed in light of the students' needs and interests and built progressively toward the attainment of challenging physical education goals. The following discussion is an overview of recommended program emphasis at the different school levels.

Preschool and Kindergarten Program

The emphasis should be on freedom of movement and basic skills, with attention centered on gross motor patterns (running, walking, crawling, climbing, pushing, pulling, and dodging), balance and stability, gross hand/eye coordination, and the development of self-awareness and expression through movement. The orientation should be toward the child as a unique individual with particular physical, mental, emotional, and social needs. Developmentally appropriate activities should emphasize spontaneous movement in an enjoyable environment that provides freedom to explore and opportunities to create movement patterns. Activities should integrate and reinforce other developmental domains, such as imagining and planning, and should offer contextual reinforcement to use these skills elsewhere. An abundance of vigorous large-muscle activities ensures motor fitness (Garcia, Garcia, Floyd, & Lawson, 2002).

Elementary School Program

The physical education program at the elementary school level should include activities with emphasis in the following areas (Stillwell & Willgoose, 2002):

> *Fitness Activities*
> *Game activities*
>   low-organized games
>   relays
>   cooperative games
>   creative games
>   lead-up games
>
> *Movement Exploration Activities*
> *Rhythmic Activities*
>   basic rhythms
>   creative rhythms
>   singing rhythms
>   traditional and contemporary dance
>
> *Self-testing Activities*

In the primary grades (K–3), emphasis should still be on fundamental or basic movement patterns. Attention should be given to (1) movement and performance awareness; (2) simple, organized activity concepts; (3) basic rules of safety; and (4) the development of flexibility, agility, balance, and coordination. Activities should concentrate on large-muscle development and movement exploration.

With the increasing popularity of middle schools, most elementary schools now are organized through fifth-grade level. However, a quarter of elementary schools still include grade six, and a small percentage of K-through-8th-grade schools remain (Clearinghouse, 1998). In the intermediate grades (4–5 or 6), the emphasis in physical education should be on the development and refinement of motor skills, physical fitness, the use of correct body mechanics in basic skills, and the development of activity-related concepts, such as rules and basic strategy. The orientation should be toward individual development with the individual as a member of the group (social interaction).

Middle School Program (Grades 5–8 or 6–8)

Middle schools were designed to bridge the gap between elementary school and high school in a way that junior high schools were not. Middle schools have now replaced junior highs by a ratio of ten to one. Junior high school is becoming a historical term both conceptually and in practice, although several school systems still use this designation. The middle school's mission is to attend to young adolescents' social, emotional, and physical needs as well as their intellectual development. In order to better accomplish the mission, some programs are exploring such practices as interdisciplinary teams of teachers, common planning time, eight-period days, flexible schedules, activity periods, and cooperative learning (NMSA, 1995).

In many instances, the physical education program has not been properly designed for middle schools but simply extends the program for grades five and six on through grade eight. The emphasis in middle school physical education should be on (1) the development and maintenance of physical fitness, (2) the development of a wide variety of specific activity skills, (3) a basic knowledge and appreciation of a variety of activities to serve as a base for intelligent choices relative to out-of-class participation, and (4) the development of self-awareness and self-confidence.

The orientation should be on the student as an emerging adult who needs broad exposure and significant challenges. The activities should emphasize the development of a diverse range of human talents and interests, which would form a base for choices and contribute toward a feeling of fulfillment and success. The student's frequent awkwardness should be de-emphasized and the need for social interaction, group involvement, and vigorous participation should receive ample emphasis. Team sports are of obvious importance. The program should also include challenging aquatic activities, graduated stunts, tumbling, and gymnastics skills, combative activities, intermediate-level dance activities, outdoor education skills, and graduated fitness activities and tests.

High School Program

Three major considerations are involved in organizing curriculum content at the high school level. The first involves designing the content area and placing various activities in appropriate categories. The second consideration pertains to time assigned to each content area. The third consideration is developing the structural framework, which incorporates the above two factors. It is often advantageous to initially organize the major content areas into blocks of time according to the degree of emphasis by grade level (Stillwell & Willgoose, 2002). Generally, the high school physical education curriculum is divided into six content areas:

aquatics
conditioning activities
gymnastics
individual/dual activities
rhythms/dance
team sports

In addition, outdoor education—or adventure education—has gained popularity in recent years. It may be taught as an extension of physical education or as a separate discipline. (See the discussion on outdoor education in a later section.) Table 13-2 shows recommended time allotments for various categories of activities in the physical education curriculum for grades 7 through 12.

Table 13–2
Time Allotment for Upper Middle and Secondary School Physical Education Content (in percent)

Content Area	Grade Level					
	7	8	9	10	11	12
Aquatics	10	10	10	10	10	10
Conditioning activities	10	10	10	15	15	15
Gymnastics	15	15	15	10	10	10
Individual/dual activities	20	20	20	30	30	30
Rhythms/dance	15	15	15	10	10	10
Team sports	30	30	30	25	25	25

Source: Adopted from Stillwell & Willgoose, 2002.

College Programs

The emphasis in college should be much the same as in high school except with more orientation toward lifetime activities, exposure to all the different areas of physical education, and the opportunity for advanced instruction. The curricular orientation should be toward the students' needs to gain insight into their futures as nonstudents in terms of wellness and enrichment of life. These two concepts should be carefully interwoven into the whole program.

Over the last few decades, the curricular emphasis on the college level has shifted from team sports to individual sports to motor fitness development, and now to health-related fitness, with a focus on promoting healthy living behaviors and lifelong activity skills. Administrators of basic instruction programs list the four main outcomes of college physical education as to:

- facilitate lifelong participation,
- help students enjoy physical activity,
- enable students to become fit and healthy, and
- to help students understand the importance of movement in their lives.

The curricular emphasis on fitness activities and individual sports reflects the continuance of a trend that began in the 1970s and 1980s. Activity courses in these two areas remain popular among students. Specific activity courses in their order of popularity are: fitness or aerobic activities, weight training, golf, tennis, bowling, and racquetball. When aggregated, outdoor activities represent the seventh most popular activity. Rock climbing was most frequently mentioned. Multidimensional, concept-based courses that combine a lecture component with activity experiences also remain popular in colleges across the nation. "Fitness for Life" is the most frequently reported title for this type course, although the terms "health" and "wellness" also are commonly used in course titles (Hensley, 2000).

If students entering college have previously had outstanding physical education experiences, they are ready to elect advanced courses and are prepared to participate successfully in intramurals and sports clubs. However, many students enter college without this background and consequently have poor fitness profiles and are deficient in basic movement and recreation skills.

Elective versus Required Physical Education

Whether to classify physical education as required or as elective is a decision that should be incorporated into the philosophy of the school or college. Physical education traditionally has been a required subject. A strong case can be made for maintaining the requirement at the elementary and middle-school levels. However, the case weakens slightly at the high school and college levels, in the sense that the students are better prepared to make sound judgments relative to electives. However, the lack of fitness of students in high school and beyond indicates a definite need for effective fitness education. At any level, an effective physical education program easily can be justified, whether required or elective.

Changing from a required to an elective program does not necessarily mean lower status or a reduction in faculty or student enrollment. It does, however, change the nature of the program. An elective program defines the activities to be offered, and students select those in which they desire instruction. In this situation, departments often have to market their offerings.

Both required and elective physical education have their proponents and opponents. Following are some pros and cons from both viewpoints.

Arguments in Favor of Required Physical Education

- Mandatory enrollment facilitates efficient scheduling.
- Physical education might not be emphasized by the particular administration or might deteriorate if not required.
- Because students take so many other required courses, they would not have time for physical education as an elective.
- All students need daily vigorous activity for emotional release, physical fitness, and total development.
- Students look at required subjects as being more important than electives.

Arguments against Required Physical Education

- A required program does not motivate teachers to meet student needs and interests.
- Physical education classes may be perceived as irrelevant or a waste of time.
- Students resent requirements, and so do parents.
- Required classes are counter to the principle of individualized education.
- Many students lack motivation toward required participation.
- Some students learn to dislike physical education when it is required and want nothing to do with it in later life.

Arguments in Favor of Elective Physical Education

- Student choice of activities increases enthusiasm for participation.

- Elective programs increase possibilities for meeting individual differences among students.

- Elective programs increase student freedom and responsibility for making their own decisions and increase the possibility of carryover into later life.

- The drive for activity will cause students to enroll voluntarily.

- Elective programs often better utilize community facilities, thus reducing expenses to school districts.

- An elective program motivates teachers to do a better job because students will enroll only if the classes are interesting and useful.

Arguments against Elective Physical Education

- Those who need physical education the most will not take it.

- Elective physical education is harder to schedule and tends to result in less efficient use of faculty and facilities.

Status of Required Programs

Physical education used to be an accepted component of the school curriculum, but this is changing. The percentage of high school students enrolled in daily gym classes declined from 42% to 29% during the 1990s. Typical of this trend is Atlanta (GA) Public Schools, which eliminated required physical education in the fall of 2000. Programs continue to be eliminated, despite a 1996 report by the Surgeon General of the United States that recommended requiring daily physical education for all students in kindergarten through twelfth grade. No federal standard requires that physical education be offered to students in U.S. schools, however. Each state and territory of the United States has legislation that describes the educational requirements. In Illinois, the only state to require daily physical education in all grades, 20% of the school districts have asked for a waiver. A district or school may enact requirements beyond those of the state—even in Colorado and South Dakota where no mandates exist, most districts provide for some physical education—but a district may not require less than a state requirement. In many states the legislated mandate requires only that physical education be offered, and local districts provide the content and guidelines. For a table summarizing state requirements, see chapter 12 (table 12-1).

At the elementary school level, state-mandated requirements for physical education range from 30 minutes a week to 150 minutes per week (NASPE recommends 150). At the middle school level, physical education time requirements range from 80 minutes a week to 275 minutes per week (NASPE recommends at least 225). The majority of high school students take physical education for only one year between ninth and twelfth grades. The time requirements range from no time specified to above the NASPE recommendation.

Oklahoma shifted from specific time requirements to a standards-based requirement.

Policy on Substitutions for Physical Education

When schools have to deal with requests for substitutions for participating in the physical education instructional program, a consistent policy should be in place. As indicated in table 12-1, requests typically include participation on athletics teams, ROTC, and marching band. None of these activities duplicate a comprehensive physical education program. Participation on an athletics squad promotes specialization in one sport, which may or may not be a lifetime sport. Likewise, marching and basic training exercises fall short of meeting the comprehensive objectives of physical education. In fielding requests for substitutions, the administrator should inquire as to what special contribution the activity makes toward student development that would justify this substitution. Unless the answer is acceptable, the request should be denied.

Exempting students from physical education because of disabilities is no longer an option, as federal law requires instructional programs to accommodate all students with special needs. Modifying physical activities because of medical problems encompasses a separate set of issues. Medical excuses from family physicians are problematic, as a few doctors will write excuses indiscriminately to accommodate patients and their families. The school should develop a modified activity form (see figure 12-6 in previous chapter) for the physician to fill out, which specifically excludes inappropriate activities for the student but discourages substituting another course for physical education.

RELATED FIELDS

This chapter has been concerned with the physical education curriculum. However, physical education closely relates to some other areas of the school and college curriculum—notably *health education, outdoor education, recreation,* and *sports management.*

Health Education

Traditionally, health and physical education have been closely associated. In the public mind the differences between them has been indistinct, often to the detriment of health education. Traditionally the two subjects have been taught by the same person, although the goal is to increase the number of teachers certified in health education to teach these classes. About forty states require the schools to teach health education. A survey by the Centers for Disease Control (CDC) conducted in 2001 found that not quite half of fifth graders, about a fifth of eighth graders, and even fewer high school students were receiving instruction in health education. Health education was integrated into other subject-area courses as often as it was taught as a separate course. Only ten percent of those currently teaching in this area hold a degree in health education. Even fewer are certified health education specialists ("AAHE Responds . . .," 2001).

Even though health and physical education should remain closely associated and strongly supportive of each other, the days are gone when health education can be viewed as part of physical education or supplemental to it. Health should be a separate phase of the curriculum under the leadership of a health education specialist.

In 1995 the National Health Education Standards were published by the American Association of Health Education (AAHE); and in 1998, their Assessment Framework was released. These standards provide a foundation for curriculum development, instruction, and assessment of student performance. The standards were designed to be compatible with the goals in Healthy People 2000 (superseded by Healthy People 2010), promulgated by the Office of Disease Prevention and Health Promotion in the Department of Health and Human Services.

The Centers for Disease Control (NCCDPHP, 2002) has identified the key elements of comprehensive health education as:

1. A documented, planned, and sequential program of health instruction for students in grades kindergarten through twelve.

2. A curriculum that addresses and integrates education about a range of categorical health problems and issues at developmentally appropriate ages.

3. Activities that help young people develop the skills they need to avoid tobacco use, dietary patterns that contribute to disease, sedentary lifestyle, sexual behaviors that result in HIV infection, other STDs and unintended pregnancy, alcohol and other drug use, and behaviors that result in unintentional and intentional injuries.

4. Instruction provided for a prescribed amount of time at each grade level; management and coordination by an education professional trained to implement the program.

5. Instruction from teachers trained to teach the subject.

6. Involvement of parents, health professionals, and other concerned community members.

7. Periodic evaluation, updating, and improvement.

In some schools, health education receives little attention beyond that provided by the physical education teachers. In such cases, the physical educators ought to make an honest effort to provide useful health instruction and encourage sound health practices. Health and physical education are too important and too closely related for us to allow either area to be neglected.

A comprehensive health education curriculum should consist of planned learning experiences that will assist students to achieve desirable understandings, attitudes, and practices related to critical health issues. These issues include (but not limited to) emotional health and a positive self-image; appreciation, respect for, and care of the human body and its vital organs; physical fitness; health issues of alcohol, tobacco and drug use and abuse; health misconceptions and quackery; effects of exercise on the body systems and on general well-being; nutrition and weight control; sexual relationships; the scientific, social, and economic aspects of community and ecological health; communicable and degenerative diseases including sexually transmitted diseases; disaster preparedness; safety and driver education; choosing professional medical and health services; and choices of health careers.

The curriculum should be related directly to the needs, problems, and interests appropriate to the growth, development, and maturity level of students. The scope and sequence of the comprehensive school health education curriculum, grades K–12, should be planned and organized so as to avoid both serious omissions and unnecessary duplication and repetition.

The elementary teacher and the teacher of health education in secondary schools should receive assistance from a health education specialist or supervisor who is an outstanding instructional leader. The supervisor should also provide the liaison with health agencies and other community groups that will tie school and community health education together in a consistent fashion.

Outdoor Education

Outdoor education should not be separated from other subjects in the school but integrated as much as possible. Physical education contributes to the skills and attitudes involved in outdoor education, and some aspects of outdoor education often are administered as part of the physical education program. Outdoor education also relates to recreation, as much of what is learned is subsequently applied to camping, hiking, mountain climbing, skiing, canoeing, kayaking, and orienteering.

Outdoor education includes two major categories: education *in* the outdoors and education *for* the outdoors. The two categories are related and complement each other. Education in the outdoors has evolved to become roughly synonymous with environmental education and usually is regarded as outside the purview of physical education. It entails direct outdoor experiences involving observation, study, and research in both public and private outdoor settings such as parks and recreation areas, camps, bodies of water, forests, farms, gardens, zoos, sanctuaries, and preserves.

Education for the outdoors entails the learning of skills and the development of appreciation for activities such as camping and survival skills, including mapping, orienteering, outdoor cooking, casting and angling; water sports (canoeing, kayaking, rafting, swimming, water skiing, skin and scuba diving); winter sports; rock climbing; hiking; bicycling; and outdoor photography. Skills supporting several of these activities are part of the physical education curriculum.

Recently, outdoor education has placed more emphasis on reflection and learning from experience, and on group dynamics. From the administrative perspective, risk management has taken on increasing importance in running outdoor education activities.

Recreation

Schools become involved in recreation in three ways: (1) education for the worthy use of leisure time; (2) the provision of recreational activities, such as playground activities in the lower grades, intramurals, and sports clubs at the higher levels; and (3) involvement in and support of community recreation programs. Instruction in physical education classes supports the development of recreational skills. School sponsorship of out-of-class recreational programs builds a foundation for continuing recreational habits of students. In addition, some schools cooperate with the surrounding community in developing the park–school concept and in co-sponsoring recreational programs during the summers. Schools have a large number of indoor and outdoor facilities, which they can make available to community groups for recreational purposes.

At the college level, recreational administration and therapeutic recreation are professional majors often housed in the same department or school as physical education. Students in these programs often enroll in courses in both disciplines. Upon graduation, they find positions in a variety of private and public agencies such as nursing homes, hospitals, city parks and recreation departments, or in private-sector fitness programs.

Sport Management

Sport management developed as a subdiscipline within physical education on the college level. It is one of the alternative career paths for those whose career goals are outside teaching. The thirty-five-year-old field of study now includes some 200 schools that offer undergraduate and/or graduate degree programs. The disciplinary basis for sport management combines sport with business. Degree programs incorporate courses in communications, interpersonal relations, business, accounting, finance, economics, and statistics. Students engage in field experiences through practicums and internships. Career paths in sport management include athletic team management, finance, sports medicine/athletic training, journalism, broadcasting, public relations, development and fundraising, sports information, facility management, cardiovascular fitness and wellness administration, and aquatics management, among others. Graduates in sport management can find jobs as entrepreneurs, representatives, or administrators (Stier, 1993).

References and Recommended Readings

"AAHE Responds to CDC School Health Data," 2001. *HExtra*, 27(1)(winter): 1.

American Alliance for Health, Physical Education, Recreation and Dance. 1986. *Guidelines for Secondary School Physical Education*. Waldorf, MD: AAHPERD Publications.

Clearinghouse on Educational Management. 1998. *Trends and Issues: School Organization—Grade Span*. Eugene, OR: University of Oregon.

Commission on Reorganization of Secondary Education. 1918. "Cardinal Principles of Secondary Education." Washington, DC: U.S. Government Printing Office.

Cooper, Shirley, 2002. "Imperatives in Education." ERIC Resumé Abstract for ED012961. Online article: <http://www.edrs.com/members/Detail.cfm?id=38228&CFID=290229&CFTOKEN=15969088>.

Garcia, Clersida, Luis Garcia, Jerald Floyd, & John Lawson. 2002. "Improving Public Health through Early Childhood Movement Programs." *Journal of Physical Education, Recreation & Dance*, 73(1): 27–31.

Goodlad, John. 1994. Chapter 3 of *What Schools Are For*. Bloomington, IN: Phi Delta Kappa.

Greenwood, M., & Jim Stillwell. 1999. "State Education Agency Curriculum Materials for Physical Education." *The Physical Educator*, 56(3), 155–58.

Hensley, Larry D. 2000. "Current Status of Basic Instruction Programs in Physical Education at American Colleges and Universities." *Journal of Physical Education, Recreation & Dance*, 71(9) 30–33.

National Association for Sport & Physical Education. 1995. *Moving into the Future: National Physical Education Standards*. Reston, VA: American Alliance for Health, Physical Education, Recreation and Dance.

National Association for Sport & Physical Education. 2001. *Shape of the Nation Report*. Reston, VA: American Alliance for Health, Physical Education, Recreation and Dance.

National Middle School Association. 1995. *This We Believe: Developmentally Responsive Middle Level Schools*. Columbus. OH: NMSA. Online article: <http://www.nmsa.org/>.

National Center for Chronic Disease Prevention and Health Promotion (2002). "School Health Defined: Comprehensive Health Education Curriculum." Online article: <http://www.cdc.gov/nccdphp/dash/about/comprehensive_ed.htm>

Oslin, Judy, Connie Collier, & Steve Mitchell. 2001. "Living the Curriculum." *Journal of Physical Education, Recreation & Dance*, 72(5): 47–51.

Stier, William, Jr. 1993. "Alternative Career Paths in Physical Education: Sport Management." *ERIC Digest*, ED362505 (August). Monograph, n.p.

Stillwell, Jim, & Carl Willgoose. 2002. *The Physical Education Curriculum*, 5th Ed. Prospect Heights, IL:Waveland Press.

Online Sources

American Association for Active Life Styles and Fitness, Council on Outdoor Education
http://www.aahperd.org/AAALF/councils_societies.html#7
The Council's mission is to support professionals who work in higher education, elementary schools, secondary schools, recreation and park programs, and clinical settings who provide programs in, for, and about the out-of-doors; particular emphasis is placed on team building through the use of ropes and adventure courses.

American Association for Health Education
http://www.aahperd.org/aahe/template.cfm
AAHE supports and assists health professionals concerned with health promotion through education and other systematic strategies. AAHE has provided leadership in organizing and administering collaborative projects resulting in: (1) development of national health education standards for grades K–12, (2) development of standards for the preparation of graduate-level health educators, and (3) development of joint terminology for health education.

American Association for Leisure and Recreation
http://www.aahperd.org/aalr/template.cfm
The mission of AALR is to promote and support education, leisure, and recreation by: (1) developing quality programming and professional training, (2) providing leadership opportunities, (3) disseminating guidelines and standards, and (4) enhancing public understanding of the importance of leisure and recreation in maintaining a creative and healthy lifestyle.

Association for Outdoor Recreation and Education
http://www.aore.org/
The mission of the Association is to provide opportunities for professionals and students in the field of outdoor recreation and education to exchange information, promote the preservation and conservation of the natural environment, and address issues common to college, university, community, military, and other not-for-profit outdoor recreation and education programs.

The Association for Supervision and Curriculum Development
http://www.ascd.org/
ASCD is an association of professional educators from all grade levels and subject areas. They hold conferences, provide resources and professional development opportunities, and maintain an online store.

Curriculum and Instruction Academy
http://www.aahperd.org/naspe/template.cfm?template=main.html#
Click on "Interest Areas" and then "Academies." The purpose of this academy under NASPE is to further the body of knowledge in sport and physical education that relates to curriculum and instruction.

North American Society for Sport Management
http://www.nassm.com/
The purpose of NASSM is to promote study, research, scholarly writing, and professional development in the area of sport management, including the applied aspects of management theory and practice specifically related to sport, exercise, dance, and play as pursued by all sectors of the population. The Society endeavors to support and cooperate with local, regional, national, and international organizations having similar purposes and organizes meetings to promote its purposes. It publishes the quarterly *Journal of Sport Management*.

14 Intramural Recreation

Educator John Holt commented, "Sports, athletics, and games are too important to be *just* for the varsity." Intramural programs embody this ideal. They should be designed to serve the recreational needs of the entire student body and not just the athletes. Even teachers and administrators can participate in intramural activities. For example, college faculty members enter teams in intramural volleyball, and public schools schedule annual student/teacher softball games.

While varsity athletics emphasize competitiveness and winning, intramurals emphasize enjoyment and maximum participation. Athletics programs reach outside the school into the community; intramurals build a sense of community within the school.

BACKGROUND, PURPOSE, AND SCOPE OF INTRAMURALS

Intramurals are one of the oldest forms of organized sports in U.S. education. Early programs were the result of students wanting to participate in leisure-time athletic activities of their own choosing. The early college-level programs were student led, student financed, and student administered. Student-organized intramurals were present on campuses before required physical education and intercollegiate athletics. They persevered alongside intercollegiate sports programs that grew rapidly in the last half of the nineteenth century. Students continued to run their own programs until after World War I, when colleges began forming intramural departments with professional directors. High school intramurals emerged later over the course of the century, but these programs have not fared as well as the college programs. At mid-century, intramurals acquired its own professional association, now known as the National Intramural-Recreational Sports Association, which grew out of the early efforts of Dr. William Wasson to organize men and women intramural directors at historically Black colleges. In the last quarter of the twentieth century, intramurals began to "catch on" in secondary schools, as colleges and universities shifted their emphasis toward the broader concept of *campus recreation*.

299

Purpose

The primary *purpose* of intramurals is to provide a variety of recreational activities, selected on the basis of their contribution to the development of the individual participants. Outcomes of intramural participation include enjoyment, health and fitness, sociability, leadership, and a sense of achievement.

Intramurals emphasize variety in activities and inclusiveness in participation, yet intramurals cannot be all things to all people. A sense of focus is needed. A *mission statement* can help provide overall direction to the intramural program. Byl (2002) lists the following criteria for creating a coherent mission statement. The statement should:

- Explain what the program is, what it stands for, and why it exists.
- Extend the program toward new, attainable heights.
- Be short and simple (something that can be read in 30 seconds or less).
- Be inspirational (by utilizing strong verbs in voicing the mission).
- Be reflective of the intentions of everyone involved in the program.
- Be visible (post copies and disseminate it).
- Be evaluated periodically and rewritten to more closely reflect the purposes of the program.

Following is an example of a mission statement written for a college intramural recreation program (Student Affairs, 2001):

> Our mission is to provide the students, faculty and staff at the University at Stony Brook with the opportunity to recreate in a positive and friendly environment. Recreation is provided through a wide variety of programs and opportunities, including an extensive intramural sports program, sports clubs, open recreation, fitness activity classes, non-credit instruction, special events and equipment rental. We strive to complement the academic goals of the university by encouraging the physical, emotional and social growth of individuals within the diverse structure of the program.

Scope of Intramural Programs

The term *intramural* is derived from two Latin words: *intra* (within) and *murus* (walls). Traditionally, the term has referred to sports activities conducted within the walls of a school. The present concept of intramurals refers to sports and recreational activities that are conducted less formally and separately from varsity athletics but may extend beyond the walls of the institution. The term *extramural* ("outside the walls") has been coined to describe informal competitive and recreational activities involving two or more schools. This can include everything from *sports clubs* (discussed later in the chapter), which compete with other clubs off campus, to *play days* that integrate students from different schools into competition/participation units.

Institutions differ with respect to the scope of their intramural programs. Some programs provide a wide range of team sports, individual and dual activities, and recreational pursuits, whereas other programs include only a few competitive sports. The scope of intramurals is often related to the size of the institution, with the large schools usually having more comprehensive programs. Public school programs generally sponsor fewer activities than college programs.

Specific types of activities and services that might be offered by an intramural program include:

1. Informal recreational opportunities (e.g., "open gym," pick-up games).
2. Special events, such as tournaments and meets.
3. Organized sports leagues: co-recreational, women's, and men's.
4. Extramural activities, including club sports (On some campuses, this may be a separate division from intramurals.).
5. Rental or loan of recreational equipment for personal use.

The selection of activities should be in keeping with sound educational objectives and the interests of the student body. The following guidelines ensure that this standard is met.

1. The program should be based on the local needs of the students at their level and on a sound philosophy of education and sports participation.
2. Activities should be available that offer opportunities for each person, regardless of gender, physical ability, race, or economic status.
3. Appropriate representatives of the school population, both students and staff, should be involved in planning and administering the program.
4. The program should encompass sufficient variety in activities.
5. Novelty and special events may be used for enhancement, but the heart of the program should be regularly scheduled recreational activities, well planned and competently conducted.
6. Whenever possible the program should be free to participants, financed from school funds.
7. The program should not serve as a substitute for physical education instruction or interschool athletics.
8. Activities should be modified to suit the needs and abilities of the participants.
9. Participation should not be an end in itself, but also the means to teach rules, sportsmanship, and values. Students should learn how to win and lose appropriately.
10. The decision-making process in the program should be student-centered with guidance from the directors.
11. Students should be offered both leader and follower roles.

Intramurals at Various Levels

Intramurals have their place in school programs at all levels from elementary school through college; however, the purposes, approaches, and program content are different at each stage of the educational system.

Elementary School

The need of elementary school students to be active, their desire to play, their willingness to accept challenges, and their desire to belong to a group create justification for a certain kind of intramural program at the elementary school level.

Some objectives of intramurals in the elementary school are: (1) to develop movement and basic sports skills, (2) to provide moderate competition in a game setting, (3) to provide opportunities for cooperative play, (4) to foster fun, (5) to provide involvement in democratic processes, and (6) to introduce students to play opportunities across a variety of activities.

Activities utilizing basic motor skills, low-organization games, lead-up activities, and modified games should be included. To sponsor a particular activity in the school, the club approach may be used, as well as interest groups or a special theme. Likewise, special events, field days, and mass activities for students are vital parts of the program.

Middle School

Excellent opportunities exist for intramural programs at this level, due to the unique needs and interests of this age group. Following are some important objectives of middle school intramural programs.

1. Provide team activities for the enhancement of esprit de corps and group closeness.

2. Present a wide range of activities to satisfy the various levels of ability and the broad range of interests.

3. Emphasize co-recreational opportunities to aid in socialization.

4. Offer vigorous activities to dissipate nervous energy and contribute to fitness.

5. Give opportunities for individual achievement and recognition.

6. Offer some interesting special events so that students can have fun, meet new friends, and enjoy both competition and cooperation.

Ideally, intramural personnel or other faculty members should serve as officials, and student helpers should secure the equipment and prepare for the contests. Supervision by adults is important at this level. Usually the activities are conducted after school, but some successful programs also utilize hours before school, during lunch, and on weekends. To some degree, transportation problems can be reduced through careful analysis and adjustments.

Units of competition may be based on homerooms, neighborhood units, physical education classes, clubs, or pickup teams. A few schools may use a classification index or skill-test results for assignment to intramural teams, but administrative limitations of time and personnel normally prohibit this approach. Students in middle school should have an opportunity to participate at least once each week.

Regarding eligibility, all students except varsity players of the particular sport should be eligible to play in intramural activities. Students should play for only one team at a time, so as not to reduce participation opportunities for other students. However, this does not mean that varsity players (if there are varsity sports at the middle school level) could not serve as timers, scorers, and even officials.

High School

The general objective of intramurals at the high school level is to provide a wide variety of activities, in order to encourage regular participation in team and individual sports and in special events. Skill development, a positive self-image, fun, friendly competition, and the development of interest in lifetime sports are all valid reasons for a high school program. Improving self-esteem and breaking down social barriers that occur through participation in intramurals can reach students who are inclined to drop out or who may not be inclined to participate in other school activities.

At the high school level, intramural activities can be conducted before school, at noontime, during activity periods, after school, or in the evening. The ideal time is after school, except that conflicts with varsity sports often arise. To some extent, facility problems can be overcome by playing the sports out of season and through cooperative use of community facilities. Widespread busing of students has diminished the opportunities to schedule intramurals, although some schools provide activity buses to accommodate students participating in extracurricular activities.

Colleges and Universities

Intramurals in higher education is evolving conceptually and administratively. "Campus recreation" has emerged as the umbrella term, as governance of these programs has shifted from physical education departments to student affairs. Some colleges report that close to half of their students participate in intramurals, with dozens of teams in popular sports like flag football and basketball. Faculty, staff, and families also participate in some recreational programs. Most programs and facilities are financed in part by student fees.

The popularity of intramurals on college campuses has triggered a construction boom in student recreation centers. By 2002, it was estimated that over 60 percent of campuses had a student recreation facility. Most of these facilities include lounge areas, aquatic centers, racquetball courts, dance exercise areas, indoor and outdoor tennis courts, basketball courts, indoor running/jogging tracks, strength and cardiovascular training areas, climbing walls, indoor/outdoor rollerblade hockey courts, and even indoor soccer areas. In 2001, Washington State University completed a recreation center with 160,000 square feet of floor space (equivalent to four football fields). Georgia Tech was building a $45 million recreation complex as this book went to press.

Technology is employed to support large programs. Intramural departments are utilizing the Internet to facilitate registration for and scheduling of activities. Intramural directors are employing palm computers to schedule, coordinate, and mange activities on the field and to record results of competition (Mihoces, 2001; Spoor, 1998).

The benefits of intramurals are the same in college as in high school. However, college programs differ from those in public schools in several aspects that affect their administration: (1) College intramural programs are relatively large, affording the opportunity for a greater variety of program offerings. (2) The facilities are more extensive and more available. (3) A large portion of the student body may live in residence halls or apartments, in close proximity to the campus, thus reducing the transportation problems that often exist with public-school students. (4) College students have more flexible schedules than high school students do. (5) The eligibility issue is more complex. (6) Many students at a university live away from home, creating an increased need for involvement in intramurals and other school activities.

ORGANIZATIONAL PATTERNS

An intramural program can be organized in several ways. Three different patterns are shown in figures 14-1, 14-2, and 14-3. The educational level, size of the program, and administrative philosophy are all influencing factors. College-level intramural programs are associated with the office of the Vice President for Student Affairs or the Student Union, or with the Physical Education or Recreation Departments, and rarely under the Athletics Department.

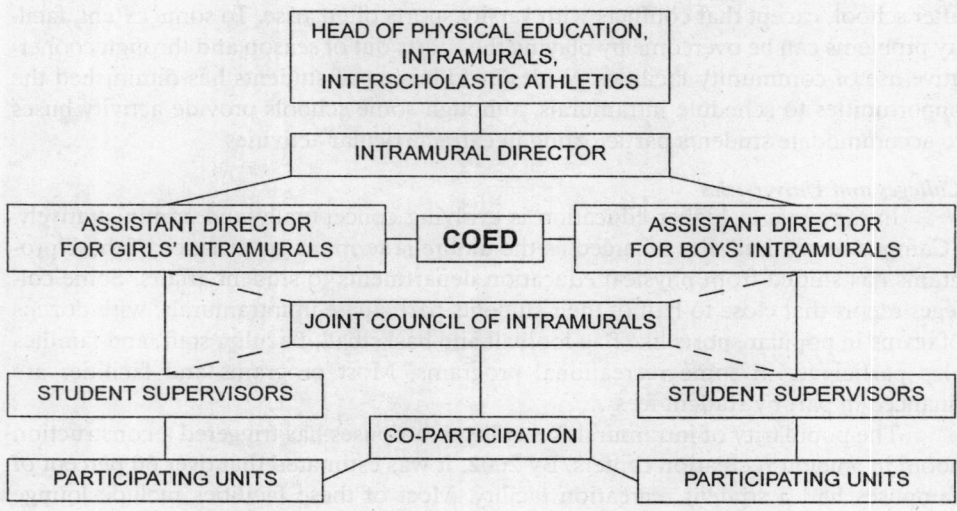

Figure 14–1. Example of a desirable intramural organizational pattern for a secondary school (courtesy of Bruce Holley)

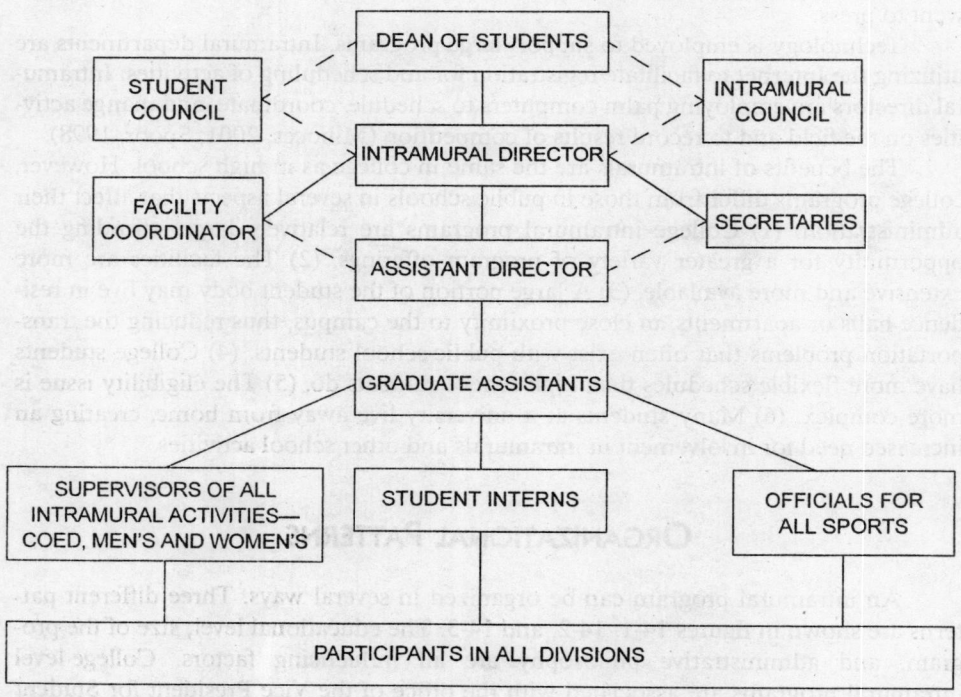

Figure 14–2 Example of an intramural organization for a college, with the program administered under student government

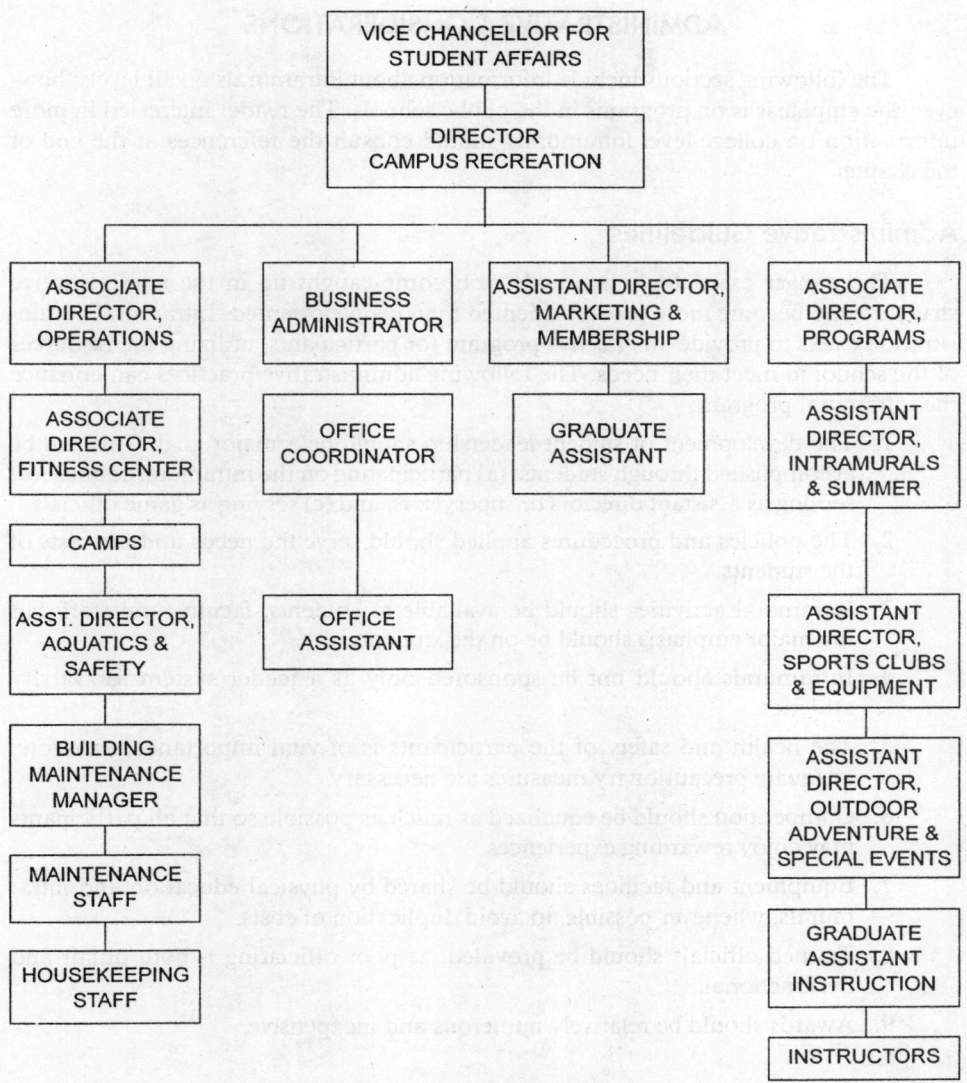

Reproduced with permission of the Department of Campus Recreation, University of Houston.

Figure 14–3.   Organizational structure for campus recreation at a large university

ADMINISTRATIVE CONSIDERATIONS

The following sections include information about intramurals on all levels; however, the emphasis is on programs in the public schools. The reader interested in more information on college-level intramurals should consult the references at the end of the chapter.

Administrative Guidelines

The danger exists for intramurals to become caught up in the administrative structure and become more process oriented than people oriented. Intramural administration exists to provide an effective program for participants, utilizing the resources of the school to meet their needs. The following administrative practices can enhance the intramural program:

1. The development of student leadership should be a major goal. This can be accomplished through students: (a) participating on the intramural council, (b) serving as assistant directors or supervisors, and (c) serving as game officials.

2. The policies and procedures applied should serve the needs and interests of the students.

3. Intramural activities should be available to students, faculty, and staff; but the major emphasis should be on the students.

4. Intramurals should not be sponsored only as a feeder system for varsity athletics.

5. The health and safety of the participants is of vital importance; therefore, adequate precautionary measures are necessary.

6. Competition should be equalized as much as possible so that all participants may enjoy rewarding experiences.

7. Equipment and facilities should be shared by physical education and intramurals, whenever possible, to avoid duplication of costs.

8. Trained officials should be provided, as poor officiating is both unfair and dysfunctional.

9. Awards should be relatively numerous and inexpensive.

Finances

The finances of an intramural program involve two broad considerations: obtaining a constant and adequate source of funding, and administering the funds wisely toward meeting the objectives of the program. Although an ample intramural budget is important, many effective programs are sponsored with limited funds, especially in the public schools. Shortage of funds should not be used as an excuse for not having a program. If the program has a history, the director should be able to forecast revenue needs and expenditures based on the past record. Intramural programs usually receive their allocation from the general budget at the beginning of the school year, so the directors know how much they have to work with. If the figures suggest a deficit budget for the coming year, then additional sources of funding must be considered.

Possible sources of additional revenue for intramurals include participation fees, facility rental, fundraising activities, and sponsorships. Participation fees usually are charged per event. The fee system has the drawback of pricing some students out of the program. When this happens, fees become counterproductive. Some schools rent their facilities and direct part of this money to intramurals. However, other schools barely have enough facilities to accommodate in-house programs, and rental opportunities are limited. Fundraising ideas include contests, dances, car washes, "a-thons" (e.g., walk-a-thons), and amateur nights. Check with local authorities and the school district office regarding regulations and restrictions on fundraising before scheduling an event (Byl, 2002).

The major expenditure for school intramurals is on recreational and sports equipment, awards, and office supplies. Sometimes the intramural director can share office supplies with other departments. If not, an adequate amount should be budgeted for this. Trophies can quickly exhaust limited budgets. Many schools resort to donated awards such as T-shirts. Travel expenses to extramural events may be covered by the school transportation budget. Likewise, school insurance should cover intramural activities. Budget management can be facilitated by using spreadsheet programs on computers.

Facilities and Equipment

While many colleges build large, modern recreational buildings to accommodate intramurals, public school programs usually have limited facilities and often have to share them with other programs like athletics. Most schools have a gymnasium, classrooms, outdoor fields, and blacktop areas. Other school facilities like hallways or the cafeteria might be utilized for some intramural activities. In addition, the director should survey the neighborhood and the community for facilities that may be rented, leased, or even used free of charge. A combination of the above tactics should be adequate to facilitate most intramural programs.

Many gymnasiums have a stage at one end that can accommodate games like table tennis or foosball (table soccer). Hallways might be used for bowling, indoor horseshoes, or shuffleboard. Solid walls outside can be utilized for handball or paddleball. Paved parking lots provide a good surface for ball hockey and similar activities. Neighborhood parks and wooded areas can be used for hiking, cross-country, or even skiing in the winter. Check with local managers of commercial recreation facilities for discounts or group rates for students. Steering students into local activity clubs organized outside the school still meets the goal of promoting participation in healthy recreation. The important point is not to get discouraged but to be creative in utilizing what is available.

Equipment needs provide a similar challenge for intramural directors. Rarely do programs in public schools have an adequate budget for their own equipment. It's worth talking to vendors, service clubs, and other groups in the community to solicit additional equipment. Sometimes intramurals can share equipment with physical education. On the other hand, most athletics coaches are reluctant to share equipment (or facilities). Often, it becomes a matter of equipment determining program. Make the most of what you have. Creative directors have made their own equipment, such as beanbags or hula-hoops constructed from rubber piping. In some instances, students share their personal sports equipment. Here again, the bottom line is to be creative in meeting program needs. Finally, the director must recognize that equipment distribution and control can be a real problem for intramural programs. Be sure to set up equipment checkout and return procedures that hold users accountable (Byl, 2002).

Intramural Council

The director should form an intramural council largely made up of students. Several methods can be considered to staff the council: students can volunteer, students can be elected, the director can select students, or the director can ask teachers to recommend students. Each method has its advantages. The goal is to obtain dedicated and competent council members. The size of the intramural council should be workable—usually ten or less members. The duties of councils vary among schools and grade levels, but generally the council plans and implements the program. More specifically, it may make policy decisions, rule on forfeits, or adjudicate disputes. Councils that meet on a regular basis can save the director a great deal of time. Council members receive valuable leadership experiences, and councils move the program away from paternalistic oversight toward participatory governance (Byl, 2002).

Volunteer Staffing

Intramural recreation departments at the larger colleges and universities have professional staffs ranging from a handful of individuals to more than a dozen. They also utilize student managers and assistants. Public school intramural programs rely almost entirely on volunteers (a term that may describe the intramural director, as well) to run their programs. This section will address briefly the administrative considerations of volunteer staffing. (Professional staffing concerns are treated at length in chapter 6.)

Student volunteers can assist with administration, promotion, equipment distribution, supervising contests, and officiating. Students benefit greatly from these experiences, which teach them leadership and planning skills, a sense of responsibility and accountability, and how to work with others. Older students can take on more responsibility. High school and college students, if properly tr.ained, can practically run the program themselves under the director's guidance. The director should schedule recruiting and training sessions for volunteers at the beginning of the school year (Byl, 2002).

Contest Officials

Most public school programs cannot afford to pay officials and must utilize student volunteers. A few schools make officiating part of the physical education course requirements. Directors may want to set up training workshops for officials after school or on Saturdays to make certain that they are knowledgeable of rules and officiating techniques. Varsity athletes are potentially good officials, as they are usually knowledgeable, are respected by their peers, and benefit from being on the other side of the whistle. Another commonly used method for obtaining officials is to have each team entered in a league or tournament to provide official(s) to officiate other contests. This way the responsibility is rotated and leads to more tolerance of officials' decisions (Byl, 2002).

Poor officiating can be one of the biggest headaches for intramural directors. It causes feelings of unfairness and can lead to acrimony. In some instances, intramural directors have found that contests can be played without officials, much like what occurs on the playground. Under this system, conflicts are resolved by the participants who usually are quick to admonish their peers with, "Do you want to play ball or argue?"

Team Captains

Team captains are an important element in leagues and tournament play. Although students can choose their own team captains, some intramural directors feel it is wise to appoint them. Participants should understand their important responsibilities before taking on this role. Some of the responsibilities of captains include (Byl, 2002):

- Signing up the team, and submitting the registration forms and fees (if any).
- Confirming the eligibility of team members and filling out game sheets before contests.
- Representing the team at meetings and coordinating scheduling.
- Distributing schedules and other communications to team members.
- Advising team members on facility, program, and contests rules, and seeing that they abide by them.
- Obtaining and distributing necessary equipment to their team.
- Checking weekly standings and scores for accuracy.

Rules and Regulations

Rules and regulations are used in intramural programs to equalize competition, provide safe playing conditions, and establish standards of conduct. The following guidelines apply:

1. State the rules and regulations in a positive manner, carrying out suggestions from the intramural council and the players.
2. Establish only those rules and regulations that are reasonable, beneficial, and enforceable.
3. State the rules in simple language.
4. Standardize the administrative procedures for handling violations.
5. Review and update the rules on a timely basis.

The Intramural Handbook

The intramural handbook is a valuable resource that should be assembled, reproduced, and distributed to program participants. It contains policies, regulations and procedures, relevant activity and event documents, necessary forms, and promotional materials. A handbook assures continuity in the program and provides a central source of information. It is a ready reference for addressing inquires. A master copy should be maintained on a computer file to facilitate updating. Many schools have placed their handbooks online for easy access by students. For example, the University of Illinois Intramural Handbook can be accessed online at <http://www.campusrec.uiuc.edu/Intramurals/h_ contents.html>.

Promotion

Promotion refers to activities that raise people's awareness of the intramural program and its benefits. Promotional media might include Web sites, bulletin boards, posters, clothing with intramural logos, live events, or anything else that draws attention to the program. Word of mouth among students may be your best source of promotion. Some programs place advertisements and announcements in the school newspaper. Other programs distribute an intramural newsletter periodically. Photographs and visual images are particularly effective, especially when students appear in them. Time your promotional activities for maximum effect. Allow enough lead time before new activities. Remove promotional material once it is outdated. Consider your audience: whom specifically are you trying to reach, and with what message? (Byl, 2002) (See chapter 8 for additional promotion and public-relations strategies.)

Point Systems and Awards

Awards are used to generate student interest, increase participation, and recognize outstanding accomplishments or service. Many intramural programs give annual awards for the individuals and/or teams who accumulate the most points based on competitive record and participation, as well as an award to the all-around champion. Following is the system used at the University of South Dakota (Office of Campus Recreation, 2001, p. 8):

> At the end of the year, winners of each division [of play] are presented awards and the team with the greatest number of points is awarded the All-University trophy. There are regulations as to the number of individuals who must participate in meet sports in order to receive team points. These meet sports include track, cross-country, golf, and swimming. Teams are awarded points on the basis of their finish in each sport. A minimum number of points are guaranteed each participating team. Additional points are awarded for playoff competition.

One practice is to keep a permanent trophy that stays in the school's intramural trophy case, with the name of each year's winner inscribed on its base. Colleges often award a traveling championship trophy, which can be displayed in the winner's residence (Greek house, dormitory floor) for the year. It is returned to the intramural department for the next year's competition.

Some intramural directors consider awarding pins, plaques, medals, certificates, and trophies as undesirable, saying that it causes students to focus on the extrinsic reward and not participate for the intrinsic values received from the experience itself. Rather than spend limited funds in the intramural budget on expensive hardware, some directors award donated T-shirts, sports equipment, or other practical items commemorating intramural participation and achievements. The main goal should be to motivate students to participate for the values and experiences inherent in the activities.

A carefully managed and conservative awards system can be an asset to most intramural programs. Such a system requires a great deal of record keeping, depending on how elaborate the factors are that go into determining awards. Points may be given for each contest entry as well as for winning or placing high. Mistakes or lost records can lead to hard feelings among highly competitive participants. Computerized record keeping can greatly facilitate recording points for awards.

Risk Management

The same precautionary measures must be taken with intramurals as with participation in other aspects of the physical education program (see chapter 9). These considerations include: (1) adequate safety measures for protection of the participants, (2) evidence of fitness to participate, (3) adequate opportunity for student insurance coverage, (4) measures to guard against potential law suits, and (5) providing adequate supervision and qualified officials.

PROGRAM STRUCTURE

Suggested Intramural Activities

Following are suggested events that could be offered during the year. Included are individual, dual, and team activities. Certain activities would be conducted separately for men and women, and others would be coed.

5K road race	handball	swim meet
arm wrestling	horseshoes	table tennis
badminton	individual fitness	tennis
basketball	racquetball	triathlon
bicycle race	road rally	water basketball
bowling	roller hockey	water polo
football (flag)	soccer	wrestling
golf	softball	volleyball

Units of Competition

Units of competition or participation are essential to a successful intramural program as these units form the basis for effective organization and administration. Individual, team, league, and all-campus are the units usually found within intramural programs. In *individual* sports competition, the individual is considered a unit because other players are not needed to form a team. *Teams* are developed by groups of interested players, and a team then constitutes a unit. A *league* is composed of a logical grouping of teams (or individuals), such as those representing a grade level or a residence hall. *All-campus* means that an event is open to everyone who qualifies under the rules of participation.

Sign-up Procedures

A variety of methods exists for intramural sign-ups. Several guiding principles apply to all methods: (1) maximize participation, (2) simplify the procedure, (3) promote equality in skill level, and (4) reorganize teams if necessary. The simplest sign-up method is *open participation* (which can be restricted by grade level or other qualifiers). Anyone who wishes to participate can sign up. This approach works best when the anticipated level of participation is uncertain. Open participation also applies to unstructured gym time for free play and pick-up games. The advantage of open participation is that it requires a minimal amount of administration beyond supervision. The major drawback is that it provides a low level of meaningful participation. The other methods are: (1) random signups (sometimes referred to as the "house" system), in which students are arbitrarily grouped; (2) sign-ups based on such units as grade level or home rooms (public schools) and residence halls, Greek houses, or academic departments (college level); and (3) sign-ups in which individuals form their own teams or competitive units. All of these approaches have specific advantages and disadvantages. The director and council will have to determine which method works the best in a given situation (Byl, 2002).

Eligibility

The question of who is eligible to participate is one that must be approached with precision and consistency. Since intramural participation is encouraged, the general guideline should be that a person is eligible unless sufficient reason exists to deny eligibility. Restricted participation in intramurals should not be used as a disciplinary measure for students who fail to meet requirements or expectations in other school programs. If the student performs poorly in mathematics, he or she is not prohibited from attending school assemblies. Likewise, a student should not be prohibited from participating in intramurals due to poor performance in academic areas. In fact, the opposite ought to be encouraged, because intramurals have their own potential for development and can be an integrating and socializing force in the school.

Questions stemming from the main issues that pertain to eligibility are:

1. Should a current member of a varsity team be permitted to participate on an intramural team in that particular sport?
2. Should a former member of a varsity team be permitted to participate on an intramural team in that particular sport?
3. Should an athlete who is trying out for a varsity team but has not yet made it be permitted to participate on an intramural team?
4. Should members of junior varsity teams be eligible for intramurals?
5. Should members of the faculty and staff participate in the same league with students, and if so, should they be on separate teams or play on teams with students?
6. Should part-time students and adjunct teachers participate?
7. Should any limit be placed on the number of intramural activities a person can enter in a given year?
8. Should a participant be able to play on more than one team at a time in a particular sport?

All of these questions concern who can and cannot participate. Other questions concerning eligibility may arise, depending on the particular circumstances. These should be resolved before competition begins and written into the intramural handbook.

Leagues and Tournaments

A tournament is organized competition in an activity, usually for a relatively short period of time, which results in determining a champion. The purposes of tournament play include:

1. Stimulating interest and developing excellence in the particular activity.
2. Providing challenging competition in wholesome activities.
3. Exposing players to higher-level competition.
4. Determining a champion.

Tournament Terminology

Bracket: The lines used in a tournament chart to show opposing players or teams.

Bye: When a player or team advances to the next bracket without opposition.

Challenge: An invitation to engage in a match or contest.

Default: A contender (e.g., team) declares ahead of time that it is unable to be present for a scheduled contest.

Forfeit: Failure of one of the contenders to begin or complete a scheduled contest.

Division: Normally, a large component of a program, containing one or more leagues grouped under a logical classification system.

Handicap: An advantage given a contender as a means of equalizing competition.

League: The grouping of teams (or organizations) of like characteristics into units for competition.

Match: A contest to declare a winner between two or more opponents.

Qualifying round: A contest that allows the winners to advance.

Round: The first round in the tournament is completed when all teams have played one match or had a bye.

Seeding: The placement of teams or individuals in different brackets so that those with superior records do not meet in the early rounds of a tournament.

Tie breaker (overtime or sudden death): A means for determining a winner/champion after the regular event has ended in a tie, based on points or on a period of time (depending on the activity). This process ensures an immediate victor.

Types of Tournaments

The particular tournaments described in this section are the round robin, Lombard, challenge tournaments (ladder, pyramid, and funnel), and elimination tournaments (single, consolation, and double). These are the more frequently used tournaments in intramural programs. Other less frequently used tournaments may fit a given situation. Descriptions of such tournaments can be found in textbooks on intramurals.

Tournaments selected for use in intramural programs ideally should: (1) provide a near equal amount of involvement for all participants (reduce early elimination); (2) provide well-matched competition; (3) be neither too short nor too long; and (4) select a champion with a few other placings also determined.

The five basic criteria for selecting the best kind of tournament are the number of participants, available facilities and equipment, participants' interest, availability of qualified officials, and the nature of the sport.

Round Robin

If sufficient time and adequate facilities exist, the round robin tournament is a good choice. It is superior to other tournaments because it produces a winner who has been fairly tested, ranks other contestants, and allows all contestants to play until the end. Each entry plays all other entries at least once; in a double round robin, each contestant plays all others twice. The winner is the contestant who wins the most games.

In scheduling a round robin tournament, the entries are arranged in two columns, as shown in figure 14-4 under "Rd 1." This provides the pairings for the first round of play. If the number of entries is uneven, a bye is added. The entry in the top left column remains stationary and the other contestants rotate one position each round. If a bye is necessary, it may be placed in the stationary position. Entries rotate either clockwise or counterclockwise. Counterclockwise rotation is used in the example below. The check round is not scheduled for play. It is the same as round one, and its purpose is to provide a check to make sure that all rounds have been completed correctly.

Rd 1	Rd 2	Rd 3	Rd 4	Rd 5	Rd 6	Rd 7	Check
A vs H	A vs G	A vs F	A vs E	A vs D	A vs C	A vs B	A - H
B vs G	H vs F	G vs E	F vs D	E vs C	D vs B	C vs H	B - G
C vs F	B vs E	H vs D	G vs C	F vs B	E vs H	D vs G	C - F
D vs E	C vs D	B vs C	H vs B	G vs H	F vs G	E vs F	D - E

Figure 14–4.   Round robin tournament involving eight entries

It is usually best to schedule rounds of play at the same time on the same day each week. This makes the schedule easy to remember. The total number of tournament games required for a round robin is determined by this formula:

Total contests $= \dfrac{N\,(N-1)}{2}$ where N represents the number of entries.

(N must be an even number. If not, add a bye to the list of participants.)

Example:

7 teams + Bye = 8 teams $\dfrac{8(8-1)}{2}$ $\dfrac{8 \times 7}{2} = 28$

Thus, there are 28 contests for a single round robin tournament with 7–8 units of competition.

When the schedule is printed, it should include the date, time, and place for each game, as shown in the following example.

January 15–Round 1			January 22–Round 2		
Bye vs. Hotshots			Bye vs. Snappers		
8:00	Snappers vs. Topnotch	Court I	8:00	Bombers vs. Hotshots	Court I
8:00	Bombers vs. Sliders	Court II	8:00	Panthers vs. Topnotch	Court II
9:00	Panthers vs. Flash	Court I	9:00	Flash vs. Sliders	Court I

A win–loss chart such as shown in figure 14-5 can be posted to report the scores and keep the team records.

If time permits, a *double round robin* provides a more valid test than the single round robin, because each entry plays each other entry twice.

Lombard

The Lombard tournament is a round robin played in one day of competition. The number of games to be played is divided into the available time to give the length of each game. As in the regular round robin, each entry plays all other entries. Winners may be determined by adding total points scored by a team and subtracting points scored against it. The entry with the most points left is the winner. More simply, win–loss records may also be used to declare the winner.

	Hotshots	Snappers	Topnotch	Bombers	Sliders	Panthers	Flash
Hotshots	✕		9–5W				
Snappers		✕					7–9L
Topnotch	5–9L		✕		3–6L		
Bombers				✕		4–5L	3–4L
Sliders			6–3W		✕		2–3L
Panthers				5–4W	3–2W	✕	
Flash		9–7W		4–3W			✕

Figure 14–5. A win-loss chart for a round robin tournament

Example: Six teams want to play a basketball tournament in which each team plays every other team. They have two courts for 3 hours (6 hours total playing time) to play 15 games. Dividing 15 games into 6 hours (360 minutes) allows 24 minutes per game. Each game could last 20 minutes, with 4 minutes between games.

Challenge

In a challenge tournament, several contestants are placed in vertical order, with the strongest entries on top. Each entry has the opportunity of improving its position (or standing) by challenging and defeating one of the entries above it. The general procedures are as follows:

1. Describe the rules, and set the starting and completion dates for the tournament.

2. Make a sign-up sheet available.

3. Place contestants on the schedule as they sign up, by drawing or reverse seeding.

4. Challenges begin. If the challenger wins, that contestant exchanges places with the defeated entry. Otherwise, no change in placement is made.

Once a challenge is placed, the challenged entry must agree to play within a specified period of time or forfeit. This keeps the tournament moving. Usually, a player or team is required to play a contestant from below if a challenge has been made before challenging another player or team above. In this type of tournament, each contestant's objective should be to get to the top by challenging and defeating the contestants above.

There are three kinds of challenge tournaments—ladder, pyramid, and funnel (inverted). These are illustrated in figure 14-6. In the *ladder* tournament, the entries are arranged in vertical sequence like the rungs on a ladder, and then the challenges begin. As a rule, contestants are limited to challenging only one or two places above their own standing. This restriction should be decided before competition starts.

The *pyramid* tournament offers a greater possible number of contests because a contestant may challenge (or be challenged by) more than one player on the lines above and below.

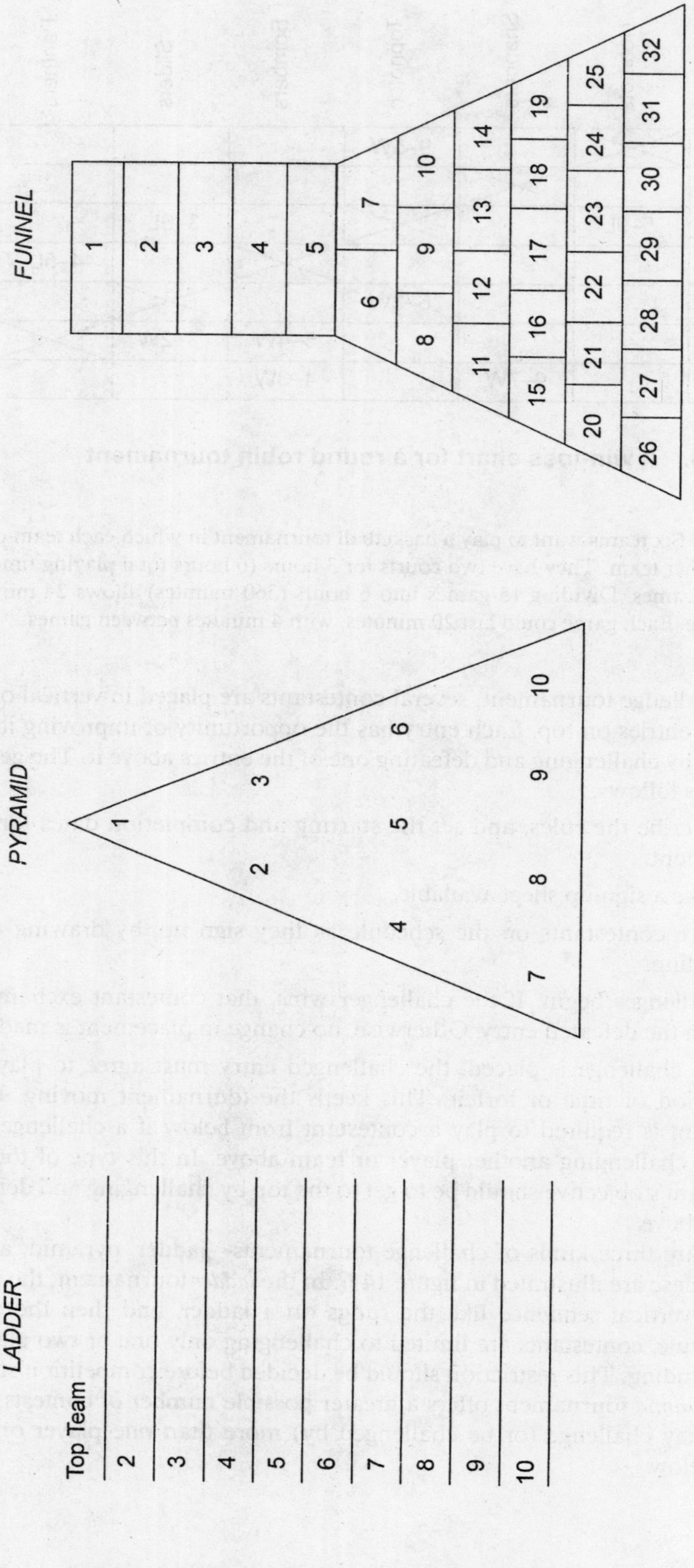

Figure 14–6. Three kinds of challenge tournaments: ladder, pyramid, and funnel

Example: In the funnel in figure 14-6, player 12 may challenge players 8, 9, or 10 on the line immediately above, or be challenged by five different players (15 to 19) on the line below.

In the case of an unusually large number of entries, the *funnel* tournament may be a good choice. It combines features of the ladder and pyramid tournaments. In the lower portion of the funnel, play is governed by the rules for the pyramid tournament, while in the upper portion play is governed by the rules for the ladder tournament.

Challenge tournaments are a good choice when limited time is available and the skill level of competitors is fairly even. One advantage of these tournaments is that no one is eliminated during play.

Elimination

In elimination tournaments, all competitors except the winner will be eliminated. Each entry is eliminated after either one or two losses, depending on whether the tournament is single or double elimination. An elimination tournament is bracketed as follows:

1. Brackets are set up to provide positions for the teams entered if the number of teams is 2, 4, 8, 16, 32, or any other power of 2.

2. If the number of teams is not equal to a power of 2, the brackets are set up to the next larger power of 2. For example, 10 teams would require brackets for 16.

3. Byes are awarded to teams without opponents, making the schedule even. (*Example*: Use a 16-player bracket for 12 players. Award four byes, two in the upper bracket and two in the lower bracket).

4. To determine the number of byes, subtract the number of teams or players from the next larger power of 2.

5. Place byes as far apart as possible so they are eliminated in the first round of play.

The seeding process places the strongest teams in separate brackets so they will not meet in early tournament rounds. Sometimes there is no seeding, but often between one-fourth and one-half of the entries are seeded. Any necessary byes should be awarded to seeded teams.

In the *single elimination* tournament, the losing contestants are dropped from the tournament and play continues until only one entry remains. There are advantages and disadvantages to this kind of tournament:

1. The tournament is short and expedient, since one-half of the entries are eliminated in the first round and half of those remaining are eliminated in each succeeding round.

2. Even though the winner is determined quickly, the best entry may not win, because injuries or an "off game" may cause upsets.

3. Each team is assured of playing only once, and this limits participation.

The formula for determining the number of games in a single elimination tournament is $N - 1$ (N is the number of entries).

Example: 25 entries will require 24 games ($25 - 1 = 24$). Figure 14-7 shows examples of elimination tournaments for 6 entries and 13 entries. Six entries require a tournament with 8 positions, thus having 2 byes. Thirteen entries require a tournament with 16 positions, thus having 3 byes. The seeded teams are placed so as to prevent them from meeting each other in the early rounds.

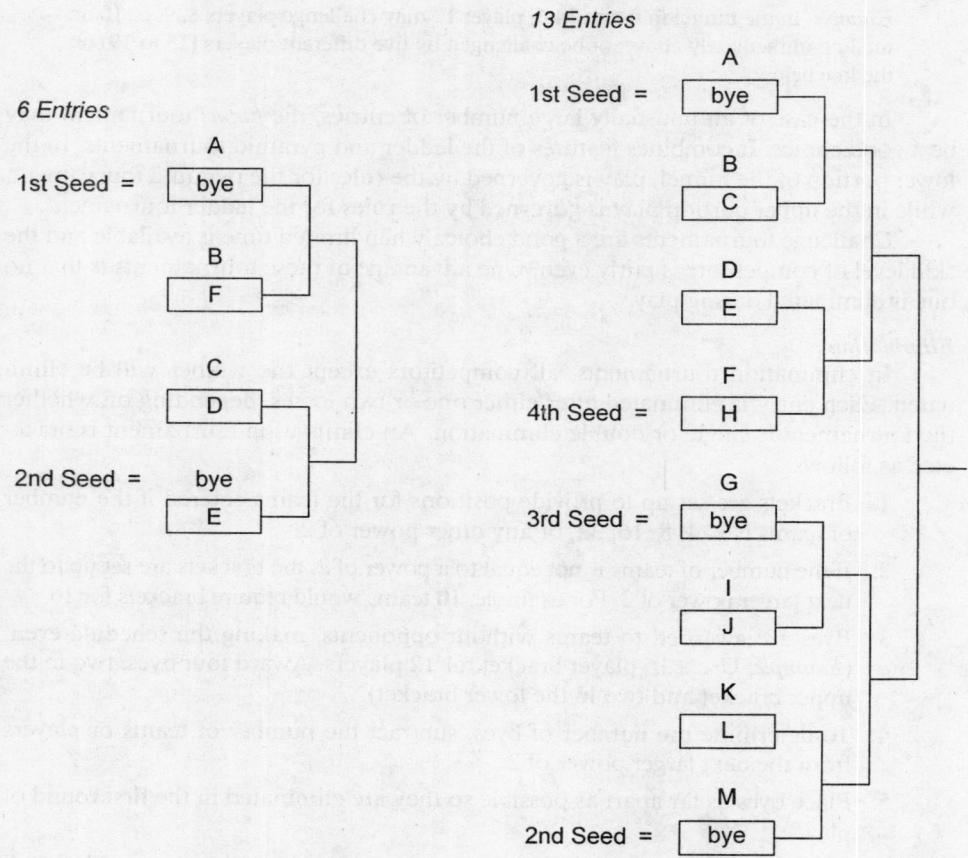

Figure 14–7. Example of two single-elimination tournaments, involving 6 and 13 entries

In a *consolation elimination* tournament, the teams are arranged in brackets in the same fashion as in the single elimination tournament, but with two sides. The first-round winners move to the right (championship side), and the first-round losers move to the left (consolation side). The final winner (A) on the right side is the tournament champion and the loser of the championship match (C in the 4-team tournament) is the runner-up. The entrant (B in the 4-team tournament) who emerges from the brackets on the left side is the consolation winner. Figure 14-8 shows three consolation tournaments involving 4 teams, 8 teams, and 16 teams. Seeding can be done in the same way as in the single elimination tournament.

In the *double elimination* tournament, each entry is assured of playing at least two contests; however, after losing twice a contestant is eliminated. The brackets are set up in the same manner as in the single elimination tournament and seeding is done in the same way. In the double elimination tournament, the formula for determining the number of contests is $2N - 1$ (with N meaning the number of entries).

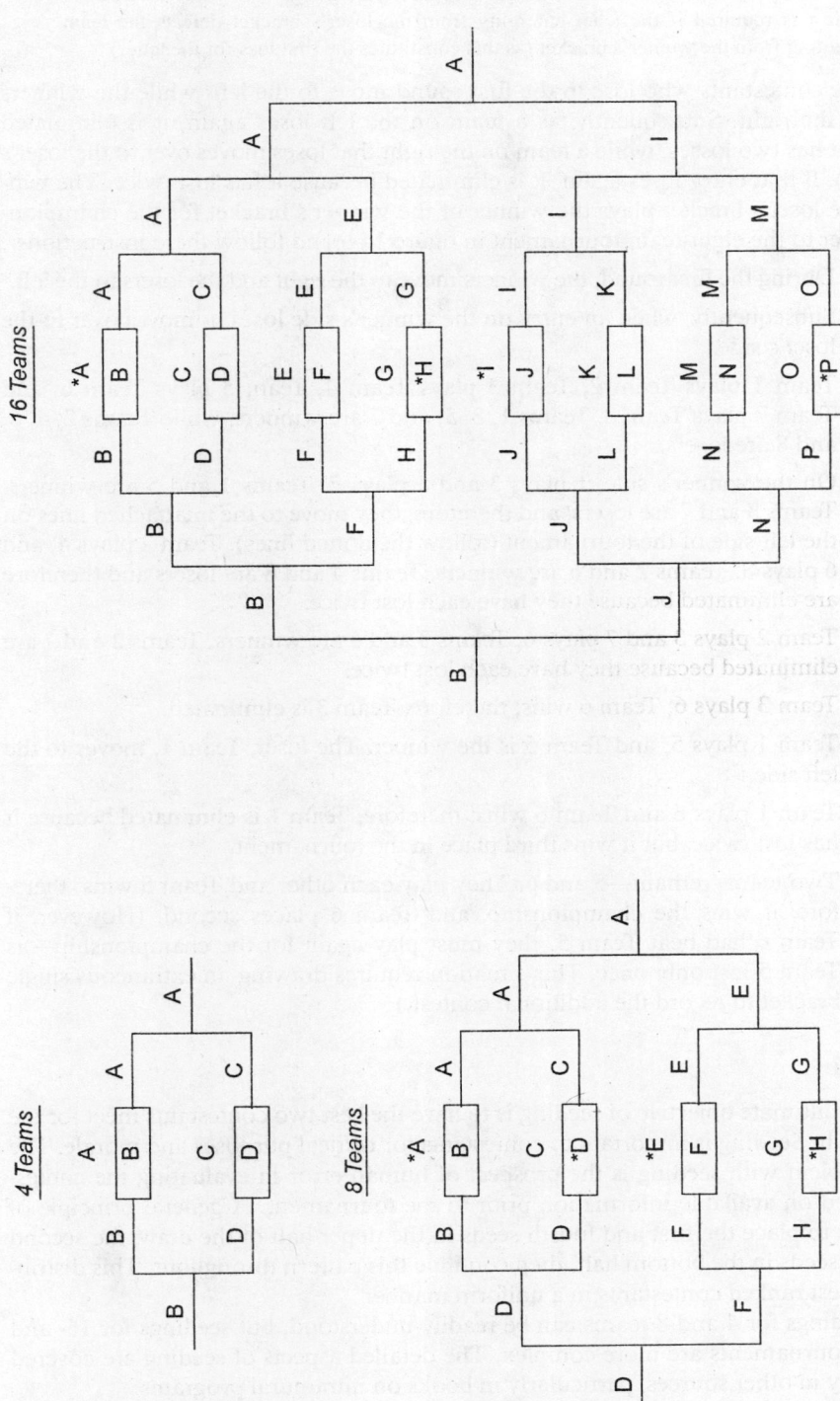

Figure 14–8. Consolation tournaments involving 4, 8, and 16 entries (an asterisk indicates seeded entries)

Example: with 12 entries, the formula would be $2 \times 12 = 24 - 1 = 23$ games. One extra contest is required if the team emerging from the loser's bracket defeats the team emerging from the winner's bracket (as this constitutes the first loss for the latter).

The contestants who lose in the first round move to the left, while the winners move to the right. Subsequently, as a team on the left loses again, it is eliminated because it has two losses, while a team on the right that loses moves over to the loser's side (left). If that entry loses again, it is eliminated because it has lost twice. The winner of the loser's bracket plays the winner of the winner's bracket for the championship. Refer to the eight-team tournament in figure 14-9 and follow these instructions:

1. During the first round, the winners move to the right and the losers to the left.

2. Subsequently, when an entry on the winner's side loses, it moves over to the loser's side.

3. Team 1 plays Team 2, Team 3 plays Team 4, Team 5 plays Team 6, and Team 7 plays Team 8. Teams 1, 3, 5, and 7 are winners, while Teams 2, 4, 6, and 8 are losers.

4. On the winner's side, 1 plays 3 and 5 plays 7. Teams 1 and 5 are winners, Teams 3 and 7 are losers, and therefore, they move to the unattached lines on the left side of the tournament (follow the dotted lines). Team 2 plays 4, and 6 plays 8. Teams 2 and 6 are winners, Teams 4 and 8 are losers and therefore are eliminated because they have each lost twice.

5. Team 2 plays 3 and 7 plays 6. Teams 3 and 6 are winners. Teams 2 and 7 are eliminated because they have each lost twice.

6. Team 3 plays 6; Team 6 wins; therefore, Team 3 is eliminated.

7. Team 1 plays 5, and Team 5 is the winner. The loser, Team 1, moves to the left side.

8. Team 1 plays 6 and Team 6 wins; therefore, Team 1 is eliminated because it has lost twice, but it wins third place in the tournament.

9. Two teams remain—5 and 6. They play each other and Team 5 wins; therefore, it wins the championship, and Team 6 places second. (However, if Team 6 had beat Team 5, they must play again for the championship—as Team 5 lost only once. This situation requires drawing an extraneous single bracket to record the additional contest.)

Seeding

The ultimate objective of seeding is to have the best two contestants meet for the final match. Seeding is important to contestants for tactical purposes and morale. The main problem with seeding is the prospect of human error in evaluating the contestants based on available information prior to the tournament. A general principle of seeding is to place the first and fourth seeds in the upper half of the draw, the second and third seeds in the bottom half, then continue this pattern throughout. This distributes the best ranked contestants in a uniform manner.

Seedings for 4 and 8 teams can be readily understood, but seedings for 16- and 32-team tournaments are more complex. The detailed aspects of seeding are covered adequately in other sources, particularly in books on intramural programs.

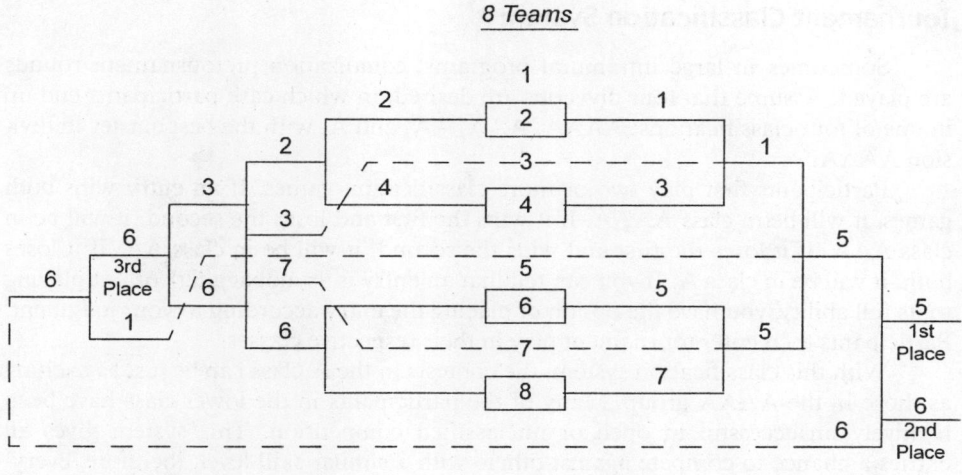

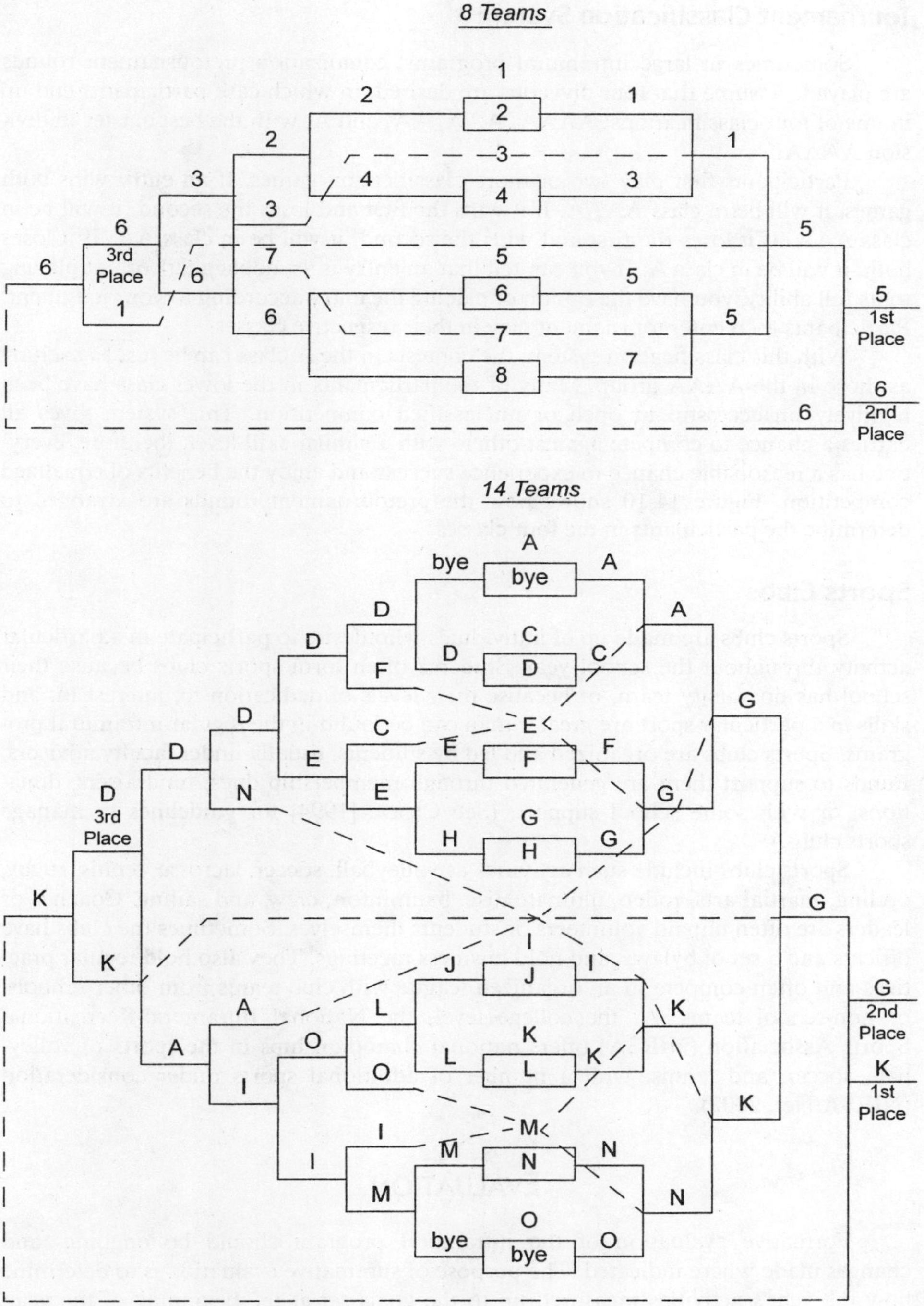

Figure 14–9. Examples of double-elimination tournaments, involving 8 and 14 teams. *Note:* Dotted lines may cross each other to different brackets, reducing the chances of playing the same team twice

Tournament Classification System

Sometimes in large intramural programs, equalization pretournament rounds are played. Assume that four divisions are desired, in which case participants end up in one of four classifications: AAAA, AAA, AA, and A, with the best entries in division AAAA.

Participants first play two or more classification games. If an entry wins both games, it will be in class AAAA. If it wins the first and loses the second, it will be in class AAA. If it loses the first and wins the second, it will be in class AA. If it loses both, it will be in class A. If you can tell that an entry is "sandbagging" or not playing to its full ability, you have the option of placing the entry according to your judgment. Participants then enter tournament play in their respective classes.

With this classification system, the contests in the A class can be just as exciting as those in the AAAA group. Many of the participants in the lower class have been relatively unsuccessful in open or unclassified competition. This system gives all entries a chance to compete against others with a similar skill level; therefore, everyone has a reasonable chance to experience success and enjoy the benefits of equalized competition. Figure 14-10 shows how the pretournament rounds are arranged to determine the participants in the four classes.

Sports Clubs

Sports clubs are made up of individuals who desire to participate in a particular activity throughout the school year. Students often form sports clubs because their school has no varsity team, or because their levels of dedication to, interest in, and skills in a particular sport are greater than can be found in the regular intramural programs. Sports clubs are organized and led by students, usually under faculty advisors. Funds to support them are generated through membership dues, fundraisers, donations, or with some school support. (See Cleave [1994] for guidelines to manage sports clubs.)

Sports clubs include such activities as volleyball, soccer, lacrosse, tennis, rugby, cycling, martial arts, rodeo, ultimate disc, badminton, crew, and sailing. Coaches or leaders are often unpaid volunteers or students themselves. Sometimes the clubs have officers and a set of bylaws, and hold business meetings. They also hold regular practices and often compete in an organized league with club teams from other schools, or non-school teams. At the college level, the National Intramural-Recreational Sports Association (NIRSA) offers national championships in the sports of volleyball, soccer, and tennis, with a number of additional sports under consideration (NIRSA.Net, 2002).

EVALUATION

Formative evaluation of the intramural program should be ongoing, and changes made where indicated. The purpose of summative evaluation is to determine how the year's activities measured up to the goals set at the beginning of the year. Intramural directors can utilize both formal and informal methods for evaluating their programs. Examples of informal evaluations would be noticing that game officials are not doing a good job or that students can't keep up with the pace of the aerobics instructor. Formal evaluation implies using more standardized procedures. Statistics

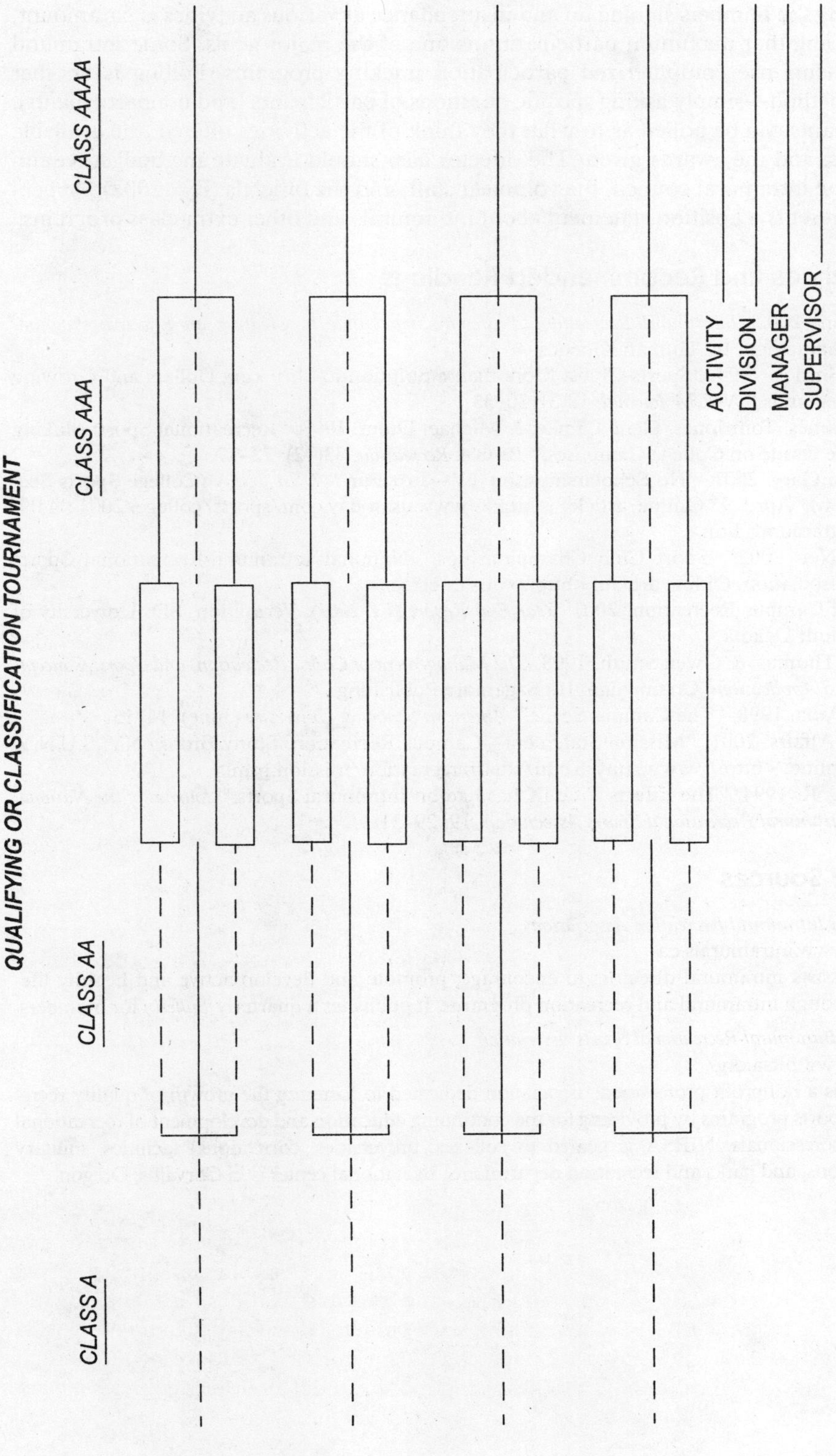

QUALIFYING OR CLASSIFICATION TOURNAMENT

CLASS AAAA

CLASS AAA

CLASS AA

CLASS A

ACTIVITY
DIVISION
MANAGER
SUPERVISOR

Figure 14–10. Pretournament for arranging the participants into four classes for league or tournament play

regarding the numbers signing up and in attendance at various activities is paramount, considering that maximum participation is one of the major goals. Some intramural departments use computerized participation tracking programs. Polling is another direct method—simply asking specific questions of participants (and nonparticipants). Participants can be polled as to what they think of the activities offered, the available facilities, and the awards given. The director also should evaluate the budget, equipment, the intramural council, the volunteer staff, and the officials (Byl, 2002). Appendix E presents a position statement about intramurals and other extra-class programs.

References and Recommended Readings

Byl, John. 2002. *Intramural Recreation: A Step-by-Step Guide to Creating an Effective Program.* Champaign, IL: Human Kinetics.

Cleave, Shirley. 1994. "Sports Clubs: More than a Solution to Shrinking Dollars and Growing Demands." *NIRSA Journal*, 18(3): 30–33.

Lewis, James, Tom Jones, Gene Lamke, & Michael Dunn. 1998. "Recreational Sport: Making the Grade on College Campuses." *Parks & Recreation*, 33(12): 72–77.

Mihoces, Gary. 2001. "No Scholarships, no TV—Just Fun." *USA Today* (College Sports Section), April 9. Online article: <http://www.usatoday.com/sports/college/2001-04-09-intramurals.htm>.

NIRSA.Net. 2002. "Sport Club Championships." National Intramural-Recreational Sports Association. Online article: <http://nirsa.net/sc/>.

Office of Campus Recreation. 2001. *Year End Report (FY 2001)*. Vermillion, SD: University of South Dakota.

Sawyer, Thomas, & Owen Smith. 1998. *The Management of Clubs, Recreation, and Sport: Concepts and Applications.* Champaign, IL: Sagamore Publishing.

Spoor, Dana. 1998. "The Campus Scene." *American School & University* (June): 14–15.

Student Affairs. 2001. "Mission Statement—Campus Recreation. Stony Brook, NY: S.U.N.Y. Online: <http://www.sunysb.edu/stuaff/recreation/mission.html>.

Tharp, L. R. 1994. "The Effects Title IX has had on Intramural Sports." *Journal of the National Intramural Recreational Sports Association*, 19: 29–31.

Online Sources

Canadian Intramural Recreation Association
http://www.intramurals.ca
CIRA assists intramural directors to encourage, promote and develop active and healthy lifestyles through intramural and recreation programs. It publishes a quarterly *Bulletin* for members.

National Intramural-Recreational Sports Association
http://www.nirsa.org/
NIRSA is a nonprofit professional association dedicated to fostering the growth of quality recreational sports programs by providing for the continuing education and development of recreational sports professionals. NIRSA is geared to colleges, universities, correctional facilities, military installations, and parks and recreation departments. Its national center is in Corvallis, Oregon.

15 The School Athletics Program

Two notable quotes illustrate the dilemma facing U.S. athletics. "Winning isn't everything. It's the only thing," was the motto of former Green Bay Packers coach Vince Lombardi. This sentiment stands in contrast with the Modern Olympic creed expressed by its founder Baron Pierre de Coubertin: "The most important thing . . . is not winning but taking part" (see Overman, 1999). Schools must put into place policies and incentives to assure that a healthy balance is struck between winning at all costs and the welfare of students who participate in athletics. Coaches should recognize that what is achieved in the win–loss column is transitory in importance, whereas athletic experiences that contribute to the development of young men and women have long-term consequences.

VALUE OF ATHLETICS

The most widely accepted justification for interscholastic athletics relates to its alleged character-building benefits. Traits associated with participation in athletics have included development of a strong work ethic, sacrifice for a cause, loyalty, dedication, perseverance, leadership, competitiveness, and cooperation. The claim that sports develop character is almost universally accepted by educators, students, and their parents, despite the fact that little empirical evidence exists to support the claim (Stoll & Beller, 2000). Virtually all the evidence for character building is anecdotal. Educators must realize that whether or not sports build character depends on the values that the community, the school, coaches, and parents choose to emphasize.

Evidence is mounting that the environment surrounding some athletics programs actually hinders the development of ethical and moral reasoning. Social scientists point out that coaches can create a "moral callousness" in their athletes if they place too much emphasis on winning. Kretchmar (1994) lists the symptoms [paraphrased below] of a climate of moral callousness in an athletics program:

- Appeals to the fact that "everyone is doing it" as an excuse for unethical practices.
- Inability to distinguish between what is part of the game and what is not.
- Inability to draw a line between shrewd strategy and blatant rule breaking.
- A sense that "not getting caught" implies that nothing unethical occurred.

The crucial question for educators is: "What can be done to insure that athletics promotes the development of positive moral and ethical reasoning skills of participants?" Professors Sharon Stoll and Jennifer Beller (2000), former athletes and coaches, offer the following guidelines to insure that school athletics programs fulfill their potential to build moral character:

1. *Develop a program philosophy.* The purpose of the program must be clearly stated and articulated to all participants including coaches, parents, athletes, the media, and the public.

2. *Develop ethical standards and codes of conduct.* Standards of deportment must be outlined and articulated in a code of conduct in order to place the philosophical principles in a behavioral context.

3. *Develop educational programs.* Workshops, seminars, and school programs must be designed to educate all participants regarding the ethical aspects of athletics.

4. *Develop a penalty structure.* There must be clearly defined consequences for failure to act appropriately. This applies to coaches, players, and fans. Punishment for unsportsmanlike behavior must be clear, swift, and certain.

5. *Evaluate the program.* Once provisions are put into place, the effectiveness for promoting positive change must be evaluated through observations and formal methods.

6. *Demand good role models.* If we expect athletics to promote ethical behavior, everyone connected with the program has a responsibility to be a good role model; actions speak louder than words.

The coach is the central role model who determines the character of his or her athletes. Schools and professional organizations have developed codes of conduct governing the behavior of coaches. Figure 15-1 reproduces the Coaches Code of Conduct developed by the Illinois High School Association.

The Coaches Council of the National Association for Sport and Physical Education (NASPE) has established a Code of Conduct to which coaches at all levels should be held accountable. It can be accessed at <aahperd.org/naspe/template.cfm?template=position-papers.html>.

The educational benefits of athletics are highly touted. However, administrators must make certain that participation in athletics doesn't preempt academic education. Several examples have surfaced recently of athletes who graduated from high school or college without having acquired fundamental skills. Extracurricular activities should supplement and complement the curriculum, not replace it. Athletes shouldn't be removed from the classroom to attend practice sessions. Games and tournaments should be scheduled to provide minimal interference with class schedules. Coaches should not ask teachers to excuse athletes from exams or to apply lower standards when determining their grades. Athletes should not be enrolled in "easy" courses that poorly prepare them for further education or the job market.

Balanced athletics programs can contribute positively to education. Reeves (1998) reported a three-year study of 240,000 North Carolina students which indi-

---

Coaches Code of Conduct

The IHSA believes that sportsmanship is a core value and its promotion and practice are essential. Coaches have a duty to assure that their teams promote the development of good character. This code of conduct applies to all coaches involved in interscholastic athletics and activities.

1. The coach will promote academics and the educational process.

2. The coach will teach, enforce, advocate, model, and promote the development of good character to include:

   a) Trustworthiness
   b) Respect
   c) Responsibility
   d) Fairness
   e) Caring
   f) Citizenship

3. The coach will respect participants, officials, opponents, parents, and all others involved.

4. The coach will promote fair play and uphold the spirit of the rules in the activity.

5. The coach will model appropriate behavior at all times.

I have read and understand the requirements of this Code of Conduct and acknowledge that I may be disciplined if I violate any of its provisions.

_____   _____
Signature   Date

Figure 15–1.   Illinois High School Association's Coaches Code of Conduct (http://ww/ihsa/org/addatude/coaches_code.htm)

cated that athletes had higher grades, better attendance, fewer behavior problems, and lower dropout rates than nonathletes. School athletics has other potential benefits. It can foster school pride and tradition, can promote cohesion within the community, has obvious entertainment value, can garner alumni loyalty, and may encourage financial support for the school. However, these are peripheral values. The only justifiable rationale for sponsoring interscholastic athletics is that it is educational.

GOALS FOR THE PROGRAM

Interschool athletics should be considered an integral part of the total educational process that has as its aim the development of physically, mentally, socially, and morally fit individuals. Athletics can serve as a vital educational medium by incorporating the following worthy goals:

1. To nurture and properly channel the competitive urge that is inherent in all of us.

2. To engender the will to win by fair and honest means.

3. To promote pride in successful effort for the sake of accomplishment.

4. To build team spirit and desire on the part of those who engage in sports, and to work with each other in a cooperative effort.

5. To teach self-discipline and self-control.

6. To build morale.

7. To develop sound minds and bodies through mental and physical preparation for competition.

8. To encourage spectators and nonparticipants to take an interest in and appreciate the values of competitive sports.

9. To stimulate a continuing institutional interest and loyalty among students, alumni, and the public.

Statements of beliefs regarding interschool athletics are reproduced in Appendix E. School and athletics administrators should be familiar with these principles.

AREAS OF RESPONSIBILITY

Coaches provide the direct leadership for the athletics teams, but certain school administrators also have important responsibilities. This section discusses the duties and responsibilities of school administrators and organizations.

School Superintendent (or College President)

The school superintendent has the following duties in connection with athletics (in an institution of higher education, the president would have similar responsibilities): (1) clarify and communicate the role of athletics in the particular school system; (2) establish and communicate the responsibilities of the various personnel: (3) evaluate the accomplishments of athletics in terms of its contributions to the educational goals and the welfare of the educational system; (4) maintain a high level of leadership in the athletics program; (5) establish a framework of policies for the provision and use of athletics facilities; (6) make sure an effective program of public relations is maintained; (7) give leadership to the budget and determine methods by which the program should be financed; (8) assure appropriate policies in connection with such matters as the safety of participants, the balance of athletics opportunities for boys and girls, and the scope of the total program; and (9) be accountable to the governing board for all aspects of athletics within the school system.

School Principal

The principal is ultimately responsible in all matters pertaining to interscholastic athletics in the school, and for assuring that the program adheres to the policies of the state activities/athletics association and the school district. The principal also is accountable for all revenue and expenditures associated with school programs. The responsibilities of principals are more specific than those of the superintendent, as they address problems and procedures in cooperation with the school athletics council (if one exists) and the athletics staff. Specific responsibilities include: (1) certifying the eligibility of players; (2) defining responsibilities of athletics personnel; (3) supervising faculty members involved; (4) giving administrative leadership in terms of crowd control and ethical behavior; (5) making and enforcing policies relative to the use and care of facilities and equipment; and (6) selecting and hiring athletics staff members. At the college or university level, a designated middle-level administrator, typically a vice president, working in conjunction with the athletics council, performs these responsibilities.

Athletics Director (and Assistants)

In school districts with several large high schools, the athletics director may be appointed at the district level. Otherwise, the director administers the program for a particular high school or middle school. In small schools, this position may be combined with that of principal or assistant principal, or it may include teaching and coaching responsibilities. Regardless of the level, this position entails the direct administration and close supervision of the athletics program. Among the director's major responsibilities are (Hoch, 1999):

- contracting and scheduling contests;
- arranging for officials;
- establishing travel schedules and enforcing travel policies;
- preparing team rosters and monitoring eligibility of athletes;
- making certain that the program is conducted in compliance with the rules;
- submitting reports to the state activities association and conference regulatory bodies;
- preparing for athletics contests and providing administrative supervision;
- providing adequate athletic training and medical services;
- preparing and administering the athletics budget;
- accounting for ticket sales and revenues;
- assessing risks, promoting safety, and properly reporting accidents and injuries;
- providing leadership to public relations, marketing;
- insuring external support by working with booster clubs, alumni groups, and the general public;
- coordinating awards banquets, pep rallies, and fundraising activities;
- acting as an advocate for the department; and
- hiring, supervising and evaluating coaches.

In some school districts, an *associate* or *assistant* director carries a portion of the administrative load. Many colleges now employ an associate athletics director for women's sports. Large programs, particularly at the university level, also employ a *business manager*, who is responsible primarily for financial operations; a *sports information director*, whose responsibilities include working with the news media; a *special events director*, who has the responsibility for promotion and management of the athletics contests, the sale of tickets, and the supervision of facilities; and *a compliance officer*, who monitors compliance with athletic association regulations.

At smaller schools where the athletics director is also a coach, the director must be objective and fair in providing balanced administrative support for the total program. Too often the administrator/coach has used the administrative position to provide disproportionate support for a particular sport at the expense of other sports.

Only an exceptionally fair-minded individual might be able to transcend the inherent conflicts in holding both positions simultaneously. Whether the athletics administrator should have a coaching assignment depends on the particular situation. In small schools, everyone's expertise is often needed to fill all the coaching positions. In larger programs, the directorship tends to be a full-time responsibility.

Successful athletics directors share three common characteristics: (1) they possess exceptional professional skills and positive personal qualities, (2) they have some formal or informal education in sport management, and (3) they have had some type of real-life experiences in terms of managing sports programs (Schneider & Stier, 2001). Traditionally, directors have been coaches who decided to become administrators when the opportunity arose. Experience as a coach may give an administrator added insight, but a successful coaching career does not guarantee success as an administrator. The two positions require distinct skills and qualifications. Recently, several large colleges have given preference to candidates with degrees in business administration when appointing directors of athletics. High schools are more inclined to look for an educator to fill this position.

Professional associations for athletics directors include the National Council of Secondary School Athletic Directors (NCSSAD), formed in affiliation with AAHPERD. The purpose of NCSSAD is to extend professional services to secondary school athletic administrators and to support and strengthen the role that secondary school athletics administrators play in helping to attain the educational objectives of secondary school education. Also, NASPE and the North American Society for Sport Management (NASSM) jointly developed curricular guidelines for careers in sport management in the 1980s. In 1998, their Sport Management Program Review Council developed new sport management curriculum core content for future sport managers and athletics directors (Schneider & Stier, 2001).

The National Interscholastic Athletic Administrator Association (NIAAA), affiliated with NFSHSA, implemented a certification program for directors in 1989. Three levels of certification are available, including that of Certified Athletic Administrator (CAA). For certification, the candidate must demonstrate a knowledge of and the ability to apply the essential elements of athletics administration. The CAA designation has been earned by more than 2,000 athletics administrators nationally.

Athletics Coach

In Division I college programs, virtually all head coaching—and most assistant coaching—assignments are full time and no longer include teaching. In the public schools, coaching and teaching responsibilities are normally combined. This dual responsibility is also quite common in community colleges and in small four-year college programs. Role conflicts between teaching and coaching are common in these settings, especially when coaches are teaching physical education. When coaches neglect their physical education classes or use them as varsity practice sessions, administrators become reluctant to allow physical education teachers to coach. However, the acute shortage of coaches in many areas has preempted this option. Moreover, physical education teachers usually are the most qualified teachers to coach athletics. Regardless of their college teaching major, coaches need to be able to diagnose and teach physical conditioning and motor skills to athletes. It's not always necessary for a coach to have played the sport, but certainly such experience can be an advantage.

In addition, coaching responsibilities require good human-relations skills. Coaches must be able to deal not only with players, but also with officials, parents, and fans. Coaches routinely are invited to speak before community groups. Consequently, they should develop good public-speaking skills. Also, coaches represent the school in the community and on other campuses. Their personal demeanor, health habits, and ethical behavior must be exemplary. Coaches take on an exceptional workload beyond the classroom. Their responsibilities may entail coming to school early and leaving late each day. In addition, they devote many evenings and weekends to coaching responsibilities. Coaches must develop time-management skills in order to balance their professional and personal lives.

Many states have had to resort to hiring coaches who are not teachers, because of a shortage of coaches. These "walk-on" coaches may have coaching experience, but some do not appreciate the educational goals of athletics. The principal should be responsible for hiring walk-on coaches and orient these individuals to the school's philosophy and objectives. The preferred scenario would be for nonteaching coaches to work alongside coaches on the teaching staff. Some school districts are requiring fingerprinting and First Aid/CPR certification for walk-on coaches. California requires walk-ons to complete a National Coaches Education Certification (NCEP) workshop. Maryland has implemented a regulation that 50 percent of the coaching staff must be teachers. Ideally, most athletics directors would prefer to see faculty coach interscholastic activities.

When a shortage of coaches exists, an assignment may be given to a person who is not competent in the particular sport. This diluted approach, which may seem the only alternative, results in the mere supervision of students in a self-teaching exercise without the benefit of a subject specialist to guide learning experiences. As an analogy, students would learn little about science in a class supervised by a nonspecialist. The same is true of athletics. If students are to develop a high level of proficiency in their sport, the coach must be a well-prepared specialist. The professional preparation of athletics coaches is just as important as the preparation of biology teachers, language arts teachers, or counselors.

The more qualified coaches often have enrolled in coaching courses. Many colleges now offer coaching minors or concentrations. In 2000, NASPE formed the National Council for Accreditation of Coaching Education, which reviews the quality of coaching education programs. These reviews are based on compliance of programs with Guidelines for Coaching Education and the National Standards for Athletic Coaches. Thirty-seven standards were identified that reflect the range of competencies required for quality athletics coaching. These standards fall into eight categories:

1. Injuries
2. Risk management
3. Growth, development and learning
4. Training, conditioning and nutrition
5. Social/psychological aspects of coaching
6. Skills, tactics and strategies
7. Teaching and administration
8. Professional preparation and development

Some states like California have developed their own list of competencies that apply to individuals assigned to coaching duties in the schools. These standards imply that a coach's success depends to a large extent on specific knowledge in identified areas and the ability to apply that knowledge effectively. In addition to these specific competencies, sound values and strong leadership are key elements in coaching success.

Athletics Councils

Athletics councils—also known as boards or committees—are prevalent in colleges and universities, and they exist in many secondary schools. Generally, they are advisory in nature, with their advice flowing to the chief administrator of the school. University athletics councils most often report to the university president or to a vice president, less often to the faculty senate, and rarely to the athletics director (who may serve *ex officio* on the council). A school or district athletics council is advisory to the superintendent, building principal, or athletics director. In some cases, the council is more than advisory and gets involved in making and enforcing policies. Whatever the case, the role of the council should be clearly defined.

The composition of athletics councils varies. On the college level they are composed primarily of faculty, administrators, and students, with athletics department personnel more likely to serve *ex officio*. At some high schools, the councils are made up almost entirely of coaches and athletics department staff. Less frequently, councils include cheerleader sponsors, student representatives, booster club representatives, alumni, school board members, or members of the community. Wide representation in membership is recommended for councils, as it helps to keep athletics in proper perspective and integrated with other aspects of the school and community.

Functions of councils at the public school level may differ among districts and states. Following are the functions of the Athletic Advisory Council of Centerburg (Ohio) High School:

- To continuously evaluate the interscholastic program of the high school.
- To control and regulate athletics by uniform policies consistent with sound educational aims and objectives.
- To aid in determining the scope of the athletic program.
- To develop a long-range program for the development and extension of facilities for each sport.
- To direct the athletic program according to the rules and regulations of the Ohio High School Athletic Association and the [athletic conference].

College athletics councils monitor gender equity (if not handled by a separate committee), assess athletes' academic progress and monitor graduation rates, often approve ticket prices and allocations, and may participate in the hiring of coaches and athletics staff. A few councils oversee budgets. Some colleges have separate athletics award committees. If not, the council may assume this responsibility, as well (Overman, 2001).

Booster Clubs

Most booster clubs have good intentions and usually operate in the interests of the athletes and the program; however, problems do occur. Boosters occasionally must be admonished to keep within the rules and regulations governing athletics. The director of athletics needs to maintain a degree of control over booster activities. Directors would be wise to act as liaison with the booster groups to screen and control the transfer of funds. Coaches should be discouraged from unilaterally requesting funds from boosters. A mandatory form for booster-generated funds should be used, on which coaches list the reasons for the request and its eventual disposition. As a further control, the school should develop a model set of bylaws for booster clubs that define and limit their areas of involvement in school activities. Koehler & Giebel (1997) provide a sample set of bylaws in their *Athletic Director's Survival Guide* that cover the following areas of club operation:

name, purpose, powers, offices	committees
membership	dues
meetings	accounting procedures
board of directors	contracts, checks, deposits, and funds
officers—duties and responsibilities	amendments to bylaws
definition of fiscal and elective year	

GOVERNING ORGANIZATIONS

State Associations, Leagues, and Conferences

For high school athletics, local/regional leagues or conferences are formed to enhance scheduling, provide championship competition, and implement useful policies and procedures. In some cases, a district or regional organization encompasses several leagues. Regional organizations serve essentially the same purpose as local leagues except on a broader scale.

In most states, high schools are classified for purposes of athletics competition based on selected criteria, with size of student body being the most influential factor. In a state with four classifications, the divisions might be A, AA, AAA, and AAAA, with the latter designating the larger schools. The divisions are aligned in various ways and given different labels depending on the state. Divisions often combine school population with geographic location. Leagues and conferences function within these divisions. State championships are sponsored in a variety of sports in most states. In a few of the larger states, it is not practical to complete all of the competition necessary to determine the state champions; therefore, regional championships are the highest level of competition.

Championships beyond the local league are sponsored under the auspices of the state high school activities/athletics association. There are two kinds of state associations: those affiliated with the State Department of Education, and those that are unaffiliated (volunteer and not established by law). Most of the state associations are of the second kind. When the state association is affiliated with the State Department of Education, membership in the association is automatic for all public schools, with private and parochial schools having the option of joining or forming their own associations. Several states have both public school and private school associations. In a voluntary state association, each school or district has the choice of whether to belong. In practically all cases, schools choose to belong because of the advantages, which include:

1. Eligibility for participation in regional and state championship athletics events.
2. Enforcement of regulations for the conduct of athletics.

3. Sponsorship of a classification plan for high schools.
4. Certification and assignment of athletics officials.
5. Enforcement of athletics standards.
6. Published bulletins, newsletters and online Web sites.
7. A final authority for the resolution of questions, controversies, and appeals.

College Athletics Conferences

Colleges and universities (including junior colleges) are organized into athletics leagues (usually called conferences), through mutual agreement of the member institutions. The conference is usually governed in accordance with a written constitution and a regulatory code. The conference is administered under the jurisdiction of one or more governing boards known as a Conference Council and/or President's Council (most conferences have both). If the conference is large enough to employ an administrator, this person is known as the Conference Commissioner. The large university conferences maintain a staff of employees. Some institutions that do not belong to conferences, remaining disaffiliated or unattached, are known as independents.

In addition to belonging to athletics conferences, colleges and universities are affiliated with the appropriate national governing organization. For intercollegiate athletics, this means affiliation with one of the following: National Junior College Athletic Association (NJCAA), National Association of Intercollegiate Athletics (NAIA), or the National Collegiate Athletic Association (NCAA). Within certain of these national organizations, the member institutions are classified into divisions according to their athletic prominence or their emphasis on athletics. These national organizations are made up of colleges and universities, not of athletics conferences, but conferences sometimes become affiliates of the national organizations.

League and Nonleague Competition

Most athletics personnel agree that the ideal size of a high school league is six to eight schools. College leagues function especially well with eight or ten schools. This size permits a complete round of competition in football and certain other sports, and it permits a home and home arrangement in sports like basketball. In addition, a league of this size enhances championship meets in individual sports such as track, swimming, and wrestling.

League schedules are usually prepared by an individual or a committee and are subject to approval by the appropriate representatives of the member schools. Nonleague contests are arranged by the officials of the two schools involved. Important considerations in selecting nonleague opponents include: (1) schools of approximately the same size and with similar athletics programs and facilities; (2) schools that provide natural rivalries; (3) schools that have positive relationships with each other; and (4) schools in locations that afford interesting trips without excessive travel.

REGULATIONS AND RULES

The athletics governing organizations establish and enforce regulations for member schools. Among the purposes of the regulations are to standardize athletics programs and procedures, to protect student athletes and school personnel against possible abuses of athletics, and to help keep athletics in the proper perspective in the educational setting. The regulations are valuable in protecting students, coaches, and administrators against exploitation.

Regulations apply to both individuals and to schools. Regulations that apply to individuals often deal with: (1) age of the athlete, (2) completion of a minimum number of units with an acceptable grade point average, (3) current registration for a specified number of units (minimum full-time student status), (4) parental permission to participate when applicable, (5) amateur status, (6) disaffiliation with nonschool teams during the athletics season, and (7) adherence to transfer rules.

Regulations that pertain to schools typically relate to: (1) conditions governing athletics events; (2) limitation of the number and value of athletics awards; (3) use of official athletics contracts; (4) exchange of eligibility lists; (5) control of postseason games; (6) limitation of the number of contests and the length of season; and (7) specified number of practices prior to the opening contest.

Eligibility

The eligibility standards or requirements for athletic participation are normally defined by governing organizations. For example, a university affiliated with an athletics conference and with the NCAA must meet the eligibility requirements of both. At the high school level, an athletics conference might have eligibility regulations in addition to those of the State High School Athletics/Activity Association.

Most states enforce an age limit for high school athletes, often based on the nineteenth birthday. Some states follow the eight-semester rule that limits the student to eight semesters of eligibility commencing with the ninth grade. The courts have ruled that schools must make reasonable accommodations to this rule for students with physical or mental disabilities covered by the Americans with Disabilities Act (ADA). However, they often disagree on what constitutes reasonable accommodations. Other common eligibility policies include the so-called "no pass, no play" rule, which requires a 2.0 or similar grade point average, enrollment in a minimal number of credit hours, and good attendance in order to be eligible to participate in athletics or other extracurricular activities. Following the movement toward initiating strict academic standards in the 1980s, a few states have rescinded the "no pass, no play" rule after reevaluating its effects.

In addition, the NCAA has added a new hurdle by defining the core-course requirements for high school athletes who intend to compete at the Division I level in college during their freshman year. The burden falls on high school principals to certify and document their courses to satisfy the NCAA's regimen. The NCAA retains the right to overrule the school's judgment.

Eligibility standards, given all their complexity, must be reasonable and functional. Individuals and institutions are expected to police themselves. Governing organizations may have a penalty system for eligibility violations. Athletics coaches and administrators must be well informed and in total compliance with eligibility rules. Some states like Delaware place eligibility rules online. School districts are encouraged to provide workshops on eligibility for their coaches and administrators. Athletes and parents should be provided with the pertinent information on eligibility. More and more athletes are challenging eligibility rules. This has become one of the most contentious areas of athletics administration, with lawsuits filed in several states. As a result, large school districts have developed computerized databases to monitor academic eligibility of athletes and those who are currently ineligible (Reeves, 1998).

The growth of the home-schooling movement has created new eligibility issues. Regulations on the eligibility of home-schooled students to participate in school extra-curricular activities vary from state to state, and courts have ruled both ways. Currently, less than two dozen states allow home-schooled students to participate in athletics. The issue to allow participation often revolves around money, as schools receive tax revenue based on average daily attendance (ADA) in some states. In addition, grades and attendance of home-schooled students are difficult to monitor. The National Federation of State High School Associations does not favor allowing home-schooled students to participate in public school extracurricular activities (Texas A&M Prof . . .," 1999).

Recruiting

Recruiting of athletes should not be practiced by high schools; most state high school activities associations have policies discouraging the practice. Typical is the Tennessee Secondary School Athletic Association regulation, which bars members from using "undue influence . . . to secure or retain a student for athletic purposes." While public schools are restricted by district boundaries, some private schools have recruited aggressively at the statewide level (Associated Press, 2001).

A recent trend among high schools has been to recruit outstanding foreign athletes. The schools have argued that athletics is a legitimate part of exchange students' education. Several notorious incidents have brought this practice into question. The National Federation of State High School Associates (NFSHSA) has recently established (optional) guidelines that prohibit schools from input into the selection of a specific student, and prohibiting coaching staffs from serving as host families. Exchange students must be part of an official program and have visas issued by the Immigration and Naturalization Service. However, the states administer their own eligibility requirements. Some state athletic associations prohibit contact with prospective athletes outside the country, while other states have required that foreign student athletes be officially adopted by families within the school district. At the same time, courts have ruled that an outright ban on foreign exchange students participating in school athletics is discriminatory (Popke, 2002).

At the college level, recruiting has become an important element in producing winning teams. Many educators, including some athletics directors and coaches, feel that recruiting has become both dominant and distasteful. They would like to de-emphasize its importance, but efforts to do so have met with resistance. The NCAA has detailed regulations governing recruiting practices. Athletics staff need to be well informed about the precise regulations associated with recruiting. Administrators outside the athletics department must also adhere to recruiting regulations. Finally, high school coaches should advise their athletes regarding appropriate and inappropriate recruiting efforts by college recruiters and professional scouts.

Scouting

The scouting of teams and players is a common practice at both the college and high school level. College coaches scout high school athletes with the intent of recruiting outstanding prospects. Recently, several private national scouting services have gone into the business of scouting high school athletes in return for a fee. This development illustrates the escalation of recruiting practices at the college level.

Scouting by high school coaches for the purpose of recruiting athletes is inappropriate. However, the scouting of upcoming opponents on one's schedule is a legitimate practice as a part of game preparation. Teams often will exchange game films or videos through mutual agreement. Most athletics departments develop scouting forms, designed to fit onto a clipboard, on which to record data and comments. PDAs may provide a high-tech alternative. Computer software is available for virtually all sports. Software packages have been developed for analyzing defensive and offensive strategies in football and certain other team sports. These programs can be used for scouting purposes.

Drug Testing

The school's right to administer random drug tests to athletes was upheld in the 1995 U.S. Supreme Court decision *Vernonia School District 47J v. Acton* (115 S. Ct. 2386). The Court held that random drug testing of student athletes does not violate their constitutional right to be free from unreasonable searches and seizures. (In 2002, the scope of this ruling was expanded by the Supreme Court in the Tecumseh School District case to include students participating in all extracurricular activities.) Students who wish to participate in athletics complete a "specimen control form" which identifies any prescription medications they are taking. Athletes are tested at the beginning of their season, and then random samples of athletes are tested on a schedule under the supervision of a monitor. If a test result is positive for banned drugs, a second test is conducted as soon as possible to confirm the result. If a second sample tests positive, the student's parents are notified. Only the superintendent, principal, vice principal, and athletics director have access to the test results, which remain on record for one year. Generally, the results of drug tests aren't turned over to law enforcement officers.

Some school districts have a policy invoking in-school suspensions for first-time offenders, which requires them to complete their class work in a separate room. Offenders also may be suspended from school activities for a period of time and/or required to undergo drug counseling. Repeat offenders usually face longer suspensions (Yardley, 2000).

ATHLETICS IN MIDDLE SCHOOLS

The question of whether middle schools should sponsor interschool athletics— and if so, to what extent and under what conditions—has received much attention and remains controversial. Many educators feel that sports days and intramurals are more appropriate programs for sixth through eighth graders. However, middle schools often receive intense pressure from parents and high school coaches to implement competitive interschool athletics. The worst abuse of middle school sports occurs when the program is used as a "feeder" system for high school teams without regard to the maturation level of students. Sports psychologists have major reservations about middle school students' ability to handle the stress of highly competitive sports.

Athletics programs in middle schools should evolve out of good basic instruction in physical education and be designed to provide opportunities for students to apply the skills learned in the classroom. This implies that the physical education staff be involved in athletics. All students who wish to participate should be allowed to play at some level. Students should be discouraged from specializing in one or two sports at this age. Team sports have the advantage of accommodating large numbers, being popular with students, and providing valuable social skills. The amount of after-school practices, length of road trips, and number of contests generally should be less than in high school programs.

Younger students are not physiologically ready for intense competition. The wide range in skeletal maturation at this age suggests caution in sponsoring contact sports. Studies show that sports injuries are disproportionately high for ten- to four-teen-year-olds, and football has the highest injury rate. Flag football is preferable to tackle football for younger students. Many middle schools that sponsor tackle football provide both lightweight and heavyweight teams. A typical maximum weight in the lightweight category is 115 pounds (McEwin & Dickinson, 1998).

In spite of the above reservations and a shortage of qualified coaches, most middle schools continue to offer some interschool sports. In doing so, they should keep the following thoughts in mind: middle school athletics ought to be less intense; more fun oriented; less restrictive in terms of participation and specialization; and less demanding physiologically, mentally, and emotionally. Educators should realize that overemphasis on highly competitive athletics at this level often backfires. Many students "burn out" and discontinue sports by age fifteen after having participated in intensely competitive programs from an early age.

ATHLETICS AWARDS

Educators would like to believe that the joy and satisfaction of participation in school athletics is sufficient reward for time and effort spent. This is what is known as an intrinsic reward. However, schools also promote extrinsic rewards. An athletics letter, certificate, or other similar award is meaningful as a symbol of achievement when it represents a standard of excellence. Even well-managed awards program experience problems, however. Schools should develop clearly defined awards criteria that are fair and equitable. Administrators should contemplate how much awards cost the school and whether this money could be better spent in other ways. Some schools are having second thoughts about purchasing expensive trophies, which require more and more floor space for trophy cases. Utilitarian awards, like printed T-shirts or inexpensive sports equipment, are practical alternatives.

The NFSHSA and its counterparts at the intercollegiate level place an upper limit on the monetary value of athletics awards. The rationale is to preserve the ideal of amateurism. These organizations are concerned about budgetary demands and problems of exploitation if limits were not imposed.

Some Guiding Principles

1. Awards should serve as symbols of achievement and have little monetary value.

2. The primary goal of student athletes should be the joy and satisfaction derived from participation and accomplishment, of which an athletics award is a symbol.

3. Opportunities should be provided for an ample number of students to earn awards, yet the criteria should be high enough to make the awards meaningful.

4. Good citizenship should be considered along with participation and achievement.

5. No major distinction should be made between awards for different sports.

6. An award should be given for recognized academic achievement, such as academic all-conference and academic all-American awards.

Specific Requirements (Criteria)

The requirements for earning an athletics award, such as a letter, should be specific but simple. The following are suggested, and they apply equally well to boys and girls in the appropriate sports. The criteria are skeletal and not all sports are included, but the list provides a sampling of recommended criteria.

Football. Participation in at least one-third of the quarters played by the team in the total schedule.

Basketball. Participation in at least one-third of the quarters played by the team in the total schedule or participation in the state tournament.

Baseball/softball. Participation in at least one-fourth of the innings or one-third of the games played by the team in the total schedule or participation in the state tournament. (Exception: Pitchers may be recommended for letters without an established requirement of participation. Base coaches may also be recommended for letters.)

Track. Scoring an average of two points per meet for the dual meets or qualifying in the district or state meet.

Swimming. Scoring an average of two points per meet for the dual meets or qualifying in the district or state meet.

Tennis. Participation in at least one-half of the matches played by the team in the total schedule.

Cross-country. Participation in at least one-half of the meets on the team's schedule and finishing among the first seven in the school in half the meets.

Soccer. Participation in at least one-third of the halves played by the team in the total schedule.

Wrestling. Scoring an average of one point per match in the dual meets, or scoring two or more points in the district or state meet.

Service letter. May be awarded to a student who has been faithful in practice and participation for at least two years and has completed the sport season during the senior year without having reached the required standards, either because of injury or lack of skill.

Manager's letter. May be awarded for a minimum of one full season of service as team manager.

SOURCES OF FUNDING

For the majority of athletics programs, the main source of funding is the appropriated budget. Supplemental sources, which vary in different situations, include gate receipts, donations, concession revenue, parking fees, and revenue from special projects sponsored by the booster club, student body, or members of athletics teams.

Despite all these sources of funding, many school districts are having difficulty finding adequate financing for athletics programs. Some middle schools and high schools have begun charging students a fee to participate in athletics. In states where this practice is legal, fees start at approximately $60 and may run as high as $500 per sport (sometimes with a family maximum for the school year). Other schools are charging athletes for incidentals such as insurance, equipment, or a fee to defray the cost of transporting athletes to sporting events in school-owned vehicles. If such a fee system is implemented, provisions should be made for students whose families cannot afford these charges. The schools might work with service clubs or charitable agencies in the community to assist low-income families. No student in tax-supported schools should be "priced out" of participating in extracurricular activities.

It is usually preferable to have all forms of revenue associated with athletics deposited in the school's general fund and, in turn, have the athletics program managed exclusively on a planned and approved budget. The alternate approach is filled with potential hazards because, if the program depends directly on a variety of non-budgeted sources, large variations can result in the amount of funding from year to year. Obviously, this would be disruptive in terms of planning and continuity. Also, if the program depends heavily on nonappropriated sources, it is easy for athletics personnel to become obsessed with producing revenue instead of providing high-quality educational leadership. Outside interests that help with nonappropriated funding sometimes become too involved in athletics policies and procedures. (More information about the sources of funding is provided in chapter 7.)

Within the basic philosophy described, sponsoring fundraising events can be desirable for two reasons: they can provide funds for some extras in the program that otherwise would not be possible, and special fundraising events can have a rallying effect for the program. Here are some ideas that have proven effective in both high school and college programs:

1. *Jog-a-thons, mini-marathons, or swim-a-thons.* Funds can be raised either from the participants paying entry fees or from soliciting sponsors who pay a specified amount per mile covered by the participant (or per lap in the case of swimming).

2. *Sports camps.* These have become common at colleges and universities, and often a portion of the revenue goes to athletics. Some high schools also sponsor camps.

3. *Celebrity golf tournaments.* Two potential sources of income exist here: (a) golf enthusiasts pay a specified fee to play in a foursome with a celebrity; and (b) if the field of participants justifies it, admission can be charged.

4. *Special food events* such as breakfasts and dinners are sponsored. The tickets are priced to produce a profit. Sponsors often can arrange to have the food contributed by local merchants.

5. *Raffles* are sometimes held in conjunction with other fundraising events. The items given away are usually donated by local sporting goods vendors and other merchants. Sometimes a large item like a set of golf clubs is given as the grand prize.

6. *Novelty athletics events* such as varsity vs. alumni games are often successful. In addition, concession stands and the sale of printed programs, souvenirs, and renting of folding-chair seats are fundraising possibilities.

Schools should sponsor fundraising in a manner that doesn't exploit students or staff. Not all educators and parents are comfortable with some of the practices employing students to solicit funds. Students too often are pressured to use their leisure time for door-to-door solicitation. Parents may object to the practice of teenage girls soliciting passing motorists on street corners for a school-sponsored car wash.

ADDITIONAL REMUNERATION FOR COACHES

Athletics coaches in the public schools put in many after-school hours. There should be a reasonable trade-off for the extra time or a sound formula for additional compensation.

In some cases, coaches are allowed free periods in exchange for the after-school hours. The arrangement may involve a combination of free periods and extra compensation. The most common procedure is to provide extra compensation. Whatever plan is used, it must be equitable in view of the professional preparation and experience, and the amount and kind of extra services performed.

A relatively unsophisticated method would be to offer the coach a flat salary for extra services. This figure would usually relate to merit and the quality of prior years' work. However, a more exact approach might involve a multiplier, which relates to base salary and also includes steps for years of service. Such a method is based on the concept that the teacher should be fairly compensated for the fractional amount of time spent and in accordance with the base salary schedule. As the teacher's base salary increases, the extra compensation should increase a proportional amount. Districts often issue separate contracts for athletics coaches and activities sponsors.

The following formula provides a method of calculating extra compensation. This method can be used with other subjects besides athletics, such as for a language teacher who works extra hours with the language club or the forensics teacher who coaches the debate team.

$$\frac{\text{Number of hours per day in the extra activity}}{\text{Number of hours in the school day}} \times \frac{\text{Length of the extra activity per year (weeks or months)}}{\text{Length of the normal school year}} = \text{Time Index}$$

For example, the wrestling coach spends 2 hours per day with the team; the workday is 8 hours. The wrestling season lasts 4.5 months, and the school year lasts 9 months.

$$\frac{2}{8} \times \frac{4.5}{9} = \text{Time Index} = .125$$

If the teacher's base salary were $35,000, the extra compensation for coaching would amount to $4375 ($35,000 × .125 = $4375).

ADDING AND ELIMINATING SPORTS

The athletics department should establish procedures for discussing the addition and elimination of sports. School administrators, coaches, students, and parents should be involved in these discussions. Eliminating a sport can be a particularly sensitive issue in the community.

The school may decide to add a sport due to Title IX requirements or in response to student interests. The following criteria need to be addressed when considering a new sport (Koehler & Giebel, 1997):

- The sport should be approved by the state athletics/activities association.
- The sport should be competitive and physically challenging.

- The sport should be offered by enough neighboring schools to permit scheduled contests.
- There should be sufficient student interest to guarantee continuing participation.
- The school should have adequate facilities, equipment, and funding to accommodate the sport.
- The sport should contribute to the goal of gender equity.

When sports are being considered for elimination, the following questions should be addressed:

- Has student interest become insufficient to continue the sport?
- Has the sport become too dangerous to permit continued participation? (Many high schools have discontinued men's interscholastic gymnastics.)
- Is the program able to compete successfully with neighboring schools?
- Are other schools dropping the sport, so that scheduling is becoming difficult?
- Has funding become inadequate?

GIRLS' AND WOMEN'S ATHLETICS

Opportunities for girls and women in high school and college athletics multiplied rapidly after Title IX was went into effect in the 1970s. By the late 1990s, 2.5 million girls and women were participating in school sports—almost ten times the number before Title IX. Athletic scholarships for women were almost nonexistent before Title IX. By 2001, there were over 10,000 college scholarships for women athletes.

In the mid-1980s, the effect of Title IX was limited because of the U.S. Supreme Court's ruling on what constituted federal funding; however, passage of the Civil Rights Restoration Act in 1988 reaffirmed Title IX's institutionwide protection. Subsequently in 1992 the Supreme Court ruled that monetary damages could be awarded in Title IX cases. As a result, many schools took Title IX much more seriously. In 2001, the courts ruled that the actions of state athletics associations fall under the provisions of Title IX and the Equal Protection Clause of the Fourteenth Amendment (Women's Sports Foundation, 2001a). In addition to federal regulations, some twenty states had passed legislation to improve equity.

Major concerns continue to be expressed regarding the coaching of girls' and women's sports. Both men and women may coach teams of either gender. However, the current shortage of qualified coaches has impacted women's sports programs disproportionately, and men are displacing women as coaches (more than half of women's intercollegiate coaches in the year 2000 were men).

Requirements of Title IX

This section explains requirements of Title IX of the 1972 Education Amendments Act, which applies to interschool athletics and intramural activities. (Chapter 12 addressed the application of Title IX regulations to the physical education instructional program.) Section 86.41 of the act pertains to interschool and intraschool athletics. It states that "an institution or a district must develop and operate athletics programs according to the following specifications."

1. *General.* No person shall, on the basis of sex, be excluded from participation in, be denied the benefits of, be treated differently from another person, or

otherwise be discriminated against in any interscholastic, intercollegiate, club or intramural athletics offered by a recipient [of federal funds], and no recipient shall provide any such athletics separately on such basis.

2. *Separate Teams.* Notwithstanding the requirements of paragraph (1) of this section, a recipient may operate or sponsor separate teams for members of each sex where selection for such teams is based on competitive skill or the activity involved is a contact sport. However, where a recipient operates or sponsors a team in a particular sport for members of one sex but operates or sponsors no such team for members of the other sex, members of that sex must be allowed to try out for the team offered unless the sport involved is a contact sport. For the purposes of this part, contact sports include boxing, wrestling, rugby, ice hockey, football, basketball, and other sports the purpose or major activity of which involves bodily contact.

3. *Equal Opportunity.* A recipient school that operates or sponsors interscholastic, intercollegiate, club, or intramural athletics shall provide equal athletic opportunity for members of both sexes. In determining whether equal opportunities are available the Director will consider, among other factors:

   - Equipment, uniforms and supplies
   - Access to weight room and training room
   - Equal practice facilities
   - Same size and quality locker rooms and competition facilities
   - Equal access to practice and games during prime time
   - Same quality coaches as boys'/men's teams
   - Opportunity to play the same quality opponents
   - The same awards and awards banquets
   - Cheerleaders and band performances at women's games as well as men's.

Directors of school athletics should keep the following interpretations in mind when administering their programs (Women's Sports Foundation, 2001b):

1. Just because a sport is revenue producing or has more spectators, these facts cannot be used as an excuse for treating male athletes better than female athletes.

2. Schools cannot use the argument that boys are more interested in sports than girls to justify providing more participation opportunities for boys.

3. Schools cannot claim that male athletes are treated better than female athletes because a booster club provides funds for the boys' teams.

4. Girls must be allowed to try out for a boys' team if there is no girls' team for that sport.

5. Boys do not have a right to try out for a girls' team if there are more boys playing sports at that school.

Unequal aggregate expenditures for men's and women's programs will not automatically constitute noncompliance with this section, but such inequality will be considered when assessing equality of opportunity for members of each sex.

Plaintiffs may seek relief from their institution's Title IX compliance officer by filing a lawsuit or though an administrative complaint. The Office for Civil Rights (OCR) within the Department of Education is the agency that fields administrative complaints concerning possible Title IX violations. Currently, Title IX guarantees equal opportunity in all aspects of education. The act covers the major areas of school and college athletics: financial assistance, accommodation of student interests and abilities, and program components (Women's Sports Foundation, 2001b).

References and Recommended Readings

Associated Press. 2001. "State High School Athletic Association Can be Sued." *USA Today* (Feb. 20). Online article: <http://www.usatoday.com/news/court/2001-02-20-school-sports.htm>.

Hoch, David. 1999. "The Many Varied Roles of the A.D." *Scholastic Coach and Athletic Director,* 68(6): 4–5.

Koehler, Mike, & Nancy Giebel. 1997. *Athletic Director's Survival Guide.* Englewood Cliffs, NJ: Prentice-Hall.

Kretchmar, Scott, 1994. *Practical Philosophy of Sport.* Champaign, IL: Human Kinetics Press.

McEwin, Kenneth, & Thomas Dickinson. 1998. "What Role for Middle School Sports?" *The School Administrator* (November): 52–56.

Overman, Steven J. 2001. "Faculty Control of Intercollegiate Athletics: A Survey of Current Practices." Unpublished study.

Overman, Steven J. 1999. "Winning Isn't Everything. It's the Only Thing: The Origin, Attributions, and Influence of a Famous Football Quote." *Football Studies,* 2(2): 77–99.

Popke, Michael. 2002. "Foreign Policy." *Athletic Business* (April): 30, 32.

Reeves, Kimberly. 1998. "Athletic Eligibility: Right or Privilege?" *The School Administrator* (November), 55(10): 6–12.

Schneider, Robert, & William F. Stier. 2001. "Recommended Educational Experiences for High School Athletic Directors (ADs)." *The Physical Educator* (early winter): 211–18.

Stoll, Sharon, & Jennifer Beller. 2000. "Do Sports Build Character?" In John R. Gerdy (Ed.), *Sports in School: The Future of an Institution.* New York: Teachers College Press, pp. 18–30.

"Texas A&M Prof Does Study On Home-Schoolers, Athletics." 1999. *Aggie Daily* (August 6). Online article: <http://www.tamu.edu/univrel/aggiedaily/news/stories/99/080699-7.html>.

Women's Sports Foundation. 2001a. "Playing Fair: A Guide to Title IX in High School & College Sports." *Geena Takes Aim.* Women's Sports Foundation. Online article: <http://www.womenssportsfoundation.org/cgi-bin/iowa/issues/geena/record.html>.

Women's Sports Foundation. 2001b. "Myth Busting: What Every Female Athlete Should Know." 2001. *An Educational Resource Kit: Title IX.* East Meadow, NY: Author, n.p.

Yardley, Jim. 2000. "Family in Texas Challenges Mandatory School Drug Test." *The New York Times,* April 17. Online article: <http://www.mapinc.org/drugnews/v00/n507/a01.html%3F111862>.

Online Sources

National Association for Girls and Women in Sport
http://www.aahperd.org/nagws/template.cfm
NAGWS is the premiere organization to address issues and promote opportunities for all girls and women in sport. The Web site includes publications, news releases, participation statistics, and a legislative action center.

National Association of Intercollegiate Athletics
http://www.naia.org/
The NAIA Web site includes association news, a list of member colleges, job postings, "coaches' corners," and student/athlete information.

National Collegiate Athletic Association
http://www.ncaa.org
The NCAA Web site includes news, in-house publications and reports, governing rules and regulations, names of staff members and committees, and a list of member institutions.

National Federation of State High School Associations
http://www.nfshsa.org
The NFSHSA Web site includes current news, a calendar of events, sports rules and materials, participation surveys, publications, and links to coaches' education programs and to member associations.

National Junior College Athletic Association
http://www.njcaa.org/
The NJCAA Web site includes news, polls, statistics, a calendar of events, forms, and awards.

Women's Sport Foundation
http://www.womenssportsfoundation.org
This organization bills itself as a resource for girls' and women's sports. The Web site includes a newsletter, a calendar of events, updates on Title IX, and an "issues and action" section.

16 ▸ Conducting Athletics Events

Of all school activities, none is more visible in the community than athletics. For better or worse, athletics may be the one facet of the educational program that most shapes public opinion and garners public support for the schools. Hundreds—sometimes thousands—of spectators, including parents, attend these contests. In addition, they often are broadcast or telecast to a local or regional audience. Consequently, it is paramount that these events are conducted in a manner that reflects positively on the school and its educational mission. These events should be well planned, well organized, and conducted professionally and efficiently.

Athletics events generate public enthusiasm and sometimes controversy because competition is dramatic in nature and, therefore, stimulates the emotions of both participants and spectators. Because athletics contests attract the attention of a relatively large segment of the public, this population will include a few individuals who become obsessed by athletics. In many small communities, school athletics is the main form of live entertainment on weekends. It's what people talk about with their neighbors and read about in the local newspaper. Athletics competition has tremendous potential for positive impact on both individuals and the community. The extent to which this potential is achieved depends on the quality of management.

SCHEDULING

Event management begins long before the event takes place, with scheduling being a particular instance. Athletics contests, especially those that attract a large number of fans, should be scheduled and announced well in advance. This can be beneficial in reserving facilities, promoting the events, and avoiding competition or conflict with other activities. In addition, scheduling in advance affords an advantage in the selection of preferred opponents. Computers have facilitated long-range scheduling of athletics contests by managing data on starting times and dates, facility availability, and conflicts with other events.

347

Schedule distribution is important. Copies of schedules, both poster size and wallet size, can be printed. Schools often contract with advertisers to print schedules at the company's expense. Many schools now place their athletics schedules on their Internet Web sites. Of course, schedules should be shared with news media.

In scheduling athletics events, realistic limits should be placed on their number and the length of the season. There is a tendency for the competitive seasons to become longer and longer and to include more and more contests. Reasonable limits must be enforced for the good of the participants and the school program. State athletics associations typically mandate limits on the number of contests for each sport, as shown in table 16-1.

Table 16–1
Maximum Number of Contests: Michigan High School Athletic Association

	Games	Scrimmage		Games	Scrimmage
Basketball	20	4	Baseball	*	4
Football	9	4	Girls Comp. Cheer	12	4
Ice Hockey	24	2	Cross Country	15	4
Skiing	15	4	Golf	16	4
Soccer	18	4	Girls Gymnastics	15	4
Swimming	16	4	Girls Softball	*	4
Tennis	16	4	Track & Field	18	4
			Girls Volleyball	18	4
			Wrestling	16	4

*Baseball and softball teams and individuals may play a maximum combination of 56 dates and contests (e.g., 36 games on 20 dates).

Courtesy of Michigan High School Athletic Association, from MHSAA *Handbook*.

Guidelines for scheduling athletics events include:

1. Maintaining membership in a reputable athletics league made up of schools similar to yours that sponsor similar athletics programs.

2. Facilitating scheduling by aligning men's and women's programs with the same athletics league.

3. Maintaining membership in appropriate state, regional, or national athletics organizations. These organizations facilitate scheduling by the classification of schools according to the nature of their athletics programs.

4. Carefully selecting nonleague opponents to give the athletics schedule variety and interest. Contests should be scheduled within a reasonable distance due to the time, expense, and hazard of travel.

5. Involving coaches in the scheduling process. The advantages: (a) it is a morale factor for the coaches; (b) their contacts with other coaches can often facilitate scheduling; and (c) coaches should be involved in determining the caliber and location of the competition.

GAME CONTRACTS

Written game contracts are recommended whenever athletics contests are scheduled months or years ahead, and if they involve spectators or financial considerations. Game contracts reduce the possibility of misunderstandings and protect against default or unwarranted changes. In some states, the state athletics/activities association provides contract forms. Increasingly, middle schools and high schools are utilizing contracts for both revenue and nonrevenue sports. This practice has supplanted the use of simple letters of agreement.

Traditionally, contracts involving athletics contests have been relatively free from litigation; however, disputes do occur on occasion. Thus, the administrator needs to understand the essential elements of a contract. When drafting these documents, it is important that the language be clear, explicit, and comprehensible, and that it cover the various contingencies that might occur prior to the final outcome of the event. Game contracts usually consist of a single form covering the following terms:

1. The parties, location, date, and time of the contest.
2. Provision for and payment of the officials.
3. The state association and conference rules governing eligibility and competition.
4. Radio and television rights pertaining.
5. Financial guarantees and complimentary tickets.
6. The termination provisions under the agreement.
7. Renewal provisions.
8. Remedies available in the event of breach of contract by one of the parties.

Schools generally cancel a contest by mutual consent; however, the game contract should address conditions under which either party will be considered in breach and provide for remedies when a breach has occurred. In order for a contract to be legally enforceable, it must be supported by *consideration*—something of value exchanged. Consideration may include money or other benefits. Some legal experts recommend that contracts include an arbitration clause to avoid excessive delays involved in civil litigation if a dispute should occur (Fleischman, 1996).

Figure 16-1 is a sample athletics contract. Contracts range in complexity, depending on the significance of the event and the particular conditions associated with it.

MANAGEMENT OF ATHLETICS EVENTS

A host of duties must be handled in connection with conducting athletics events. The best way to identify and keep track of these duties is to prepare a checklist (see table 16-2) and review it to make sure that all the duties are handled effectively and on the proper timetable. It is desirable to identify the person responsible for each duty; however, since this varies from one situation to another, such identification does not appear on the sample checklist that follows.

Some athletics administrators prefer to have a separate checklist specific to each kind of athletics event, such as one for basketball games and another for football or track. If this method is preferred, the checklist presented here can serve as a useful guide.

Normally, the athletics director does not perform all the functions on the checklist but is responsible for seeing that they are handled properly.

ATHLETIC CONTRACT

THIS AGREEMENT, Made and entered into this _____ day of

_____ , 20 ___ , stipulates:

FIRST: That the _____ teams representing the below

named institutions shall play a game of _____

at _____ on _____ , 20 ___ at _____ p.m.

SECOND: That in consideration of playing the above named game, the administrator

of the _____ program shall pay the administrator of

the _____ program the sum of _____

_____

_____

THIRD: That the officials for games shall be settled at least _____

_____ before the contest and the expenses of the

same shall be _____

_____

FOURTH: If either institution refuses to play except for some breach of this contract,

the administrator of the program of the institution refusing to play shall forfeit to the

administrator of the other program the sum of _____

_____

FIFTH: That this game shall be played under the rules of the _____

_____ Conference, and the _____

(national governing organization).

For _____   For _____

(institution)   (institution)

_____   _____
Faculty Representative   Faculty Representative

_____   _____
Director Athletics   Director Athletics

Figure 16–1. Example of a contract form for an athletic contest between two schools

Table 16–2
Before-Game Preparations

Check when Completed	Responsible Person (Optional Information)

_____Contract completed

_____Facilities officially scheduled

_____Eligibility records checked

_____Medical examinations completed

_____Permission slips obtained

_____Game officials scheduled and contracted

_____Equipment available and in good repair

_____Facility prepared (lined, cleaned, etc.)

_____Media coverage arranged

_____Courtesy arrangements made for visiting team

_____Tickets printed and ready for distribution

_____Programs printed and ready for distribution

_____Arrangements made for concessions

_____Ushers assigned

_____Game statisticians assigned

_____Police arrangements made

_____Parking areas available and properly prepared

_____Parking area personnel assigned

_____Scoreboard in working order

_____Public address system ready

_____Spectator area properly prepared

_____Halftime arrangements made

_____Decorations completed

_____Scorers, timers, judges, and announcer selected and scheduled

_____Physician's presence arranged (game format)

_____Written time schedule and contest procedures distributed to those who need them

Game Responsibilities

_____Supplies and equipment on hand

_____Tickets and ticket takers available

_____Ushers present

_____Contest programs on hand

_____Officials' quarters ready

_____Visiting team's quarters checked and courtesies provided

_____Flag-raising ceremonies arranged

_____Intermission (halftime) program set

_____Players' benches in position

Table 16–2   *(continued)*

_____ Physician present
_____ Scoreboard working
_____ Guards for dressing rooms on duty
_____ Concessions open
_____ Cheerleaders and pep band ready
_____ Police personnel on duty
_____ Rest rooms open, clean, and policed as necessary
_____ Extra equipment guarded
_____ Media personnel arrangements completed
_____ Game statisticians ready
_____ Game management personnel prepared (announcers, timers, scorers, etc.)

After-Game Responsibilities

_____ Payment of officials
_____ Payment to visiting school
_____ Equipment repair and maintenance
_____ Storage of equipment
_____ Deposit of ticket receipts and concession income
_____ Financial statement prepared
_____ Proper use of game statistics
_____ Proper attention to any lingering problems

Event Management Technology

Communication among key personnel is crucial immediately before, during, and following an athletics event in order to deal with last-minute details and unexpected occurrences. Cell phones or two-way radios, along with pagers and PDAs, are indispensable. Excellent computer software is now available for planning and conducting most athletics events and is highly recommended for track and swim meets. The software makes it possible to have the meet organized in the computer prior to the date of the meet, including time schedule, entries, heat and lane assignments, records for each event, and so on. Coaches or meet directors can e-mail the seeding information, which may be placed directly into the computer program. At the time of the meet, the names of athletes are simply entered into the template. Immediately after the completion of each event, reliable and detailed results are run and readily available to all concerned, including the media. In both track and swimming, automated timing is also available that will interface with the computer programs. At the conclusion of the meet, results can be e-mailed to designated recipients (Denney, 2002).

CROWD CONTROL

Athletics events that charge admission fall into two basic categories. *General admission* permits patrons to sit in any available seats on a first-come, first-serve basis;

reserved seating provides patrons with specific seating locations. The latter approach usually requires the assistance of ushers. Gates should be opened well enough in advance of starting time (at least an hour for large crowds) to allow orderly progression to seating. General admission requires more lead time than reserved seating. Provide an adequate number of ticket takers (one per 1000 patrons is recommended). Turnstyles provide for control at entrances with minimal interference to traffic flow. If tickets are sold on site, be sure that this function is adequately staffed and secured (Russo, 1998).

Athletics events require special monitoring of both participants (including coaches) and fans. For school sports, this responsibility lies in the hands of coaches, athletics directors, and security directors. The coaches have the primary responsibility for the conduct of the athletes and themselves, whereas the conduct of spectators is primarily a responsibility of administrators. Coaches and administrators would be well advised to hold a hard line against misconduct of both participants and fans, as one group can influence the other. Problems can escalate rapidly if not strictly controlled, and can lead to bad publicity, injuries, and even litigation. The AAHPERD guidelines for crowd control appear below:

1. Provide separate seating areas for the supporters of the two teams.
2. Have uniformed police officers inside and outside the facility when needed and appropriate.
3. Ban alcoholic beverages, mechanical noisemakers, and undesirable signs.
4. Provide free faculty passes to encourage faculty to attend athletics events; and have the faculty interspersed throughout the spectators.
5. Place continuous emphasis on sportsmanship and proper self-control among both athletes and coaches.
6. Provide adequate lighting of parking lots and areas adjacent to the athletics facilities.
7. Carefully plan spectator flow in and out of the seating areas.
8. Have an adequate number of ushers or supervisors located throughout the facility.
9. Provide adequate divisions or barriers between spectators and playing area.
10. Play restful music over the public address system after the contest.

Sometimes, brief remarks by a school administrator, coach, or student body officer can be helpful in communicating that athletics events will be conducted contingent on desirable behavior of participants and spectators. Those unwilling to accept this standard should be discouraged from attending these events.

Make sure that contest venues are well lighted and have adequate space, and amenities such as rest rooms and concession areas are adequate in number, strategically located, and clean. A clean facility has a positive effect on behavior. Many sports venues restrict food and drink consumption to designated areas. Well-planned barriers can prevent potential crowd-control problems by routing traffic and isolating groups. Low walls in indoor areas, and shrubbery and fencing outdoors are options. More schools are utilizing portable, freestanding fencing, with the advantage of being able to reposition it according to need. Fencing can be purchased or rented. Fence sections are available that can be connected with hidden "bungee" cords or staked to the ground. Fences and barriers should be able to withstand formidable stress from bulging crowds without collapsing.

Where there has been a history of crowd-control problems, schools have resorted to nontraditional scheduling including weeknight instead of weekend games, scheduling contests only during daylight—and in extreme situations contests without the public admitted, unannounced contests, or (as a final resort) cancellation of contests.

In response to recent threats of terrorist attacks at public venues, more stringent security measures have been put into place at sports stadiums and arenas. Measures have included rigorous pregame facility inspections; use of metal detectors at access points; a ban on coolers, bottles, and cans; most large bags and backpacks, as well as searches of all other bags brought in by spectators. Entry gates to stadiums and arenas now have designated bag-search lanes and express lanes. Fans wearing coats may be asked to open them. An increased presence of police and security personnel and restrictions on parking near the stadium are evident. Security needs to be positioned both inside and outside the athletics facility.

Three concerns must be addressed in implementing security procedures at school athletics events: cost, legality, and public relations. Both high-tech and low-tech options should be explored when implementing security measures. Regarding legality, public schools are considered agents of the state. Procedures carried out at their sports events are subject to the Fourth Amendment of the U.S. Constitution, which protects citizens against unreasonable search and seizure. Check with the school attorney before implementing new procedures. A growing number of sports fans have objected to increased security measures, including security cameras and invasive searches. Increased security must be justified to the public through a coordinated educational campaign (Miller, Stoldt, & Ayers, 2002).

Finally, develop an emergency procedures manual for the facility and distribute it to key staff members, with their individual roles spelled out. This manual should include an emergency evacuation plan, which should be rehearsed and then evaluated.

ADA REQUIREMENTS FOR FACILITIES

The Americans with Disabilities Act (ADA) requires that public accommodations are usable by all people. School facilities (including sports stadiums, arenas and fields) must make "reasonable accommodations" for participants and other patrons who have disabilities. Spectator areas in sports facilities must provide integrated wheelchair seating with a "line of sight" to the contest area. Barriers such as steps or curbs that would impede wheelchair access must be removed when "readily achievable" in existing facilities. Federal regulations supplement local and state building codes. New construction must comply with all ADA accessibility standards. All fixtures, such as public telephones, drinking fountains and hand dryers, must be positioned within reach of wheelchair occupants.

The first step in compliance is to conduct a complete facility review. Designate a competent staff member to be the ADA resource person. The best approach is to work from the outside to the inside of the facility, following the same travel path as a disabled user. Compliance requires that particular attention be given to parking area, sidewalks and ramps, entranceways, stairs/elevators, and rest rooms. Following the review, prepare a written evaluation of needed repairs that includes priorities, cost estimates, and completion schedule. For extensive renovations or repairs, facility mangers may consider hiring an ADA consultant (Fried, 1998).

GAME OFFICIALS

Poor officiating can be one of the triggers for negative behavior of sports fans and unethical conduct by coaches and players. Good officiating, therefore, should be viewed as important in itself and also as an effective method of controlling related problems. Of course, no official is capable of doing a perfect job, particularly in the eyes of partisan fans, coaches, and participants. A strong effort should be made to obtain well-qualified officials and get the players, coaches, and fans to recognize that the officials deserve their understanding and support. To gain experience, a new official often starts with middle school, freshman, and junior varsity games. Another starting place is preseason practice games. This approach provides game experience with less pressure than league games. A prospective official is usually required to pass written and practical certification tests.

Game officials should be selected and assigned by the appropriate representative of the athletics association, conference, or league, not by a representative of either school. A sample contract for game officials is shown in figure 16-2. The National Federation Officials Association represents sports officials and publishes *Officials' Quarterly*, a national publication for high school officials.

```
Date _____ 20____

This is an agreement to officiate as _____ by
_____ at the _____ school
         (name)

on _____ 20___. The _____
        (date)                              (sport)

contest between _____ and
_____ will begin at _____.
                                                  (time)
The fee for this activity will be _____. Mileage will be paid
at _____ ¢ per mile both ways. Other officials who will be working with you are:

_____        _____
        (name)                                  (position)

_____        _____


You will be expected to report _____ minutes prior to the contest. In
case of postponement, you will be informed of such by _____.
                                                              (time)

_____        _____
(Director of Athletics)                 (Official)

_____        _____
(Address)                               (Address)

_____        _____
(Telephone number)                      (Telephone number)

Return one copy and keep one for your records.
```

Figure 16–2. A sample form for contracting game officials

CHEERLEADING SQUADS

Cheerleaders play an important role at athletics events. Like athletes, they are representatives of the student body, chosen as a result of their skills and ability to lead. By nature of their positions, they have an excellent opportunity and important responsibility to promote goodwill and sportsmanship during athletics contests, pep rallies, student assemblies, and many other school settings. Cheerleaders can exert a significant positive influence on other students and on fans in general, thus playing an important role in crowd control at athletics contests. Cheerleaders should be trained with the following objectives:

1. To develop a wholesome and enthusiastic school spirit.
2. To develop loyalty to the school and team regardless of the outcome of the game.
3. To promote a cooperative spirit between the student body, faculty, and school administration.
4. To promote the kind of sportsmanship that will help students acquire the basic attributes of good citizenship.

Appropriate policies should govern the selection process for cheerleading squads. Tryouts must be conducted in a fair and open manner. Legitimate criteria for selection include poise, enthusiasm, and athletic ability. Occasionally, controversy accompanies selection. The ideal is to appear inclusive rather than exclusive. Squads should be representative of ethnic groups in the school and preferably include both boys and girls.

Uniforms also can be problematic. Some schools have incorporated cheerleading uniforms into their athletics budgets, but squads often have to engage in fundraisers to defray the cost. Students should not be "priced out" of cheerleading when the cost of uniforms is borne partially or completely by parents. Uniforms should be selected to facilitate free movement but should not be immodest to the point that they distract from the purpose of cheerleading or encourage ogling of female cheerleaders by spectators. What is deemed acceptable for cheerleaders in professional sports should not be used as a model for young women.

Hundreds of cheerleading camps and venues for competition currently exist, sponsored by groups such as the National Cheerleaders Association and the Universal Cheerleaders Association. Attending camp is not an essential element to having a successful cheerleading program; however, camps can provide cheerleaders and coaches with an opportunity to learn new material; build leadership, discipline, and team unity; and become educated about safety techniques and regulations. Most camps are held in the summer and now promote intersquad competition. Cheerleading competition is becoming more popular at all levels. As with athletics competition, cheerleading competition can be overemphasized. Schools need to tailor these experiences to the students" maturity level and justify participation within budget limitations. *American Cheerleader Magazine* provides information on national and regional competitions.

The National Federation of Interscholastic Spirit Associations (NFISA) regulates cheerleading under the organizational umbrella of the National Federation. NFISA publishes safety rules and guides for cheerleading and drill-team sponsors. The American Association of Cheerleading Coaches and Advisors (http://www.aacca.org) instructs school personnel about safety and liability. AACAA offers the following administrative guidelines to minimize risks to cheerleaders:

Knowledgeable Coach
- Attends training camps with team
- Participates in local, state or national coaches' conferences
- Completes safety certification course

Appropriate Practice Facilities
- Adequate matting
- Adequate space
- Adequate height
- Safety procedures
- Written emergency plan
- Access to athletic trainers
- Staff member certified in CPR/first aid

Travel
- Cheerleading coach should be aware of and follow school travel policies for safety and proper insurance coverage.

Legal Issues
- Pre-participation physicals should be required in accordance with policies on all student athletes.
- All forms should be completed for each participant:
  Medical release
  Informed consent/liability waiver
  Insurance information
  Parental contact information

Following instances of catastrophic injuries, some school districts have banned off-the-ground stunts for cheerleaders. Administrators and sponsors should familiarize themselves with school district policies and state liability statutes.

Cheerleading may be athletic and encompass competition, but it is not considered an athletic team or sport for purposes of satisfying provisions of Title IX, according to the Office of Civil Rights. Consistent with this interpretation is The Women's Sports Foundation position, stated in part:

> Danceline, cheerleading, drill teams, baton twirling and marching band . . . usually exist to entertain or educate a spectating audience, or, in the case of cheerleading, to coerce audience enthusiasm. . . . These exhibition activities are secondary . . . to the existence of an athletic team or the purpose of an athletic activity (Women's Sports Foundation, 2001).

ATHLETICS TRIPS

In scheduling trips off campus, the following considerations should be taken into account. A trip should not involve excessive distance or time; transportation arrangements should be adequate and safe; and an ample amount of responsible supervision should be provided. The mode of transportation to use—vans, school bus, or commercial carrier—has both safety and legal implications with respect to the choice of vehicle and driver (see risk management section below). Occasionally, sufficient reasons exist to cancel an athletics event, such as adverse weather or prevalent illness among team members. Leagues should establish clear guidelines as to what justifies cancellation and the procedures that should be followed.

The best approach to thorough planning and effective arrangements for an athletics trip is the utilization of a checklist, such as the example that follows.

Check when Completed

_____ Confirm game details with the host school by letter or telephone.

_____ Make travel arrangements by common carrier or other approved mode.

_____ Plan a travel itinerary, including food and lodging arrangements.

_____ Draw and cash a check from the school fund to cover the expenses.

_____ Secure and distribute or sell the tickets provided for the visiting school.

_____ Send itinerary to parents and secure the signed permission slips for all athletes, as required by school district policy.

_____ Inform the athletics groups, student groups (band, cheerleaders), and adult groups (athletics booster club, parents of players) of itinerary.

_____ Pack game equipment and assign a student manager to guard the baggage. Include athletics training supplies.

_____ Load equipment; inventory the items.

_____ Load athletics team personnel; check roster.

_____ Check team personnel before leaving each stop.

_____ Unload personnel and equipment at the contest site.

_____ Contact the host staff immediately on arrival.

_____ Assign the student manager to guard clothing and equipment during the contest.

_____ Have a travel vehicle prepared to leave immediately after the contest.

_____ Secure the contract payment.

_____ Account for equipment and personnel before departing.

_____ Account for equipment and personnel on arrival.

_____ Write a courtesy follow-up letter to the host school.

_____ Confirm all matters that apply relative to insurance.

Risk Management in Transporting Student Athletes

Student athletes should be transported in a manner that insures their safety and doesn't compromise the school's liability. Schools owe a legal duty of care when they provide transportation. The laws governing the transportation of students differ widely from state to state. Teachers, coaches, and administrators should be well informed about applicable state laws. If a statute is violated, the standard of care is irrelevant if proximate cause can be shown. The school then would be held absolutely liable. Thus, it is imperative that educators provide transportation in an informed and safe manner to avoid charges of negligence. Liability usually applies from the point of departure until return to that point, or from the time students enter the transporting vehicle until they disembark, depending on the jurisdiction.

Schools have four options to consider when transporting students: (1) use of independent contractors, (2) use of organization-owned vehicles, (3) use of employee vehicles, and (4) use of nonemployee (parents, volunteers, participants) vehicles. If a school can afford to hire an independent contractor, this is the best legal option since the contract shifts most liability to the contractor. Investigate before hiring to make sure that the contractor has liability insurance and meets other state requirements, as well as requirements of the Interstate Commerce Commission. For school districts that can't afford to hire a private contractor, school-owned vehicles are the next best option. This is the most common method of transporting students. The school has a duty to assure that its vehicles are in safe operating condition and that drivers are properly qualified and licensed. Equip all school vehicles with the necessary emergency equipment. The National Highway Traffic Safety Administration has cautioned schools to carry no more than ten students in midsize vans, as a greater number of passengers increases the vehicles' likelihood to roll over.

Before a school allows students to be transported by employees or nonemployees in their private vehicles, stringent risk management policies should be established to assure that both the vehicles and drivers conform to safety standards. Where a *principal-agent* relationship is established by law, the school may be liable for the negligent action of a driver. Transporting students in privately owned vehicles creates the greatest liability for schools and their employees. Students under the age of eighteen should never be allowed to transport other students (Pittman, 1996).

Traditionally, some states defined the duty owed by an automobile operator to a nonpaying passenger as less than that of ordinary care, through so-called "*guest statutes.*" Guest statutes prevent nonpaying passengers from suing the driver or owner of a car for accidental injuries except in cases of gross negligence or willful or wanton misconduct. An increasing number of states have ruled that guest statutes are unconstitutional, denying due process, and now hold that drivers and owners of automobiles owe a duty of care to transported guests.

USE OF FACILITIES

Facilities used for athletics practices and competition are also used for other activities, such as physical education classes, intramurals, and free play. In view of this, at least three important guidelines pertain to facilities:

1. Enter requests for facilities for both contests and practices well in advance, and make certain that the requests are complete and clearly communicated. Early requests are especially important in connection with interschool athletics contests.

2. Be reasonable and considerate in terms of others' needs for the facilities. Unfortunately, some athletics coaches become greedy and inflexible about facility use. Practically all school sports facilities should be multipurpose in nature, and each of the different uses should be held in proper perspective and correct priority.

3. Teachers and coaches have a responsibility to make sure the facilities are used appropriately. In addition, users are responsible for leaving the facility in as good a condition as possible for subsequent users.

The best way to avoid conflicts in scheduling facilities is to have well-defined priorities and clearly described scheduling procedures that are strictly enforced. In addition, teachers and coaches should be encouraged to avoid infringing on the scheduled time of others and to promote a congenial and mutually supportive relationship in terms of facility use. It is a good practice to prepare a schedule form for each teaching station prior to the beginning of each term.

ATHLETICS ELIGIBILITY

Athletics administrators and coaches should make sure that all their athletes are eligible prior to their participation in athletics contests, and that lists of eligible athletes are kept on file and updated (see section on eligibility in chapter 15). Figure 16-3 is a sample of a high school eligibility certificate.

SAFEGUARDING THE HEALTH OF ATHLETES

The National Federation of State High School Associations and the American Medical Association's Committee on the Medical Aspects of Sports have issued a joint statement that the athlete has the right to optimal protection against injury to the extent this can be assured through proper conditioning, careful coaching, good officiating, proper equipment and facilities, and adequate medical care. Periodic evaluation of each of these factors should help provide a safe and healthy experience for players. The five major areas of the joint statement follow.

1. *Proper conditioning.* This helps to prevent injuries by hardening the body and increasing resistance to fatigue. Coaches and athletics administrators should:

- Give prospective players directions to follow and activities to perform for preseason conditioning.
- See that there is a minimum of three weeks of practice before the first game or contest.
- Take precautions to prevent heat exhaustion and heat stroke. This is particularly important in preseason practice for football, soccer, cross-country, and field hockey.
- Require players to warm up and stretch thoroughly before participating in rigorous activity.
- Make substitutions without hesitation when players show signs of disability, injury, or fatigue.

2. *Careful coaching.* Athletes who are well coached and more skilled have a reduced chance of injury. Coaches and other athletics department personnel should:

- Stress safety in performance techniques and elements of play.
- Analyze injuries to determine their cause and suggest ways to prevent them.
- Discourage the use of tactics and techniques that increase hazards on the field of play.
- Plan practice sessions carefully and see that they are neither too long nor too short.

3. *Good officiating.* Qualified officials who know what they are doing promote enjoyment of the game and protection for players. In this respect:

- Players and coaches should be thoroughly schooled in the rules of the game.
- Roles and regulations should be strictly enforced in practice sessions as well as in games. If they are not, the risk of injury will increase in practice, and players will get into bad habits that will carry over into games.
- Officials should be qualified both emotionally and technically for their responsibilities on the field or court.

ATHLETIC ELIGIBILITY CERTIFICATION

This is to certify that the students listed below are eligible to represent the
_____ High School in the following
athletic contest _____
to be played at _____ on the
_____ day of _____.

FACULTY REPRESENTATIVE _____

PRINCIPAL _____
                              (Signature)

COACH _____
                              (Signature)

Names of Contestants	Uniform Number		Birth Date	Class
Last Name First Initial	Home	Away	Mo. Day Yr.	
1.				
2.				
3.				
4.				
5.				
6.				
7.				
8.				
9.				
10.				
11.				
12.				
13.				
14.				
15.				
16.				

Figure 16–3.   Sample form for certifying eligibility of student athletes

- Players and coaches should respect the decisions and ruling of officials. If they don't, tempers will flare and the game may get out of control.

4. *Proper equipment and facilities.* Student athletes should not be expected to participate in athletics programs unless they are protected by safe equipment and play on safe fields and courts. Specifically, schools should see that:

- The best protective equipment is provided for contact sports.
- Careful attention is given to the proper fitting and adjustment of equipment.
- Equipment is properly maintained; worn and outmoded items are discarded. The temptation to try to make equipment last one more year should be resisted when the safely of student athletes is at stake.
- Proper and safe areas of play should be provided, and these areas should be carefully and continuously maintained.

5. *Adequate medical care.* (See figures 16-4 and 16-5.) Proper medical care is a necessity in the prevention and control of injuries. Athletics educators and school administrators should:

- See that a thorough preseason medical examination is provided for each athlete and that a detailed health history is taken. Screening should be repeated every two years.
- Arrange to have a physician present at games when possible, and see that a doctor is available during practice sessions.
- Adopt a policy to the effect that the physician will make all decisions as to when an athlete should resume participation following an injury. This applies to both competition and practice.
- Adopt policies on exactly how much medical attention a trainer can perform without a doctor present.

The Preparticipation Physical Examination

The great majority of state high school athletics/activities associations require some type of physical examination prior to participation in sports. However, few associations provide specific guidelines for the exam or for excluding athletes for medical reasons. Regardless of state association requirements, all schools should conduct exams by a physician with expertise in sports medicine, as a matter of course. Several physicians' groups, including the American Academy of Family Physicians and the American College of Sports Medicine, have established guidelines and recommendations for determining what conditions warrant restrictions in sports participation. Athletes with disabilities may have a right to participation under Section 504 of the Rehabilitation Act of 1973. State and federal laws appear to support schools' right to require preparticipation exams as qualification to participate in athletics programs. If the athlete fails the exam, he or she generally has the right to a second opinion by another competent physician (Block, 1998).

Annually a small number of students die while engaged in physical education or athletics. Most of these deaths can be attributed either to heat stroke or to an undetected heart condition. It is crucial that teachers and coaches are trained on how to prevent heat-related deaths. Regarding heart conditions, some states have begun to include echocardiogram tests as part of physical exams for athletes. Physicians often will volunteer their time to conduct these tests. Some state athletics associations are

ATHLETIC EMERGENCY INFORMATION FORM

To be completed by Parent or Guardian
(please print)

Name of Student _____

NAME _____
(parent or guardian)

ADDRESS _____

Signature _____ Date _____

PHONE (home) _____

(business) _____

FAMILY DOCTOR (1) _____ Phone _____

(2) _____ Phone _____

CLOSE RELATIVE (1) _____ Phone _____

(2) _____ Phone _____

In the event parent, family doctor, or relative cannot be reached, indicate your hospital preference:

(1) _____

(2) _____

IF CONTACT CANNOT BE MADE WITH ANY OF THE ABOVE, THE COACH WILL USE HIS BEST JUDGMENT TO PROTECT AND ASSIST THE INJURED ATHLETE IN ACCORDANCE WITH SCHOOL POLICY.

Figure 16–4.   Emergency medical form to be placed on record at the school

---

SUGGESTED HEALTH EXAMINATION FORM

(Cooperatively prepared by the National Federation of State High School Associations and the Committee on Medical Aspects of Sports of the American Medical Association.) Health examination for athletes should be rendered after August 1 preceding school year concerned.

(Please Print)   Name of Student _____   City and School _____

Grade_____ Age_____ Height_____ Weight_____ Blood Pressure_____

Significant Past Illness or Injury _____

Eyes _____ R 20/ ____; L 20/ ____; Ears _____ Hearing R ____/15; L ____/15

Respiratory _____

Cardiovascular _____

Liver_____ Spleen _____ Hernia _____

Musculoskeletal _____ Skin _____

Neurological _____ Genitalia _____

Laboratory: Urinalysis _____ Other _____

Comments _____

Completed Immunizations: Polio_____ Tetanus _____
                                                    Date   Date

Instructions for use of card   Other _____

"I certify that I have on this date examined this student and that, on the basis of the examination requested by the school authorities and the student's medical history as furnished to me, I have found no reason which would make it medically inadvisable for this student to compete in supervised athletic activities, EXCEPT THOSE CROSSED OUT BELOW."

BASEBALL	FOOTBALL	ROWING	SOFTBALL	TRACK
BASKETBALL	HOCKEY	SKATING	SPEEDBALL	VOLLEYBALL
CROSS COUNTRY	GOLF	SKIING	SWIMMING	*WRESTLING
FIELD HOCKEY	GYMNASTICS	SOCCER	TENNIS	OTHERS _____

*Estimated desirable weight level: _____ pounds

Date of Examination _____ Signed _____
                                                                    Examining Physician

Physician's Address _____ Telephone _____

Figure 16–5.   Health examination form for athletics participation (reprinted by permission of the National Federation of State High School Associations and the American Medical Association)

convening medical advisory committees to establish policies in these areas (Popke, 2002). Often, local hospitals and clinics assist school districts by offering free medical services to student athletes for sports-related injuries and other services such as preparticipation physicals or injury evaluation.

Procedures in the Absence of a Physician

When an athlete is injured and no physician is present, the following procedures should be carried out by the trainer (or the coach, if a trainer is not available). All athletics staff should be familiar with these important procedures.

1. Make an immediate preliminary examination of the injury.

2. Send for a physician and/or ambulance immediately if the injury is beyond the realm of your ability.

3. Give first aid as needed.

4. Notify the parents or, if they are not available, a designated relative or friend, and inform them of the athlete's condition, what treatment you have administered, or what action you suggest.

5. Determine whether the condition is such that the athlete should be removed from the area. First, ascertain whether such removal would be sanctioned by a physician. Unconscious players or players who are unable to move comfortably should be moved only if necessary and then only with adequate assistance and by proper stretcher or backboard techniques.

6. Have an ambulance available for all games involving contact sports. Keep it out of view of both spectators and players. Have the attendants readily available.

7. Use a standard accident report form to record all important information.

Athletic Trainers

Currently there is a trend toward providing certified athletic trainers at all levels of school sports programs. However, less than half of the nation's high schools currently employ athletic trainers. A few states like West Virginia won't allow high schools to participate in football practices or games without a certified trainer. Well over half the states now have certification, registration, licensure, and exemption regulations for athletic trainers. States usually require trainers to be certified through state trainer certification boards. Schools may require trainers to be certified by the National Athletic Trainers Association Board of Certification (NATABOC). Waivers or temporary exemptions may be granted, but schools are expected to employ certified trainers (who may also be teachers). In addition, schools use student athletic trainers under professional supervision to assist with taping and other duties. Student trainers should be certified in first aid and CPR.

Duties of athletic trainers fall into three basic categories: prevention of injuries, treatment of injuries, and rehabilitation after injuries. Trainers give massages and basic first aid, tape or wrap limbs or muscles, and prescribe supportive equipment to allow athletes to continue competing. Trainers also assist team doctors in deciding when players can compete following an injury.

The trainer has numerous specific duties; the more prominent ones are included in the following list:

1. Assists the physician with the arrangements for preparticipation exam and record keeping.

2. Works cooperatively with the coaches in setting up and carrying out a program of conditioning for athletes, especially those aspects that help in the prevention of injury.

3. Administers appropriate emergency treatment to injured athletes on the field, in the gymnasium, or in the training room.

4. Applies protective or injury-preventive devices such as strapping, bandaging, or bracing.

5. Works cooperatively with and under the direction of the physician in respect to:

   a) Reconditioning procedures following injury.

   b) Use of therapeutic devices and equipment.

   c) Fitting of special braces, guards, and other protective devices.

   d) Referrals to the physician, health service, or hospital.

6. Works cooperatively with the coaches and the physician in selecting and checking safety equipment.

7. Supervises the training room, which includes requisitioning and storage of supplies and equipment, keeping records, maintaining an accurate inventory, and managing the training budget.

8. Supervises and instructs assistant trainers.

9. Counsels and advises athletes and coaches on matters pertaining to conditioning and training, such as diet, rest, and reconditioning.

10. Keeps accurate and detailed records of athletics injuries and treatments.

Student Managers

Defining the role of a student manager is difficult because it can vary considerably among different sports and with the preference of coaches. Probably the most accurate definition is *an assistant to the coach who carries certain responsibilities in accordance with the agreement between the coach and the manager.* Many responsibilities of the student manager could be correctly labeled as errands, but certain responsibilities involve much more. For example, the manager's duties often include caring for equipment, keeping inventory, compiling, analyzing, and reporting game statistics and preparing equipment and supplies for team practices and games.

A coach should add a reasonable amount of dignity to the student manager's position by including some significant duties and treating the manager with respect. In this way, being a student manager can be a rewarding experience, and the manager can make a useful contribution.

METRIC MEASUREMENT IN SPORTS

Because of easy conversion between units of the same quantity, the metric system has become the internationally accepted system of measurement units. The United States is the only industrialized country in the world not officially using the metric system. Despite federal legislation requiring its use in some sectors of the economy, we remain a long way from full-scale conversion to metric measurement. However, science and industry are rapidly adopting the metric system, and it is being taught in many school systems.

As the scale of international sports competition has increased, the need for a common set of standards has also increased. The 1996 Atlanta Summer Olympics was utilized to promote awareness of the metric system in the United States, with mixed results. In American college athletics, the metric system is now in use in sports like track and field and in swimming, but no one sees any advantage to changing a football field to metric measurement. The NCAA rulebooks for the various sports include metric conversions for the traditional imperial measurements. Progress toward the use of metric measurements has been made in other levels of athletics competition not regulated by the NCAA, including high school, junior college, NAIA, and open competition.

Coaches and meet officials should familiarize themselves with the metric units of measurement and, where necessary, the methods of conversion. Most are fairly simple, but some can be confusing. For example, a high-jump bar set at 7 feet would be at 2.13 meters. Once the metric system is adopted, internal conversions are simple because metrics are established on a "base ten" system, whereas the imperial system lacks this element of consistency.

Resources are available to assist coaches and athletics administrators with metric measurements. *Sports Rules On File* (ISBN 0-8160-4117-2), published by Facts On File Reference Library Series, lists the major sports played at the U.S. high school and college level such as tennis, track and field, wrestling, basketball, baseball, volleyball, football, soccer, swimming, and gymnastics. When rules of the sport specify U.S. measurements, they are given without conversion to metric. When the metric system is required, U.S. conversion measurements are also noted. CD-ROM and online versions of this resource are available. A free online resource is The IFP Reference library (http://www.french-property.com/ref/convert.htm). This Web site allows the user to calculate conversions between the imperial system and metric system on the screen, as well as providing conversion tables.

References and Recommended Readings

Block, Martin. 1998. "The Preparticipation Physical Examination." In Herb Appenzeller (ed.), *Risk Management in Sport: Issues and Strategies*. Durham, NC: Carolina Academy Press, pp. 169–86.

Denney, Maurie. 2002. Athletics Director, Jefferson High School, Lafayette, IN. Personal correspondence, January 21.

Fleischman, Robert. 1996. "Game and Event-Related Contracts." In Doyice Cotton & T. Jesse Wilde (eds.), *Sport Law for Sport Managers*. Dubuque: Kendall/Hunt, pp. 230–35.

Fried, Gil. 1998. "ADA and Sports Facilities." In Herb Appenzeller (ed.), *Risk Management in Sport: Issues and Strategies*. Durham: Carolina Academic Press, pp. 253–65.

Hardy, R. 1981. "Checklist for Better Crowd Control." *Journal of Physical Education, Recreation & Dance*, 52: 70.

Miller, Lori, G. Clayton Stoldt, and Ted Ayers. 2002. "Search Me," *Athletic Business* (January): 18–21.

Pittman, Andy. 1996. "Transportation." In Doyice Cotten & T. Jesse Wilde (eds.), *Sport Law for Sport Managers*. Dubuque: Kendall/Hunt, pp. 113–21.

Popke, Michael. 2002. "Doctors on Duty." *Athletic Business* (March): 26, 28.

Russo, Frank Jr. 1998. "Events Crowd Management." In Herb Appenzeller (ed.), *Risk Management in Sport: Issues and Strategies.* Durham: Carolina Academic Press, pp. 237–52.

Shaul, Marnie S. 2000. "Interscholastic Athletics—School Districts Provided Some Assistance to Uninsured Student Athletes." FDCH Government Account Reports, Sept. 12.

Women's Sports Foundation. 2001. Position Paper. "Drill Teams, Cheerleading, Danceline and Band as Varsity Sports." *An Educational Resource Kit: Title IX.* East Meadow, NY: Authors, n.p.

Online Resources

National Athletic Trainers Association
http://www.nata.org/
The NATA Web site includes news and publications, information on certification, a calendar for trade shows, and educational material for athletic trainers.

National Federation of Interscholastic Spirit Associations
http://www.nfhs.org/nfisa.htm
Supports coaches of middle and high school cheerleading, pom, dance, and drill teams, by assisting them in risk management, good sportsmanship and citizenship, and professional development. The Web site includes resources, rules, current news and events, a discussion forum, and a section on sports medicine.

National Federation Officials Association
http://www.nfhs.org/nfoa.htm
The NFOA Web site includes membership applications, insurance information, publications, playing rules, and an officials' education link.

National School Safety and Security Services
http://www.schoolsecurity.org/resources/Links.html
A national consulting firm specializing in school security and crisis-preparedness training, security assessments, and related safety consulting for K–12 schools. The Web site includes links to news, resources, and training.

Sydex Software
http://www.sydexsports.com/default.htm
Sydex Computer Systems offers software packages for football, track and field, cross-country, basketball, volleyball, baseball, softball, and soccer.

Appendix A
Professional Resources

Professional organizations and publications make important contributions to the lifeblood of a particular discipline, so they need to be supported and utilized effectively. These resources help students and professionals to become involved and understand the current issues and trends. They keep us abreast and instill the realization that few things are more interesting and stimulating than to be on the leading edge of a profession.

Members of a profession should know which organizations relate to their area and be at least generally familiar with their purposes, programs, and services. Knowledge about these organizations can serve at least two purposes:

1. Helping the administrator better understand how the organization can contribute to the profession as a whole and to the programs of a particular institution or school system.

2. Helping the administrator better know with which organizations to affiliate and for what reasons.

The same points are true for professional literature. Students, teachers, coaches, and administrators need to keep up to date on what is being written by the leaders of their profession. They need to stay abreast of technological changes, of the norms, the issues, and professional trends. Since much of professional literature today is available via the Internet, this appendix lists the Web sites of all organizations, journals, and newsletters mentioned.

PROFESSIONAL AND SERVICE ORGANIZATIONS

Numerous professional and service organizations at the national level focus on physical education and school athletics. A brief description of each of these organizations is presented in alphabetic order. Some of the organizations have district and state subdivisions or affiliates.

American Alliance for Health, Physical Education, Recreation and Dance
1900 Association Drive, Reston, VA 20191
<http://www.aahperd.org/>
AAHPERD is the largest organization of professionals supporting and assisting those involved in physical education, leisure studies, fitness, dance, health promotion, and all specialties related to achieving a healthy lifestyle. It is made up of six national associations and a research consortium and is divided into six district associations, along with 54 state and territorial affiliates. These associations provide members with a comprehensive and coordinated array of resources, support, and programs to help practitioners improve their skills and to further the health and well-being of the American public. Founded in 1885, today the membership network reaches into more than 16,000 school districts, over 2,000 colleges and universities, and 10,000 community recreation units.

The following associations comprise the Alliance:

American Association for Leisure and Recreation (AALR) promotes school, community, and national programs of leisure services and recreation education.

Association for the Advancement of Health Education (AAHE) works for continuing comprehensive programs of health education in schools, the workplace, and the community. Position papers are developed on such health topics as certification, drug education, and sex education.

American Association for Active Lifestyles and Fitness (AAALF) has as its mission the promotion of active lifestyles and fitness for all individuals by facilitating the application of diverse professional interests through knowledge expansion, information dissemination, and collaborative efforts.

Research Consortium. The purpose of the Consortium is to promote research in health education, physical education, dance, athletics, exercise, and recreation. Membership is comprised of any of AAHPERD's six association members who conduct, or have an interest in, research.

National Association for Girls and Women in Sport (NAGWS) is the primary educational organization for professional development of girls and women as sports leaders and advocates for programs of sport and physical activity for all females.

National Association for Sport and Physical Education (NASPE) provides leadership opportunity in physical education and sports development, competition, consultation, publications, conferences, research, and a public information program.

National Dance Association (NDA) promotes the development of sound policies for dance in education through conferences, convention programs, special projects, publications, and cooperation with other dance and art groups.

The AAHPERD holds a national convention each year. In addition, the six districts each hold a convention annually, and the 54 state and territorial affiliates conduct annual conventions. The Alliance and its affiliates sponsor a large number of workshops, conferences, and clinics each year, all of which are aimed at professional development of its members or the solution of particular issues pertaining to the profession.

The Alliance publishes five regular periodicals: *Update; Journal of Physical Education, Recreation and Dance; Journal of Health Education; Research Quarterly for Exercise Sport;* and *Strategies.* More than a dozen newsletters are prepared and distributed regularly to provide information in specific areas of the associations. The Alliance also publishes books and films. The titles are available in the AAHPERD publications catalog.

American College of Sports Medicine

401 W. Michigan St., Indianapolis, IN 46202-3233

<http://www.acsm.org/>

The American College of Sports Medicine is a nonprofit association with an international membership of more than 18,000. Unlike most associations whose memberships are joined together by a single professional specialty, this organization represents over 40 different specialties. ACSM has twelve regional chapters throughout the United States. Membership is open to any individual interested in sports medicine who holds a graduate degree or who has become specialized in a field related to health, physical education, athletics, or the medical sciences.

ACSM promotes and integrates scientific research, education, and practical applications of sports medicine and exercise science to maintain and enhance physical performance, fitness, health, and quality of life. The specific objectives are to promote and disseminate medical and other scientific studies dealing with the effects of sports and physical training on people's health; to initiate, promote, and correlate research in these fields; and to establish and maintain a sports medicine library.

The College sponsors an annual national convention and several workshops in various geographic areas, as well as lecture tours by selected sports medicine specialists; and it takes position stands on important sports medicine topics and issues. The official monthly journal is *Medicine and Science in Sports Exercise.* Other publications include *Health and Fitness Journal* (for practitioners), the *Sports Medicine Bulletin*, and a *Yearbook of Sports Medicine.*

American Council on Exercise

4851 Paramount Drive, San Diego, CA 92123

<http://www.acefitness.org>

ACE is the world's largest nonprofit certifying organization for fitness professionals in the following four categories: personal trainer, group fitness instructor, lifestyle and weight-management consultant, and clinical exercise specialist. The certification exam is offered periodically in cities across the United States and Canada.

American Physical Therapy Association

1111 North Fairfax Street, Alexandria, VA 22314-1488

<http://www.apta.org/>

Founded in 1921, APTA has a membership of more than 65,000. It is divided into 52 chapters and 19 special-interest sections. APTA provides a variety of services to its members and to the profession. These services include the following: sets standards of practice and adopts and enforces ethical principles to assure high-quality physical therapy services for the public; through the Commission on Accreditation in Physical Therapy Education, accredits educational programs to assure the preparation of competent practitioners; guides and assists in the development of continuing education and professional development opportunities for physical therapists; promotes the exploration of new knowledge in physical therapy; holds two annual conferences, conducts continuing education programs, and provides for the production and dissemination of publications to help transmit pertinent information; speaks for physical therapists in legislative, regulatory, and other health policy matters; conducts a public-relations program; fosters the recruitment of minorities into the profession; and supplies information and advice to members, educational institutions, private and governmental agencies, and other associations.

The American Physical Therapy Association is a member of the National Health Council and the World Confederation for Physical Therapy. It publishes two monthly periodicals: *Physical Therapy, The Journal of the American Physical Therapy Association*; and *PT—Magazine of Physical Therapy*. It also publishes the *Guide to Physical Therapist Practice*, the *APTA Book of Body Maintenance and Repair*, and a variety of patient education and information brochures listed in its resource catalog (available upon request).

Canadian Association for Health, Physical Education, Recreation and Dance
403-2197 Riverside Dr., Ottawa, Ontario CANADA K1H 7X3
<http://www.cahperd.ca/e/>
The Canadian counterpart of the AAHPERD, CAHPERD is a national, nonprofit voluntary organization which, since its inception in 1933, has been engaged in providing and extending the benefits of physical activity to the citizens of Canada. Its particular goals are to:

- Act as a strong national advocate on issues pertaining to physically active lifestyles.
- Create a network of practitioners and researchers.
- Exercise leadership and collaborate in the forum of allied organizations that relate to physically active lifestyles.
- Promote quality programs in educational settings from kindergarten to university.

CAHPERD issues a quarterly periodical, *The Physical and Health Education Journal*, and a newsletter titled *In Touch*. It also publishes numerous pamphlets and booklets on various topics for the profession.

International Association of Athletics Federations
17 rue Princesse Florestine, BP 359, MC 98007 MONACO
<http://www.iaaf.org/>
The IAAF is the highest authority for athletics in the world. It includes 210 affiliated National Member Federations who gather once every two years for a Congress which is the IAAF's decision making body. Executive authority rests with a 26-member Council. A worldwide Coaches Education and Certification System (CECS) has been established with basic-level courses taught in each member country. Historically, IAAF has been closely affiliated with the Olympics and with track and field sports.

National Association for Physical Education in Higher Education
<http://www.napehe.org/>
The National Association for Physical Education in Higher Education is an organization for professionals in higher education. Its purpose is to provide a forum for interdisciplinary ideas, concepts, and issues related to the role of physical education in higher education with respect for social, cultural, and personal perspectives.

NAPEHE holds annual conventions at which a Young Scholar Award and Doctoral Student Poster Competition are sponsored, to recognize scholars new to the profession. Affiliated organizations include the Western Society of Physical Education of College Women and the American Academy of Kinesiology and Physical Education. NAPEHE publishes *Quest,* a scholarly journal for professionals in kinesiology and physical education in higher education, and the *Chronicle of PEHE* (see Selected Periodicals section below.) (NAPEHE doesn't maintain a home office, but officers may be contacted through the Web site.)

National Association of Collegiate Directors of Athletics

P.O. Box 16428, Cleveland, OH 44116

<http://nacda.fansonly.com/>

This organization serves intercollegiate athletics as a forum for its members in their efforts to establish common standards and educational objectives among athletics directors in colleges and universities throughout the United States and Canada. NACDA disseminates information and guidelines pertinent to the immediate problems and issues confronting athletics directors.

NACDA has over 6,000 individual members representing 1,600 institutions. It sponsors an annual convention for the major governing bodies in intercollegiate athletics—NJCAA, NAIA, and NCAA institutions. This is the only meeting held regularly with the major governing bodies in one place at one time. This yearly convention serves as a forum for discussing current issues affecting athletics and athletics administration.

NACDA publishes a journal six times annually entitled *Athletics Administration.*

National Association of Intercollegiate Athletics

23500 W. 105TH St., P.O. Box 1325, Olathe, KS 66051

<http://www.naia.org/>

The National Association of Intercollegiate Athletics is an autonomous athletics association made up of the athletics programs of 500 four-year colleges and universities. The fundamental tenet of the NAIA is that intercollegiate athletics is an integral part of the total educational program of the institution. This belief is strongly reflected in the governing documents, activities, and organizational structure of the Association.

Established in 1940 as the National Association of Intercollegiate Basketball (NAIB), it converted to the NAIA in 1952. It expanded to include national championships in golf, outdoor track and field, and tennis, and later to football and most other college sports. The initial purpose of the NAIA—to provide national championship opportunities to colleges and universities competing below the so-called major level—has been continued.

NAIA organizes and administers all areas of intercollegiate athletics at the national level, including rules and standards, and district and national sports competition.

National Collegiate Athletic Association

700 W. Washington St., Indianapolis, IN 46204

<http://www.ncaa.org/>

The NCAA is the organization through which U.S. colleges and universities act on matters of men's and women's athletics at the national level. It formed in 1906 as the Intercollegiate Athletic Association and took its present name in 1910. NCAA established divisions in the 1970s and began administering women's athletics in 1980. In 1997, the NCAA changed its governance structure to provide more control by the presidents of member colleges and universities. The association's membership exceeds 970 institutions, conferences, and affiliated organizations.

The particular purposes of the NCAA are:

- To uphold the principle of institutional control of, and responsibility for, all intercollegiate athletics in conformity with the Association's constitution and bylaws.

- To serve as an overall national discussion, legislative, and administrative body for the universities and colleges of the United States in matters of intercollegiate athletics.

- To recommend policies for the guidance of member institutions in the conduct of their intercollegiate athletics programs.
- To legislate upon any subject of general concern to the membership in the administration of intercollegiate athletics.
- To study all phases of competitive athletics and establish standards therefor, to the end that colleges and universities of the United States may maintain their athletics activities on a high plane.
- To encourage the adoption by its constituent members of eligibility rules in compliance with satisfactory standards of scholarship, amateur standing, and good sportsmanship.
- To establish and supervise regional and national collegiate athletics contests under the auspices of the Association and establish rules of eligibility therefor.
- To stimulate and improve programs to promote and develop educational leadership, physical fitness, sports participation as a recreational pursuit, and athletic excellence through competitive intramural and intercollegiate programs.
- To formulate, copyright, and publish rules of play for collegiate sports.
- To preserve collegiate athletics records.
- To cooperate with other amateur athletics organizations in the promotion and conduct of national and international athletics contests.
- To otherwise assist member institutions as requested in the furtherance of their intercollegiate athletics programs.

The services of the NCAA include the following:

- Maintains a central clearing house and counseling agency in the area of college athletics administration.
- Enacts legislation to deal with athletics problems.
- Conducts research as a means of developing solutions to athletics problems.
- Provides financial and other assistance to various groups interested in the advancement of intercollegiate athletics.
- Represents the colleges in legislative and regulatory matters.
- Administers insurance programs available to member institutions.
- Provides a film/television production department.
- Promotes and participates in international sports planning and competition by working with numerous national and international organizations.
- Sanctions national intercollegiate postseason competition and sponsors national college championship meets.
- Compiles and distributes official statistics for intercollegiate sports.
- Sponsors a national rules committee for the various NCAA sports.
- Provides other services to member institutions as requested.

National Education Association

NEA building, 1201 16th Street NW, Washington, DC 20036
<http://www.nea.org/>
The National Education Association is the largest professional association in the United States, with a membership of approximately 2.6 million. First organized in 1857 as the National Teacher's Association, the name of the association was changed in 1870 to the National Educational Association. In 1907, the NEA was incorporated under a special act of Congress.

The expressed goals of the NEA are to promote an independent, united teaching profession, professional excellence, and economic security for educators, and to provide leadership in solving educationally related social problems.

Any person who is actively engaged in the profession of teaching or other educational work, or any other person interested in advancing the cause of education, is eligible for membership in the association. The Association disseminates numerous publications in the form of periodicals, pamphlets, and books, and distributes a weekly e-mail newsletter.

National Federation of State High School Associations

11724 N. W. Plaza Circle, P.O. Box 20626, Kansas City, MO 64195
<http://www.nfhs.org/>

NFHS was formed in 1920 and has grown to encompass a membership that serves all 50 state high school associations and the District of Columbia, plus 14 affiliate members. High schools, coaches, sponsors, and officials and judges are represented by the association.

The purpose of the National Federation is to coordinate the efforts of its member state associations toward the ultimate objectives of interscholastic activities. The legislative body of the National Federation is the National Council, made up of one representative from each member state association. The administration is vested in an elected Executive Committee consisting of 12 representatives of eight geographical sections of the National Federation.

The National Federation represents high school activities at the national level. The Federation incorporates several national organizations for athletics directors (NIAAA), coaches (NFCA), contest officials (NFOA), speech and debate coaches, music adjudicators, and coaches of pep squads.

Publications and materials printed in the national headquarters include rulebooks in 15 sports, a monthly newsletter, and a quarterly magazine for athletics administrators.

Through its TARGET program, the National Federation is committed to helping students cope with tobacco, alcohol, and other drug problems.

National Intramural-Recreational Sports Association

4185 S.W. Research Way, Corvallis, OR 97333-1067
<http://www.nirsa.org>

Formed in 1950 as the National Intramural Association, this was the first national group entirely devoted to intramural sports. In 1975, the NIA changed its name to the National Intramural Recreational Sports Association. This name change reflects the expansion of the association to include intramurals, sport clubs, and other forms of recreational sports.

The mission of the NIRSA is to foster the growth of quality recreational sports programs by providing for the continuing growth and development of recreational sports professionals. It offers members the opportunity for professional interaction through state, regional, and national conferences and workshops.

The membership of the NIRSA includes professionals, students, and institutions in the United States and several foreign countries. The majority of the association's membership is represented by individuals involved in programming at the collegiate level. However, a growing number of members are employed in the military, industry, community, and elementary/middle/secondary-level school sectors.

Communication within the membership is maintained through periodic publication of the NIRSA *Journal*, a *Newsletter*, conference proceedings, and the annual *Recre-*

ational Sports Directory. Additional services offered to the membership include job placement, sports officials' development, program certification, research grants, a media center, numerous publications, and over 40 conferences and workshops each year.

National Junior College Athletic Association
P.O. Box 7305, Colorado Springs, CO 80933-7305
<http://www.njcaa.org/>
The NJCAA was formed in 1937 when a small number of junior college representatives met to organize an association that would promote and supervise a national program of junior college sports consistent with the educational objectives of junior colleges. In 1949, the Association was reorganized by dividing the nation into regions, which total 24 today.

The Association was concerned only with men's sports until 1975, when it added a women's division. This division grew rapidly, and now approximately 500 institutions are enrolled in the women's division and 530 in the men's division. The NJCAA sponsors championships in a large variety of sports for male and female junior-college athletes. Its purposes and general functions are essentially the same as those described for the NCAA, as the junior-college counterpart.

National Recreation and Park Association
22377 Belmont Ridge Road, Ashburn, VA 20148-4501
<http://www.nrpa.org/>
The NRPA is a nonprofit service, research, and education organization dedicated to the improvement of the quality of life through effective utilization of natural and human resources. NRPA sponsors various publications, workshops, and institutes. It provides direct consultation and technical assistance for the improvement of programs, leadership, and facilities. The Association is also dedicated to building public understanding of the recreation and park field.

The NRPA was formed in 1966 by the merger of five pioneer organizations in the park and recreation field: the American Institute on Park Executives, the American Recreation Society, the National Conference on State Parks, and the National Recreation Association. The NRPA is presently subdivided into the American Park and Recreation Society, Armed Forces Recreation Society, Commissioners-Board Members, National Student Recreation and Park Society, National Therapeutic Recreation Society, and Society of Park and Recreation Educators. The NRPA publishes *Parks and Recreation* (monthly), *Journal of Leisure Research* (quarterly), and several books and pamphlets.

Phi Epsilon Kappa
901 W. New York Street, Indianapolis, IN 46202
<http://www2.truman.edu/pek/>
Phi Epsilon Kappa was founded in 1913 by the students and faculty at the Normal College of the American Gymnastics Union in Indianapolis, Indiana. Membership includes those persons preparing for or engaged in the professional areas of physical education, health, recreation, dance, human performance, exercise science, sports medicine, and sports management. More than 40,000 members have been initiated into local chapters since its conception.

The purpose of the fraternity is to bind members together and to promote a friendly atmosphere within the profession. Among its objectives are "to further the individual welfare of its members and to foster scientific research in the fields of

health education, physical education, recreation education and safety education."
Publications include the *Physical Educator*, and membership is available through the
various collegiate chapters located in colleges and universities throughout the
United States.

Society of State Directors of Health, Physical Education and Recreation
1900 Association Drive, Reston, VA 20191-1599
<http://www.thesociety.org/>
The Society of State Directors of Health, Physical Education and Recreation was first
organized in 1926 in New York City. The Society is composed of state directors of
health, physical education, and recreation, and their associates. Associate memberships are available for those who have duties in one of these disciplines or who otherwise support the purposes of the Society. Over the years, the Society has issued
numerous proclamations, resolutions, and proceedings that have had an impact on
school, college, and community programs.

The Society's basic purposes are to:

- Help shape national and state policy defining and supporting comprehensive
  school health and physical education programs.

- Link state health, physical education, and recreation leaders with their counterparts in other states.

- Work to forge school/family/community linkages in support of school
  health, physical education, and recreation programs.

- Foster professional growth and the development of leadership and advocacy skills.

- Help resolve complex issues in education and health reform.

- Provide leadership in the effort to link postsecondary institutions to school
  districts for improvement in curriculum, instruction, and assessment.

- Provide a supportive network of professional and social relationships among
  members.

- Provide training and workshops for members to help them increase capacity
  to improve comprehensive school health education and programs within
  their states.

COACHES' ASSOCIATIONS

National coaches associations exist for all the well-established school sports.
They are supplemented by state coaching associations. These associations advance
their sports, update procedures for conducting sports events, oversee rule modifications, and serve a fraternal purpose for the coaches. Some of the associations publish
monthly or quarterly publications for their membership.

American Baseball Coaches Association
108 S. University Avenue, Suite 3, Mt. Pleasant, MI 48858-2327
<http://www.abca.org/>

American Football Coaches Association
100 Legends Lane, Waco, TX 76706
<http://www.afca.com/>

American Hockey Coaches Association
7 Concord Street, Gloucester, MA 01930
<http://www.ahcahockey.com/>

American Volleyball Coaches Association
1227 Lake Plaza Dr., Suite B, Colorado Springs, CO 80906
<http://www.avca.org/ >

College Swimming Coaches Association of America
P.O. Box 63285, Colorado Springs, CO 80962
<http://www.cscaa.org/>

Golf Coaches Association of American [Men]
1225 West Main Street, Suite 110, Norman, OK 73069
<http://collegiategolf.fansonly.com/index-main.html>

National Association of Basketball Coaches
9300 W. 110th St., Suite 640, Overland Park, KS 6621
<http://nabc.fansonly.com/>

National Fastpitch Coaches Association
409 Vandiver Drive, Suite 5-202, Columbia, MO 65202
<http://www.nfca.org/>

National Field Hockey Coaches Association
11921 Meadow Ridge Terrace, Glen Allen, VA 23059
<http://www.eteamz.com/nfhca/>

National Golf Coaches Association [Women]
180 North LaSalle Street, Suite 1822, Chicago, IL 60601
<http://www.ngca.com/index2.html>

National Soccer Coaches Association of America
6700 Squibb Road, Suite 215, Mission, KS 66202
<http://www.nscaa.com/>

National Wrestling Coaches Association
P.O. Box 254, Manheim, PA 17545-0254
<http://www.themat.com/nwca/default.asp>

United States Cross Country Coaches Association
c/o Don Kopriva, 5327 Newport Drive, Lisle, IL 60532
<http://www.usccca.org/>

U.S. High School Tennis Coaches Association
<http://www.ushstca.org/>

United States Track Coaches Association
1330 NW 6th Street, Suite C, Gainesville, FL 32601
<http://www.ustrackcoaches.org/>

Women's Basketball Coaches Association
4646 Lawrenceville Highway, Lilburn, GA 30047-3620
<http://www.wbca.org/>

The addresses of other coaches' associations are available through the NCAA or National Federation of State High School Associations. (See above for contact information.)

SPORTS ORGANIZATIONS

While descriptions have been given of organizations concerned with physical education and school athletics, sports organizations with broad interests transcending school settings are listed below with brief descriptions, their current mailing addresses, Internet addresses, and periodicals they publish.

Amateur Athletic Union
P.O. Box 22409, Lake Buena Vista, FL 32830
<http://www.aausports.org/mytp/home/aau_index.jsp>
Founded in 1888, AAU is one of the largest, nonprofit volunteer sports organizations in the United States dedicated exclusively to the promotion and development of amateur sports and physical fitness programs. AAU is divided into 57 distinct associations, which annually sanction more than 34 sports programs, 250 national championships, and over 10,000 local events. In 1996, the AAU joined forces with Walt Disney World and relocated its national headquarters to Orlando, Florida. More than 40 AAU national events are conducted at the Disney's Wide World of Sports Complex. The AAU publishes brochures and rulebooks for each of its sports and an online newsletter.

National Alliance of African American Athletes
P.O. Box 60743, Harrisburg, PA 17106-0743
<http://www.naaaa.com/>
Goals of the NAAAA are to facilitate the development of African-American men and boys in American society. The association is active in programs and public education, the Black male-athlete economic cluster, and the African-American male information initiative. Each year the Alliance presents the Watkins Memorial Trophy Award to the Premier African-American Male High Senior Scholar-Athlete in the United States.

National Alliance for Youth Sports
2050 Vista Parkway, West Palm Beach, FL 33411
<http://www.nays.org/about/index.cfm>
The NAYS, a nonprofit organization based in West Palm Beach, Florida, was founded in 1981 as the National Youth Sport Coaches Association (NYSCA) with the mission of improving out-of-school sports for the more than 20 million youth participants under the age of 16. The Alliance believes that participation in youth sports develop important character traits and values, and that the lives of youths can be positively impacted if the adults caring for them have proper training and information.

National Association of Collegiate Women Athletic Administrators
4701 Wrightsville Ave., Oak Park D-1, Wilmington, NC 28403
<http://www.nacwaa.org/>
The goal of the NACWAA are to promote gender equity and ethnic diversity in intercollegiate athletics. In conjunction with HERS, the association offers an Institute for Administrative Advancement, an annual week-long residential program designed to offer women coaches and administrators intensive training in athletics administration.

President's Council on Physical Fitness and Sport

Dept. W, 200 Independence Ave., SW, Room 738-H, Washington, DC 20201-0004
<http://www.fitness.com>

The PCPFS serves to promote, encourage, and motivate Americans of all ages to become physically active and participate in sports. Assisted by elements of the U.S. Public Health Service, the PCPFS advises the President and the Secretary of Health and Human Services on how to encourage more Americans to be physically fit and active. The Council publishes *PCPFS Research Digest* in March, June, September, and December.

Special Olympics, Inc.

1325 G Street, NW / Suite 500 Washington, DC 20005
<http://www.specialolympics.org/>

Special Olympics is an international program of year-round sports training and athletics competition for more than one million children and adults with mental disabilities.

U.S. Olympic Committee

USOC Headquarters, One Olympic Plaza, Colorado Springs, CO 80909
<http://www.usoc.org/>

The U.S. Olympic Committee consists of 31 national sports-governing bodies responsible for the organization and selection of team members for the sports in the programs for the Olympic and Pan-American Games, 11 multisport organizations, and the Armed Forces of the United States, from which directly or indirectly have come substantial numbers of members of the U.S. Olympic and Pan-American Teams. The Committee includes representatives from the 50 states and District of Columbia. The U.S.O.C. represents the United States on the International Olympic Committee and is the U. S. representative on the Executive Board of the Pan-American Sports Organization.

Women's Sports Foundation

Eisenhower Park, East Meadow, NY 11554
<http://www.womenssportsfoundation.org/cgi-bin/iowa/index.html>

Founded in 1974 by professional tennis player Billie Jean King, the WSF is a charitable educational organization dedicated to increasing the participation of girls and women in sports and fitness and creating an educated public that supports gender equity in sport. WSF publishes *Women's Sports Experience* for adults and *Sports Talk* for children.

YMCA of the United States of America

National Council, 101 N. Wacker Drive, Chicago, IL 60606
<http://www.ymca.net>

The YMCA is the largest not-for-profit community service organization in America, working to meet the health and social service needs of men, women, and children in 10,000 communities. Local Ys serve people of all faiths, races, abilities, ages, and incomes. No one is turned away for inability to pay.

Young Women's Christian Association of the United States of America

Empire State Building, 350 Fifth Avenue, Suite 301, New York, NY 10118
<http://www.ywca.org>

Over 300 local YWCAs are located in all 50 states, under the umbrella of the YWCA of the U.S.A. YWCA provides sports and physical fitness programs for women and girls.

Note: A comprehensive list for associations (including contact information) representing *specific* sports appears on Learning Network's Web site, *Infoplease.com* <http://www.infoplease.com/ipsa/A0111532.html>.

SELECTED PERIODICALS IN PHYSICAL EDUCATION AND ATHLETICS

AAHPERD Update
1900 Association Drive, Reston, VA 20191
<http://www.aahperd.org/aahperd/update_template.html>
A newsletter supplement published six times a year by AAHPERD. *Update* is distributed to all AAHPERD members ($4 of annual dues is for *Update*), and an online edition is available to members. The publication is devoted to Alliance news and features in the broad fields of sport, physical education, recreation, health, dance, and safety, and includes a professional job exchange. "Physical Activity Today" is bound into four of the six annual issues for subscribers. *PAT* provides the latest and most critical research findings about sport, physical education, and exercise.

Athletic Business
4130 Lien Road, Madison, WI 53704
<http://www.athleticbusiness.com/>
This magazine is distributed free to 42,000 owners, operators, and directors of sports, recreation, and fitness facilities at colleges and universities; high schools, private schools, and school districts; park and recreation departments; sports and health clubs. The journal is published monthly, with special issues in February (*Buyers Guide*) and June (*Architectural Showcase*). A searchable version of the *Buyers Guide* is available on the Web site.

The magazine covers a wide variety of facility planning, marketing, equipment, liability, operations, and management topics. Monthly columns include Sports Law Report, College Beat, High School Sports, For Profit, and Recreation. In addition, each issue features a Product Focus, highlighting different types of products for the sports, recreation, and fitness market.

Athletics Administration
P.O. Box 16428, Cleveland, OH 44116
<http://nacda.fansonly.com/athletics-admin/nacda-athletics-admin.html>
The official publication of the National Association of Collegiate Directors of Athletics (see above). *Athletics Administration* is circulated six times annually to all two- and four-year colleges and universities in both in the United States and Canada. It is written primarily by NACDA members for NACDA members. The journal discusses current issues and trends in school athletics administration. An individual subscription is $15/yr.

Athletic Journal
(Incorporated by *Scholastic Coach* in May, 1987. Name changed to *Scholastic Coach and Athletic Director* with August, 1994 issue. See *SCAD* below.)

Chronicle of Physical Education in Higher Education
Published by the National Association for Physical Education in Higher Education (see NAPEHE above). The *Chronicle* is published and distributed to members three times a year in February, May, and September. It includes articles on a range of issues in physical education at the college level including administration.

Dance Chronicle
<http://www.dekker.com/servlet/product/productid/DNC>
Published by Marcel Dekker, Inc., 270 Madison Avenue, New York, NY 10016.

Three print issues a year. Studies in dance and related arts (critiques of ballet dances, themes, etc.), as well as music, theater, film, literature, painting, and aesthetics. A print subscription is $595/yr. Online subscriptions also available.

Journal of Athletic Training
<http://www.journalofathletictraining.org>
Published quarterly by the National Athletic Training Association (see above). A subscription is $32/yr. ($40/yr. outside the U.S.) Directed at professionals interested in healthcare for the physically active through education and research in prevention, evaluation, management, and rehabilitation of injuries. Reports on research in athletics training, including the latest in common sports injuries and rehabilitation procedures.

Journal of Physical Education, Recreation and Dance (JOPERD)
<http://www.aahperd.org/aahperd/joperd_main.html>
Published monthly (except July, combined May/June and November/December issues) by AAHPERD. Subscription Rates: Professional and supportive members pay annual dues of $125, $25 of which is for JOPERD; student members pay annual dues of $45.00, $25 of which is for JOPERD. Subscription rate for institutions and libraries is $125. *JOPERD* publishes an array of articles, materials, and announcements of interest to students and professionals within physical education, sport, dance, and leisure studies.

Journal of Sport Management
<http://www.unb.ca/web/SportManagement/jsm.htm>
Published by Human Kinetics Publishers, P.O. Box 5976, Champaign, IL 61820. A subscription is $33/yr. For professionals, scholars, and students in sport management in order to keep abreast of developments in the field. *JSM* publishes articles that focus on the theoretical and applied aspects of management related to sport, exercise, dance, and play.

Journal of Teaching in Physical Education
<http://www.humankinetics.com/products/journals/journal.cfm?id=JTPE>
Published quarterly by Human Kinetics Publishers, P.O. Box 5976, Champaign, IL 61820. An individual subscription is $49/yr; an institutional subscription is $159/yr. *JTPE* features research articles based on classroom and laboratory studies, descriptive and survey studies, summary and review articles, and discussion of current topics of interest to physical educators at every level.

The Physical Educator
<http://www2.truman.edu/pek/public.html>
Published quarterly by Phi Epsilon Kappa Fraternity, 901 W. New York St., Indianapolis, IN 46202. The journal is distributed to members as part of their annual dues. It is also available on a subscription basis to nonmembers and libraries. Subscription price is $37.50 for one year. Publishes general manuscripts in the area of physical education, recreation, and safety. Includes sections on comparative physical education, dance, teaching techniques, book reviews, and bibliographies.

Referee Magazine
Referee, P.O. Box 161, Franksville, WI 53126
<http://www.referee.com/>
This magazine of some 80 pages is read by more than 70,000 officials, coaches, and administrators, and is the only monthly publication devoted to sports officiating. The magazine contains interviews, feature articles, late-breaking news, personality pro-

files, investigative reports, health, legal and tax tips, and a wide range of technical information for many sports. A subscription to *Referee* is included with membership in the National Association of Sports Officials <http://www.naso.org/>.

Research in Dance Education
<http://www.tandf.co.uk/journals/carfax/14647893.html>
Published by Carfax Publishing, 11 New Fetter Lane, London EC4P 4EE. Aims to inform, stimulate, and promote the development of research in dance education relevant to both learners and teachers. The goal of the journal is to improve the quality and provision of dance education through lively and critical debate, and the dissemination of research findings. Two issues a year. Subscription rates: institutional $175, individual $45.

Scholastic Coach and Athletic Director
<http://www.scholastic.com>
Published by Scholastic Inc., 555 Broadway New York, NY 10012-3999. A magazine for athletics directors and coaches. Articles provide information regarding training, administration, women in athletics, and on different sports throughout the school year. Ten issues annually (subscription fee of $17/yr.).

Spotlight on Dance
<http://www.aahperd.org/nda/nda_main.html>
Newsletter of the National Dance Association/AAHPERD. Members may access archives online (click on Publications).

Strategies
<http://www.aahperd.org/aahperd/strategies_main.html>
A peer-reviewed professional magazine published by NASPE (See AAHPERD above). It offers practical, how-to articles for sport and physical education professionals at all levels. *Strategies* is published six times a year in January, March, May, July, September, and November. Annual personal subscriptions are $80, except for AAHPERD members, who receive *Strategies* as part of their membership, or as an additional journal for the special member rate of $25 annually. Annual institutional rate is $80.

Teaching Elementary Physical Education
<http://www.humankinetics.com/products/journals/journal.cfm?id=TEPE>
Published by Human Kinetics Publishers, P.O. Box 5976, Champaign, IL 61820. TEPE focuses on the topics that are important to K–8 physical educators: Each bimonthly issue includes tips, reviews, and resources that will help teachers create a program that's fresh, fun, and dynamic.

OTHER RELATED SCHOLARLY JOURNALS

Adapted Physical Activity Quarterly (Human Kinetics Publishers)

Aethlon (Sports Literature Association)

Dance Notation Journal (Dance Notation Bureau)

Dance Research Journal (Congress on Research in Dance)

Football Studies (Football Studies Group)

Health and Fitness Journal (American College of Sports Medicine)

International Journal of Sports Medicine (Thieme Medical Publishers)

International Journal of Sports Psychology (International Society of Sports Psychology)

Journal of Applied Biomechanics (Human Kinetics Publishers)

Journal of Applied Physiology (American Physiological Society)

The Journal of Applied Sport Science Research (National Strength & Conditioning Association)

Journal of Leisure and Recreation (National Recreation and Parks Association)

Journal of the Philosophy of Sport (Human Kinetics Publishers)

Journal of Sport and Exercise Psychology (Human Kinetics Publishers)

Journal of Sport History (North American Society for Sport History)

Journal of Sport Psychology (Human Kinetics Publishers)

Journal of Sport and Social Issues (Center for the Study of Sport in Society)

Journal of Sports Sciences (Taylor & Francis Ltd.)

Measurement in Physical Education and Exercise Science (Erlbaum Associates)

Medicine and Science in Sport (American College of Sports Medicine)

Motor Control (Human Kinetics Publishers)

Physical Education Digest (formerly *Coaching Digest*)

The Physician and Sports Medicine (McGraw-Hill)

Quest (NAPEHE/Human Kinetics)

Science and Sports (Elsevier)

Sociology of Sport Journal (NASSS)

Sport, Education & Society (Carfax Publishing)

Sport History Review (Human Kinetics Publishers)

The Sport Journal (United States Sports Academy)

Sport Psychologist (Human Kinetics Publishers)

The Virginia Sports and Entertainment Law Journal (University of Virginia)

ONLINE JOURNALS

Athletic Insight—The Online Journal of Sport Psychology
<http://www.athleticinsight.com/>
A quarterly forum for discussion of topics relevant to the field of sport psychology. Topics covered include performance enhancement and social issues relating to the field of sports. Current and back issues available. Links, recommended books, and conferences are also noted.

AVANTE
<http://www.cahperd.ca/e/avante/index.htm>
CAHPERD's bilingual research periodical designed to stimulate and communicate Canadian research and critical thought on issues pertaining to the fields of health, physical activity, sport, physical education, recreation, leisure, dance, and active living. AVANTE is published three times per year in both print and electronic format.

European Journal of Sport Science
<http://www.humankinetics.com/products/journals/journal.cfm?id=EJSS>
Published by Human Kinetics Publishers and the European College of Sport Science, this scholarly, refereed publication covers biological, behavioral, and social sciences as they pertain to sport and exercise. It is available online by individual or institutional subscription.

The Pulse Online Newsletter
<http://www.aahperd.org/aha-pulse/about.html>
Focuses on the Jump Rope for Heart and Hoops for Heart programs. It is co-sponsored by AAHPERD and the American Heart Association. Articles and information are passed on from the headquarters of both national organizations to the state and local coordinators that conduct the actual events. This newsletter also focuses on how money is raised by the Jump and Hoops programs.

The Sport Journal
<http://www.thesportjournal.org/>
The journal is published by the U.S. Sports Academy in Mobile, AL, in cooperation with the International Coaches Association and the International Association of Sports Academies. The first issue was published in 1998. It succeeds the discontinued *Journal of Sport Behavior.*

Note: An increasing number of hardcopy journals, such *as Measurement in Physical Education and Exercise Science*, now include online access to their contents as part of subscription benefits.

Appendix B
AAHPERD Code of Ethics

Preamble

Believing that the strength of our American democracy and its influence upon the course of events everywhere in the world lies in the physical, mental and moral strength of its individual citizens; believing that the schools of America possess the greatest potential for the development of these strengths in our young citizens; believing that the teachers of physical education have a unique opportunity, as well as a responsibility, to contribute greatly to the achievement of this potentiality, believing that all teachers of physical education should approach this great responsibility in a spirit of true professional devotion, the AAHPERD proposes for the guidance of its members the following:

Principles of Ethics

1. Inasmuch as teachers of physical education are members of the teaching profession, the AAHPERD endorses without reservation the Code of Ethics for Teachers, adopted by the National Education Association.

2. The aim of physical education is the optimum development of the individual. To this end teachers of physical education should conduct programs and provide opportunities for experiences which will promote the physical development of youth and contribute to social, emotional, and mental growth.

3. In a democratic society every child has a right to the time of the teacher, to the use of the facilities, and to a part in the planned activities. Physical education teachers should resist the temptation to devote an undue amount of time and attention to the activities of students of superior ability to the neglect of the less proficient.

4. The professional relations of a teacher with pupils require that all information of a personal nature shall be held in strict confidence.

5. While a physical education teacher should maintain a friendly interest in the progress of pupils, familiarity should be avoided as inimical to effective teaching and professional dignity.

6. The teacher's personal life should exemplify the highest ethical principles and should motivate children to the practice of good living and wholesome activities.

7. To promote effective teaching, the teacher of physical education should maintain relations with associates which are based on mutual integrity, understanding, and respect.

8. The physical education teacher should cooperate fully and unselfishly in all school endeavors which are within the sphere of education. He [or she] should be an integral part of the school faculty, expecting neither privileges nor rewards that are not available to other members of the faculty.

9. It is an obligation of the teacher of physical education to understand and make use of proper administrative channels in approaching the problems encountered in education and in schools.

10. It is the duty of the physical education teacher to strive for progress in personal education and to promote emerging practices and programs in physical education. The teacher should also endeavor to achieve status in the profession of education.

11. Professional ethics imply that altruistic purpose outweighs personal gain. The teacher, therefore, should avoid using personal glory achieved through winning teams for the purpose of self-promotion.

12. It is considered unethical to endorse physical education equipment, materials and other commercial products for personal gain or to support anything of a pseudo-educational nature. Nor should a teacher profit personally through the purchase of materials for physical education by the school.

13. It is the responsibility of the teacher of physical education to acquire a real understanding of children and youth in order that he or she may contribute to their growth and development. To achieve this understanding, it is essential that an earnest effort be made to foster and strengthen good school-home-community relationships.

14. It is the duty of every teacher of physical education to become acquainted with and to participate in the affairs of the community, particularly those concerned with making the community a better place in which to live. The teacher should take an active interest in the work of the various child- and youth-serving agencies, participating as a citizen, and as a leader of children, youth and adults.

15. Inasmuch as physical education will progress through strong local, state, and national organizations, the teacher of physical education is obligated to membership and active participation in the proceedings of professional organizations, both in general education and in the specialized field of physical education.

16. Institutions preparing teachers of physical education have an ethical responsibility to the profession, the public, and the nation for the admission, education, and retention of desirable candidates for teaching. To meet this obligation, curriculum offerings must be in harmony with the highest standards of professional education.

17. Teachers of physical education should render professional service by recruiting qualified men and women for future teachers of America. The physical education teacher also has a professional obligation to assist in the learning, practice, and understanding of student teachers in the field.

Appendix C
Facility Planning Checklist

To use the following checklist, place the appropriate letter in the space indicated opposite each statement:

A = the plans meet the requirements *completely*.
B = the plans meet the requirements *only partially*.
C = the plans *fail* to meet the requirements.

The material for this checklist is from *Planning Facilities* (1986), a publication of the American Alliance for Health, Physical Education, Recreation and Dance, 1900 Association Drive, Reston, VA 20191. It is used by permission.

General Considerations

_____ 1. The facility is planned as a part of a well-conceived and well-integrated master plan.

_____ 2. The facility has been planned to meet the total requirements of the program as well as the special needs of those who are to be served, both now and in the foreseeable future.

_____ 3. The plans and specifications have been carefully checked by all agencies and individuals whose approvals are required or desired.

_____ 4. The plans conform to state and local regulations and to acceptable standards and practices.

_____ 5. The plans have been prepared to accommodate possible future additions and expansions in the best manner possible.

_____ 6. Administrative officers, faculty offices, staff offices, conference rooms, workrooms, teaching stations, and service facilities are all properly interrelated and properly located for effectiveness and function.

_____ 7. Facilities for health services, emergencies, and first aid are adequate and suitably located.

---- 8. The special needs of the handicapped have received adequate consideration.

---- 9. The building is compatible in design and adequate in control as compared with the other buildings with which it relates.

----10. Adequate attention has been given to making the facilities as durable, main-tenance-free, and vandalproof as possible.

----11. Low-cost maintenance features have received the special consideration that is deserved.

----12. Drinking fountains, rest rooms, hair dryers, and other convenience facilities are adequate in number and properly located.

----13. Provision is made for the repair, maintenance, replacement, and off-season storage of equipment and uniforms.

----14. Noncorrosive metals are used in all damp areas such as dressing rooms and laundry areas.

----15. Fire regulations and fire prevention have received adequate consideration. Also, fire alarm systems, both audio and visual, are adequate.

----16. The facility contains adequate and well-placed bulletin boards, trophy cases, and display areas.

----17. Adequate shower, dressing, and locker room facilities are furnished for stu-dents, staff members, guests, and game officials.

----18. The proportion of space for different uses and the location relationships have received proper attention in the planning.

----19. Adequate attention has been given to quiet areas for study and concentra-tion.

----20. The plan includes an adequate class signal system (bell or buzzer).

----21. Adequate provision has been made for telephone service, intercommunica-tion (both oral and written), mail service, and stereo.

----22. The lock and key plan is suitable in terms of security and convenience.

----23. The traffic-flow plan is well conceived and adequate, especially at stairways and exits.

Teaching Stations

---- 1. Teaching stations are of optimum size to accommodate the desired number of students and the particular activities.

---- 2. Teaching stations are designed with adequate safety zones between separate activity areas, and in connection with walls and fixed objects, smooth sur-faces and padding are in locations where needed.

---- 3. Each gymnasium has at least one smooth unobstructed wall for ball rebounding activities and adequate ceiling height (minimum of 18 feet for elementary schools and 22 feet for secondary schools).

---- 4. Each teaching station has the desirable accessories, such as scoreboard, seat-ing space, chalk board, bulletin board, floor markings, game standards, power-operated partitions (where useful), public address system (if needed), adequate spectator space for the activities planned for the area, and the stan-dard equipment such as nets, mats, and the like.

—— 5. Adequate storage closets and cabinets properly located and adequately secured.

—— 6. Special instructional provisions as needed, such as televisions, instant video replay, recorders, wall screens, and projection equipment.

—— 7. Other instructional items are included where needed, such as wall mirrors, exercise bars, floor diagrams, acoustical drapery, and appropriate performance surfaces. (Different surfaces are needed for different kinds of activities such as basketball, tennis, weight training, modern dance, ballet, social dance, or gymnastics.)

Walls

—— 1. Provisions are made for wall surfaces to be of attractive, durable and maintenance-free materials.

—— 2. Movable and folding partitions that are power operated and controlled by switches.

—— 3. Hooks and rings for nets and other attachments are properly placed, highly stable, and recessed for safety reasons.

—— 4. Where moisture is prevalent, such as in a natatorium, shower and dressing rooms, and laundry area, wall coverings are of moisture-proof or resistant materials. Special consideration is given in the case of walls that will be hosed down, such as shower rooms, steam rooms, and possibly rest rooms and dressing rooms.

—— 5. Walls are free of protrusions and obstacles that interfere with traffic flow or create safety hazards. Corners are rounded and surface irregularities eliminated where desirable.

—— 6. The wall texture and color of painted surfaces is the best it can be for attractiveness, lighting, and maintenance.

Ceilings

—— 1. Ceilings in all instructional areas, offices, and meeting rooms are of the material that will provide the optimum acoustical effect.

—— 2. Adequate consideration has been given to texture and color of ceiling materials.

—— 3. Moisture-resistant ceiling materials are specified for moisture-prevalent areas.

—— 4. Consideration has been given to adequate ceiling lighting, both direct and indirect, and both artificial and natural.

—— 5. Overhead supported apparatus and equipment is secured to beams engineered to withstand stress.

Floors

—— 1. Floors are made of appropriate materials for the particular activities to be conducted in each location. Performance effectiveness, safety, and maintenance are all considered.

—— 2. Floor plates are placed where needed, and they are flush mounted and stable.

—— 3. Appropriate consideration has been given to flooring material, texture, slope, and drains in shower and locker rooms, swimming pool decks, steam rooms, and other areas that will be hosed.

—— 4. Nonskid surfaces and strips are placed where needed for safety, especially in wet areas and on stairways and ramps.

—— 5. Highly durable and easily maintained baseboards are specified.

Electrical

—— 1. Lighting intensity is in accordance with recommended standards.

—— 2. Electrical control panels are properly placed and adequately secured.

—— 3. Where appropriate, lights are controlled by key switches.

—— 4. Adequate numbers of electrical outlets are properly placed.

—— 5. Dimmer units are placed in instructional areas and elsewhere, as needed.

—— 6. Natural light is properly planned in accordance with recommended standards.

—— 7. Lights are shielded for protection from balls and other game implements.

—— 8. Waterproof outlets are specified for moisture-prevalent areas and areas that will be hosed.

Temperature Control

—— 1. Adequate provisions are made in all areas for heating, air conditioning, and ventilation.

—— 2. Special consideration has been given to ventilation and individual temperature controls in locker rooms, dressing and shower areas, swimming pool, and other moisture-prevalent and odor-prone areas.

—— 3. Proper ventilation is provided for such places as the wrestling room, equipment storage areas, research-animal kennels, and other appropriate areas.

Other Considerations

—— 1. Throughout the facility, adequate storage spaces are well located and properly designed.

—— 2. Locker rooms are arranged for ease of supervision.

—— 3. Provision is made for adequate control of access.

—— 4. Adequate faculty accommodations are provided: shower and locker room, equipment and supply access, and lounge.

—— 5. Food dispensary areas have received proper attention.

—— 6. A well-defined program for laundering and handling of uniforms, towels, and equipment is included in the plan.

—— 7. Showerheads are placed at appropriate heights for the ages of those being served.

—— 8. In the locker room, adequate attention is given to locker ventilation, placement of benches for convenience and ease of cleaning, height and size of lockers, and width of aisle between lockers.

Specialized Areas

In checking the plans, special consideration should be given to such specialized areas as the following:

―――― 1. Athletics training facilities.

―――― 2. Visiting team dressing room.

―――― 3. Dance studio.

―――― 4. Facility for ice activities.

―――― 5. Bowling alley.

―――― 6. Rifle range.

―――― 7. Indoor tennis courts.

―――― 8. A facility to accommodate a large number of spectators.

―――― 9. Research facilities.

Appendix D
Criteria for Evaluating Physical Education Curriculum, Grades 9–12

I. CURRICULUM

Evaluative Criteria

A. *Content*
Program provides for a variety of developmentally appropriate learning activities that will maintain physical fitness, refine or increase expertise in performance skills, and personalized values to provide for lifelong health and pleasure in moving.
1. Fitness Remediation and Maintenance
   a) Health related
      Cardiovascular endurance
      Muscular strength
      Muscular endurance
      Flexibility
      Body composition
   b) Motor fitness
      Speed
      Agility
      Coordination
2. Psychomotor Learning
   a) Variety of team sports (list below)

_____ _____ _____ _____
_____ _____ _____ _____

   b) Variety of individual and dual sports (list below)

_____ _____ _____ _____
_____ _____ _____ _____

Reprinted with permission of the Executive Director, Illinois Association for Health, Physical Education, Recreation and Dance.

    c) Variety of dance types (list below)

_____ _____ _____ _____

_____ _____ _____ _____

    d) Adventure/challenge activities
3. Cognitive Learning
    a) Advanced movement concepts
    b) Scientific concepts relating to:
      Exercise physiology
      Psychosocial
      Humanities
      Motor learning; development
      Biomechanics
      Game rules and strategies
      Advanced dance concepts
4. Attitude/Value Learning
    a) Engagement in self-directed physical activity
    b) Displaying appropriate participant and spectator behavior

B. *Program Administration*
Program is guided by the following:
1. A daily program of instructional physical activity is provided per the Illinois School Code; Section 27-6. If a school uses blocked scheduling, time equivalency is permissible if the total minutes is comparable with the time allocated for other major subject areas.
    a) Allowable exemptions (may not exceed that allowable by ISBE; must follow all criteria for curriculum development):
      Health education
      Leader's physical education
      Adapted physical education (may not replace inclusion program; student generally has IEP including input of physical educator)
      Driver's education
    b) Not allowable:
      Marching band
      Athletic physical education
      Athletic participation
      Academic course work
      ROTC
2. A written curriculum guide is available which allows for daily lessons based on specified units of study and yearly plans that are sequenced for progressive learning.
The curriculum must:
    be explicit (written)
    align with state learning standards
    be outcome/objective driven
    be sequential, usually by grade level or other appropriate classification/ grouping
    be developmentally appropriate, linking units of instruction with intended outcomes/objectives

enable student assessment/evaluation relative to intended outcomes/ objectives

be part of local assessment plan

be operational (explicit curriculum is being implemented)

3. The curriculum is relevant to the student interests, needs, and school facilities.
4. The curriculum provides for gender-fair access and experiences.
5. Students are grouped by ability for competitive activities.
6. The curriculum accommodates developmental needs of all students, including consistency with state and federal legislation.
7. The curriculum provides for gender equity practices as provided by state and federal legislation.
8. Proper attire is required for activity.
9. Appropriate activities are substituted if religious objections are raised to specific activities.
10. The curriculum provides for elective/selective courses of study.
11. The curriculum provides sufficient time for skill acquisition.

C. *Program Evaluation*
1. Provisions are made for assessment of the following components of learning:
    a) Psychomotor learning
    b) Fitness development
    c) Attitude/value learning
    d) Cognitive learning
2. Provisions are made for identification of learning disabilities and/or other problems related to effective learning.
3. Records are kept and maintained on the physical fitness, motor-skill and growth accomplishments of the students.
4. Outcomes and/or standards have been identified for each grade level as per:
    a) State learning standards
    b) Local school improvement plan
5. Assessment results are used for:
    a) Diagnostic purposes for both students and program
    b) Improving learning/teaching
    c) Describing to parents and community, student achievement

Note: The above criteria are judged according to three standards: (1) minimum, (2) adequate, (3) exceeds expectations. A column for evaluator's comments is incorporated in the instrument.

(Evaluation Criteria for Section II., Instruction, is not included here. The entire form can be accessed at <http://www.iahperd.org/textpages/programs/blueRibbon/ IAHPERDcriteria9_12.pdf>.

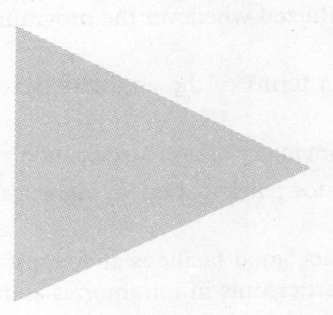

Appendix E
Position Statements

Following are statements of belief from the Society of State Directors of Health, Physical Education and Recreation about intramurals and other extra class programs, and about interscholastic athletics, with which the authors concur and with which all physical education teachers, intramural directors, and administrators ought to be familiar.

EXTRA CLASS ACTIVITIES
INTRAMURALS, EXTRAMURALS, SPORTS CLUBS AND
RECREATION ACTIVITIES

Every school should provide ample opportunity for participation in intramural, extramural and recreational activities for all students. These programs not only supplement physical education instruction but also afford opportunities for social exchange, enrichment of individual personal resources, and wholesome competition below the varsity level of excellence. Extra class activities should be consistent with the content and philosophy of the physical education program of instruction. Such activities, if handled properly, can be particularly helpful in promoting good human relationships and community involvement, and in minimizing problems of racial integration, student unrest, and the like.

About Administration

- Extra class activities should be provided for and financed by boards of education.
- School administrators should ensure ample opportunity for all students to participate in individual, dual and team sports, and other suitable physical activities. School transportation should be provided when needed.

Reprinted with permission of The Society of State Directors of Health, Physical Education, Recreation and Dance.

- Extra class activities should be administered by school officials, but community facilities and resource persons should be utilized whenever the program can be improved through such use.
- Extra-class competition should be equalized in terms of the age, skill, size and strength of participants.
- Policies should be established to protect pupils from injury or overparticipation.
- Health examinations should be a prerequisite for participation in vigorous activity.
- Protective, well-fitting equipment for participants, good facilities and equipment, and capable personnel are essential to participants in intramurals and other extra-class activities.
- Boxing should not be included at any grade level.
- Sponsoring of body contact sports below the ninth grade places the school on tenuous grounds, unless the school accepts full responsibility for providing excellent leadership, facilities, equipment, and policies for promoting.and protecting the health and welfare of participants.

Awards for extra-class activities are unnecessary. If given at all, they should be simple and of little monetary value, providing only token recognition of an activity satisfactorily completed. Students should be motivated to participate through schoolwide emphasis on the values and opportunities inherent in such participation and through the joy afforded by well-planned and well-administered programs.

About Program

Education and development of the individual through satisfying and enjoyable participation in selected activities should be the program.

Opportunity to participate in a wide variety of extra class activities should be provided for both boys and girls of all grades. Activities should be based on the interests, needs, capacities, and level of maturity of the pupils. The program should be broad enough to help meet the needs of pupils who are mentally or physically handicapped or who have social or emotional problems.

Leadership opportunities should be made available to pupils regardless of their sex and skill level.

About Personnel

Since extra-class program participation is voluntary, the program should be staffed with enthusiastic teachers who possess high-level leadership skills and organizational competencies.

INTERSCHOLASTIC ATHLETICS

A well-directed interscholastic athletics program should be made available to all boys and girls who are interested and sufficiently skilled. This area of physical education allows students to participate in competition characterized by a variety of teams, leagues, conferences, regular season schedules, specialized coaching, and high quality of performance.

About Administration

- The interscholastic athletics programs should be provided for and financed by boards of education with funds accountable through the general school budget. The program should be administered by school officials.
- Interscholastic athletics leagues or conferences should be confined to pupils in grades 9–12.
- If elementary school, middle school, or junior high interscholastic sports programs are deemed necessary, AAHPERD guidelines for such programs should be followed closely.
- The rules and regulations of the National Federation of State High School Associations should be used as guidelines in administering the interscholastic athletics program.
- The standards and guidelines of the National Association for Girls and Women in Sport may be used in administering the interscholastic sports program for girls, where considered appropriate.
- State Departments of Education should be represented on the boards of control of all state high school athletics associations.
- The health and welfare of pupils should be the first consideration in planning an interscholastic athletics program. To protect and promote the health and welfare of competing athletes, the following conditions should be met:
  - Policies for the protection of athletes should be based on the best medical knowledge available.
  - Policies should include procedures to assure adequate medical examination plus ongoing medical supervision of all athletes. A minimum of one medical examination should be required each year. Preferably this should be given immediately prior to the sport season. Reexamination should be required after any serious injury or illness before the pupil returns to participation.
  - Coaches should understand and recognize the physical capacities of individual players. Coaches should be ever alert to note signs of undue fatigue or injury, thereby guarding the health and well-being of all participants both in practice sessions and in contests.
  - High-quality protective equipment should be provided and carefully fitted for every participant.
  - Equitable competition between teams or individuals should be assured by the use of standardized classifications and eligibility requirements.
  - Playing seasons should be of reasonable length and should be preceded by an adequate period of conditioning and instruction in fundamentals. No postseason or all-star games should be permitted.
  - Contests should be confined to small geographic areas with no interstate competition except between schools located near state borders.
  - Practice periods should be of reasonable length and geared to the physical condition of participants.
  - Awards should be simple, of little monetary value, and provided by schools rather than outside agencies.

About Program

The interscholastic athletics program in all high schools should provide maximum opportunity for both boys and girls to participate in a variety of individual, dual and team activities and should provide for differing levels of size and ability.

The high school athletics program should provide full opportunities for girls to compete in interscholastic athletics in accordance with their needs, abilities and interests. Interschool games and contests should be administered and conducted to enhance wholesome social relationships among community members and students, rather than to foster conflicts or extreme behavior based on unreasonable rivalries.

Interscholastic boxing should not be permitted.

About Personnel

Coaches should be certificated teachers, members of the school staff, and well prepared for assuming their coaching responsibilities. Certification of coaches is highly recommended.

Athletics directors should be competent administrators and educators with a rich background in interscholastic athletics. It is recommended that athletics directors acquire specific preparation for their position through formal courses or regular in-service work.

Games and contests should be officiated by qualified personnel who are certified by the appropriate governing body or officials association.

Index